MAVA Math: Middle Reviews Solutions

Marla Weiss

MAVA Books and Education Company
www.mavabooks.com

authorHOUSE®

AuthorHouse™
1663 Liberty Drive
Bloomington, IN 47403
www.authorhouse.com
Phone: 1-800-839-8640

Published by AuthorHouse   04/16/2013

ISBN: 978-1-4817-3987-0 (sc)

Library of Congress Control Number: 2013906984

CONTENTS

ABBREVIATIONS

Use common sense in decoding abbreviations. For example, in division problems, R means remainder, yet R means radius in geometry and common ratio in sequences.

General
CBD	can't be done
GCF	greatest common factor
LCM	least common multiple
NTS	not to scale
T&E	trial and error

Money
D	dime
H	half dollar
N	nickel
P	penny
Q	quarter

Coordinate Plane
D	down
L	left
Q	quadrant
R	right
U	up

Percent Change
D	decrease
I	increase

Sequences
d	common difference
r	common ratio

Properties
APA	Associative Property of Addition
APM	Associative Property of Multiplication
CPA	Commutative Property of Addition
CPM	Commutative Property of Multiplication
DPMA	Distributive Property of Multiplication over Addition
IdPA	Identity Property of Addition
IdPM	Identity Property of Multiplication
InPA	Inverse Property of Addition
InPM	Inverse Property of Multiplication
ZPM	Zero Property of Multiplication
RPE	Reflexive Property of Equality
SPE	Symmetric Property of Equality
TPE	Transitive Property of Equality
APE	Addition Property of Equality
MPE	Multiplication Property of Equality

Logic
T	true
F	false

Triangles
A	acute
E	equilateral
I	isosceles
O	obtuse
R	right
S	scalene

Geometry
A	area
B	base
C	circumference
D	diameter
E	edge
H	height
L	length
M	midline of a trapezoid
P	perimeter
R	radius
S	side
SA	surface area
V	volume
W	width

Functions
ABS	absolute value
SGN	1 for > 0, −1 for < 0, 0 for = 0
TRUNC	truncate decimal part
INT	greatest integer contained in
SIGMA	number of divisors
SQR	square
SQRT	square root

Sets
N	Natural numbers
W	Whole numbers
Z	Integers
Q	Rational numbers
R	Real numbers

An Important Message To Teachers, Parents, and Students

What are middle reviews?
The word "middle" refers to middle school (grades 6–8), although this book is adaptable to junior high. The word "review" suggests repetition, although certainly practice may include some extension of knowledge. The word "middle" parallels the word "grade" in *MAVA Math: Grade Reviews* which is a collection of reviews at the elementary school level.

What is the importance of cumulative review?
Children cannot learn without cumulative review. All children, regardless of math ability, have difficulty remembering skills and concepts unless practice occurs within a school year as well as from year to year.

On which curriculum is this book based?
A Curriculum Guide corresponding to the problems in this book appears after the two sample tests. Curricula vary somewhat from school to school and book to book. This book uses a blend of common approaches for basics as well as enriched material for insight.

Does this book use a developmental approach?
Yes, this book's curriculum advances each topic from level to level. All 97 topics progress from Level 6 to Level 7 and from Level 7 to Level 8. Careful checking of the problems in this book has produced a logical plan of concepts and skills within each topic.

Are these topics and problems all that exist in a typical curriculum?
No, these topics and problems comprise a representative sample within the framework of the book--namely, 10 problems per page in a 2-column format. Some valuable exercises, such as statistical graphs, cannot fit in the allotted space. Creating a totally comprehensive math book is virtually impossible due to the richness of mathematics. Moreover, a curriculum that is too broad does not permit mastery.

Why does this book have so many problems?
People need practice to master any skill--not only in math but also, for example, in music and sports. The MAVA Math series of textbooks provides plentiful problems, so needed by students but often unavailable.

How is this book best used?
At one school students may study percents in the fall, whereas at another school students in the same grade may study percents in the spring. Thus, to make this book universally appropriate, the review problems cover the entire grade level rather than follow the sequence of one school or another. This book is best used by starting a grade level in January of that grade, completing about one third during the second semester, completing another third during the summer, and then completing the final third during the first semester of the next school year. Sixth graders would complete the last third of Level 5 of *MAVA Math: Grade Reviews* in the 1st semester of grade 6, the 1st third of Level 6 of this book in the 2nd semester of grade 6, the 2nd third of Level 6 of this book over the summer, and the final third of Level 6 of this book during the 1st semester of 7th grade. In the 2nd semester of 7th grade, students, having had a half year of 7th grade math, would begin Level 7. Then the cycle repeats. The supervising adult, whether teacher or parent, should spread the reviews over the available number of weeks.

Does this book's curriculum vary dramatically in any way from those typically seen?
This book's breadth, depth, and quantity of problems are not commonly found in a one comprehensive volume.

What is the grade level of this book?
While this book aims at grades six through eight, some students may begin earlier. Also, students in high school lacking certain concepts and skills may find this book useful.

What do the headers "Level" and "Number" mean?
"Level," a synonym for "grade," is a more flexible word because children may work above or below grade level. "Number" simply counts the review pages. Within a level, a lower number does not imply that the review is easier than one with a higher number.

If a problem type is listed in one level, does it also appear in later levels?
The Curriculum Guide shows where a problem type first appears. While problem types are cumulative within this book, the significant increase in math from level to level prohibits thorough repetition. "None" in the list means no new problem types.

Does this book have a corresponding Number Sense book?
No. Mental math exercises are among the many problems in the reviews. Students should always be conscious of using number sense rather than using a calculator or doing a full operation algorithm.

Why does this book list over 500 vocabulary words?
Students cannot do math problems without understanding the words contained therein. Unfortunately, many math words have multiple meanings. Consider base--e.g., base of a triangle, base two arithmetic, and a number (base) raised to a power (exponent). Students learn math vocabulary when they continually hear the words used correctly.

Should students complete a page before starting another page?
Because curricula vary among publishers and schools, students may omit occasional problems from a page and then return to them later. Also, students need not do the pages in order within a grade level.

Was this book field tested?
Many problems in this book were used as part of comprehensive worksheets written by Marla Weiss for classroom settings. This material yielded students who loved math, performed high on standardized tests, and earned countless awards at math competitions. The collection, with additional problems, is unified for the first time in this book.

May parents help with the reviews?
Students who receive continual help from their parents often show significantly less growth in math than students who learn to work independently. Moreover, most parents have forgotten middle school math or don't know the best ways to approach many problems. Parents should only monitor a child's work, determining weak areas needing further help.

Why are occasional problems quick or easy?
An occasional quick or easy problem on a problem set is a welcome respite for a student and keeps both the pacing and progress moving forward.

Does this book prepare for future math instruction?
Yes. This book looks ahead to skills needed for classic algebra word problems. Moreover, this book provides a solid foundation for high school math, including the college admission SAT, for which a rich math education is important (fictitious operations as one example), and the ACT, for which broad retention of topics is essential.

Should students use a calculator with this book?
Students should not use a calculator. While starting calculator use in 6th grade is often appropriate, this book encourages practice of basic skills and number sense/mental math.

Why are the decimal points bold?
Some children do not see decimal points in normal font. Similarly, some students do not write decimal points darkly enough. A happy medium exists between a light dot and a wart.

How are answers distinguished from work in MAVA Math: Middle Reviews Solutions?
Work is shown in the same font as the problem. Answers appear on the answer lines or in the charts. Otherwise, answers are circled to distinguish them from the text or from work. Students should be encouraged to circle answers not on answer lines or in charts.

What happens if a student has trouble with a problem type in this book?
Cumulative reviews help to find topics that children have not truly learned. Remediation, by backing up in grade level as much as necessary, should occur.

Why do some answers abbreviate words?
Math time should not be treated as an opportunity to teach language arts, whether spelling or handwriting. Too many students have difficulty with math or learn to dislike the subject. Attaching verbal skills to the study of math handicaps many students otherwise talented in math. Ideally, students will learn to correctly spell and neatly print math vocabulary words. However, requiring these skills while learning math is self-defeating. Most abbreviations used in this book are found on page vi.

Why are the answers to some measurement problems a number without a label following?
Consider the question: How far do you live from school? The answer could be 2 miles or 2 turtle steps. A label is needed for accuracy. Now, consider the question: How many miles do you live from school? The answer may be 2 without ambiguity because a label, namely miles, is built into the question. Requiring a label at all times is incorrect.

What does the word "unit" mean?
Unit is a general term. Regarding distance, "unit" may mean many different measurements such as inches, feet, miles, or centimeters. Understanding that the label must be square units for area and cubic units for volume is more important than what the actual unit is. By using the generic "unit," students may focus on specific skills.

Should all improper fractions be converted to mixed numbers?
No, the term "improper fraction" is a misnomer. Converting from a fraction greater than one to a mixed number is a valuable skill, but it need not always be done. For example, as a solution to an equation, the fraction is better because it may be plugged in directly to check its validity. However, measurements are best as mixed numbers. For example, one and three fourths cups flour is more helpful than seven fourths cups.

Why do some problems have charts and diagrams pre-drawn while others do not?
In Level 6, this book helps students organize their thinking by presenting an outline for some answers in the form of a blank chart or diagram. However, by Level 8, students should be able to create proper work in a blank space.

Which are more valuable–fractions or decimals?
Decimal math may be easily done on a calculator. Fractions are more important in higher math. For example, a student who does not understand how to add 1/2 + 1/3 cannot possibly add 1/x + 1/y (diagonal fraction lines for ease of typing only). Furthermore, fractions yield an exact answer when decimals sometimes yield an approximation.

Why are some answers in bold font?
In charts, number lines, and the like, some numbers come with the problem and some numbers are answers. To distinguish between the two situations, the bold numbers are answers. The numbers in regular font appear in the student book as part of the problem.

Do some problems or skills have more than one method of solution?
To find the perimeter of a rectangle, should one add the length and width and then double, or should one double each measurement and then add? Students should understand both methods and decide based on the numbers in the problems. To find the slope of a line given 2 points, which point should be considered the 1st point and which the 2nd? Again, students should decide based on the coordinate numbers, always seeking fast and accurate calculation. The inherent richness and beauty of math yield multiple approaches to many problems. Moreover, some students prefer one way of thinking and some another.

Why do some of the 97 topics have "see" following them?
Due to the richness of math, placing a concept or skill into a category may be difficult. For example, GCFs may be included with fractions or with factors. Math topics should not be thought of as residing in closed cubbyholes but in porous receptacles. When problem types in this book are fully or partially covered in another category, the references appear in the topic heading.

Why does this book contain 2 pre-algebra sample tests?
The 2 sample tests, not comprehensive of the over 500 skills in this book, offer a means to test retention. No calculator and 90 minutes duration are suggested guidelines to evaluate highest performance. A student who scores low on the 1st test should study further before taking the 2nd test. The problems are straightforward to enable students to work quickly.

How can one learn more about various problem types and solution methods such as GCF, LCM, prime factorization, cross-tabulation charts, and Multiplication Principle?
Because MAVA Math books are more for practice than teaching, www.mavabooks.com offers public service chalkboard slide shows giving instruction on various topics. Visit the website often because new material appears continually.

Does MAVA Math have another middle school level text?
MAVA Math: Middle Reviews is designed for all middle school students, intended to supplement the daily textbook. *MAVA Math: Enhanced Skills*, with topics arranged alphabetically, is more advanced for 6th, 7th, and 8th grade students who wish to study math in depth or to enter math competitions.

What are the words in chalkboard font in MAVA Math: Middle Reviews Solutions?
Work is shown in the same font as the book. Chalkboard font is used for commentary such as an insightful method.

Why are equilateral triangles also labeled isosceles?
The definition of isosceles triangle is a three-sided polygon with at least 2 congruent sides. Therefore, an equilateral triangle is isosceles, but the converse is not true.

Why are negative signs, opposite signs, and subtraction signs all represented by the same symbol?
Some math texts use a smaller, higher line as distinguished from a longer, mid-level line. Because all three signs operate equivalently, this book uses just the longer, mid-level line for simplicity.

What should a student do if MAVA Math: Middle Reviews is too difficult?
If this book is too difficult for a student, then the message is clear: either the student has not been taught adequate math in elementary school and/or the student has not retained the math taught. Such a situation is common with students who have not done cumulative review regularly. Backing up to *MAVA Math: Grade Reviews* is recommended.

Why do some geometry problems use upper case and lower case abbreviations?
When a problem involves two circles, using capital R for the radius of the larger one and lowercase r for the radius of the smaller one is helpful. Moreover, a trapezoid always has two bases; use B for the larger one and b for the smaller one. Clarity of notation leads to improved focus and accurate answers.

Does this book cover a complete course in Algebra I?
No. This book covers most topics in pre-algebra but is not in any way intended to cover a full Algebra I course.

Why is functional notation used in situations that do not involve functions?
Functional notation is precise and concise. For example, P(even) is neater than writing "probability of tossing an even." Similarly, GCF(35, 49) is neater than writing the "greatest common factor of 35 and 49."

May students and teachers write diagonal fraction lines?
Never! For correct fraction work, students must clearly see numerators and denominators, only accomplished by writing horizontal fraction lines. This book uses diagonal fraction lines only in paragraph form or in the hints due to limited space.

What happens if MAVA Math: Middle Reviews Solutions contains an error?
MAVA Math: Middle Reviews Solutions was thoroughly proofed. However, any needed corrections will be posted on www.mavabooks.com. If a correction does not address your concern, please send a concise and precise e-mail to info@mavabooks.com.

Should math be fun?
Of course, math should be fun. However, teaching math solely as a game does not lead to growth. Students who complete weekly cumulative reviews gain enough practice to truly learn math. Understanding in turn leads to natural enjoyment. Competence is pleasurable.

NOTES

Level 6	Number 1

1. Find the 6-digit mystery number.
Clue 1: The number is divisible by 11.
Clue 2: The digits are in descending order.
Clue 3: Each digit appears exactly twice.
Clue 4: The digit sum is 38.
Clue 5: All digits are odd. (997,733)

Digit sum is too small for 997,711, 775,533, and all others.

6. Find the area in square units of a square with perimeter 20 units.

P = 20
s = 5
(A = 25)

2. Given two opposite vertices of a rectangle with sides parallel to the axes. Find the other two vertices.

(2, 8) and (−3, −7)   (2, −7) (−3, 8)

(−1, 9) and (4, −6)   (−1, −6) (4, 9)

(5, −6) and (10, 7)   (5, 7) (10, −6)

7. Name the number of degrees, if any, of rotational symmetry for the figure.

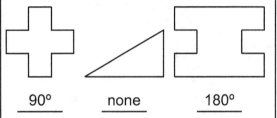

90°   none   180°

3. Complete with the vocabulary word (not congruent).

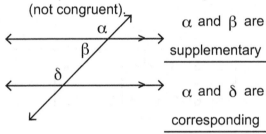

α and β are supplementary

α and δ are corresponding

8. Find the probability when tossing a die.

P(even) $\dfrac{1}{2}$ 2, 4, 6

P(odd) $\dfrac{1}{2}$ 1, 3, 5

P(prime) $\dfrac{1}{2}$ 2, 3, 5

P(composite) $\dfrac{1}{3}$ 4, 6

4. Find the percent.

35% of 80	28	95% of 60	57
65% of 60	39	30% of 120	36
70% of 90	63	15% of 200	30
40% of 35	14	55% of 40	22

9. Add.

$\dfrac{9}{70} + \dfrac{6}{35}^{2}$

$\dfrac{9 + 12}{70}$

$\dfrac{21 \div 7}{70 \div 7}$ $\left(\dfrac{3}{10}\right)$

$\dfrac{7}{20}^{4} + \dfrac{7}{40}^{2} + \dfrac{3}{80}$

$\dfrac{28 + 14 + 3}{80}$

$\dfrac{45 \div 5}{80 \div 5}$ $\left(\dfrac{9}{16}\right)$

5. Ginger bought a shirt for $21.90 and a skirt for $23.60. Find the total cost with 5% sales tax.

21.90
+ 23.60
45.50

45.50
× .05
2.2750

45.50
+ 2.28
($47.78)

Or multiply by 1.05 and omit addition.

10. Solve.

| x | = 9
x = ±9

| x | = −9
x = No Sol

| −9 | = x
x = 9

| 9 | = −x
x = −9

Level 6	Number 2

1. Find the 2 missing numbers of the pattern. Show in 2 different ways.

2, 6, 12, 20, 30, 42, __56__ , __72__

1 x 2
  2 x 3   7 x 8   8 x 9
    3 x 4
      4 x 5   OR add 4, add 6, add 8,
        5 x 6   add 10, add 12, add 14,
          6 x 7   add 16.

6. Complete the divisibility rules.

If __2__ and __3__ divide a number, then 6 divides the number.

If __3__ and __5__ divide a number, then 15 divides the number.

If __2__ and __9__ divide a number, then 18 divides the number. Not 3 and 6 (overlapping factors).

2. Identify whether the figure is an Euler Graph--can be drawn without lifting the pencil and retracing a line.

An Euler Graph must have exactly 0 or 2 vertices with an odd number of impinging segments (one to start, one to stop).

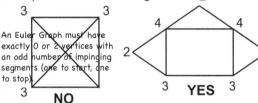

NO   **YES**

7. Name the property exemplified.

$5x + -(5x) = 0$ InPA

$-5(x + 2) = -5x - 10$ DPMA

$-9.4 + x$ is a real number   ClPA

$6y \cdot 1 = 6y$ IdPM

3. Find the simple interest on $4800 at 7% annually for 1 month.

Using a calculator reinforces the formula but not the use of fractions.

$I = PRT$

$I = \dfrac{\overset{4}{\cancel{4800}}}{1} \cdot \dfrac{7}{\cancel{100}} \cdot \dfrac{1}{\cancel{12}}$

$I = \boxed{\$28}$

8. Complete the table of values for $y = 2x + 4$.

x	−6	−5	−3	−2	0	1	3	4	7	9
y	−8	−6	−2	**0**	**4**	6	10	12	18	**22**

4. Answer true (T) or false (F).

Circles are polygons.   F

Squares are rectangles.   T

Rhombuses are parallelograms.   T

Parallelograms are trapezoids.   F

9. Calculate the area in square units of the trapezoid. (NTS)

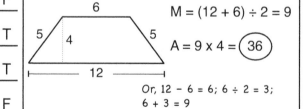

$M = (12 + 6) \div 2 = 9$

$A = 9 \times 4 = \boxed{36}$

Or, 12 − 6 = 6; 6 ÷ 2 = 3; 6 + 3 = 9

5. Operate.

$2\dfrac{4}{5} \times 3\dfrac{5}{7}$

$\dfrac{\overset{2}{\cancel{14}}}{5} \times \dfrac{26}{\cancel{7}_{1}}$

$\boxed{\dfrac{52}{5}}$

$5\dfrac{1}{3} \times 4\dfrac{1}{2}$

$\dfrac{\overset{8}{\cancel{16}}}{\cancel{3}_{1}} \times \dfrac{9}{\cancel{2}_{1}}^{3}$

$\boxed{24}$

10. Operate.

$(8 + 7) \div 3 + 7 \times 3$
$5 + 21$ $\boxed{26}$
Do 2 groupings concurrently.

$2 \times (5 − 16 \div 4 \times 2)$
$2 \times (−3)$ $\boxed{−6}$
The ÷ and x have equal weight; do L to R.

$3 \times 8 \div 6 − (9 − 1)$
$4 − 8$ $\boxed{−4}$
24 ÷ 6 = 4. 9 − 1 = 8.

$(4 + 11) \div 5 − 2 + 3$
$3 − 2 + 3$ $\boxed{4}$
2 and 3 have no ( ).

Level 6	Number 3

1. A lawn is 25 by 15 units. A bag of mulch will cover 40 square units. How many bags of mulch are needed to cover this lawn?

$25 \times 15 = 250 + 125 = 375$

$\dfrac{375}{40} = \dfrac{37.5}{4}$ $36 \div 4 = 9$ Must buy a whole number of bags.   (10)

6. Find the digit d so that the 4-digit number 67d1 is divisible by 9.

$6 + 7 + 1 = 14$ $14 + 4 = 18$ (4)

Find the digit x so that the 4-digit number 6x87 is divisible by 11.   (7)

$6 + 8 = 14$ $7 + 7 = 14$ $14 - 14 = 0$

---

2. Graph on the number line.

$x \geq 3$

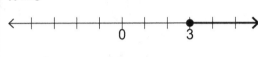

$x < -1$

7. Operate.

$5 - 10$	-5	$-10 + -30$	-40
$-15 + -15$	-30	$-50 + 25$	-25
$-10 - -10$	0	$20 - -10$	30
$-20 - 30$	-50	$-25 + -5$	-30

---

3. Solve by equivalent fractions.

$\dfrac{3}{4} = \dfrac{12}{x}$ (x = 16) $\dfrac{x}{7} = \dfrac{20}{35}$ (x = 4)

$\dfrac{5}{8} = \dfrac{x}{24}$ (x = 15) $\dfrac{2}{x} = \dfrac{14}{49}$ (x = 7)

8. Convert to scientific notation.

87,000	8.7×10^4	200	2.0×10^2
123,000	1.23×10^5	4000	4.0×10^3
7 million	7.0×10^6	7,300	7.3×10^3
2 billion	2.0×10^9	478	4.78×10^2

---

4. Find the area of the triangle. One box equals one square unit.

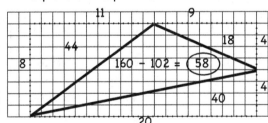

$160 - 102 =$ (58)

9. Write as an algebraic expression.

one more than twice a number $2x + 1$

two less than double a number $2x - 2$

half a number plus three $\dfrac{1}{2}x + 3$

triple a number minus four $3x - 4$

---

5. Complete with CONGRUENT (C), SIMILAR (S), or NEITHER (N).

All equilateral triangles are   S

All right triangles are   N

All squares with sides 2 cm are   C

10. Complete the table.

Decimal	Percent	Fraction
1.25	125%	$\dfrac{5}{4}$
2.0	200%	$\dfrac{2}{1}$
$0.\overline{3}$	$33.\overline{3}\%$	$\dfrac{1}{3}$

Level 6	Number 4

1. Ali scored 82, 70, and 98 out of 100 on 3 of 5 math tests. What is the minimum score she must earn on the 4th test to have an average of 87 on the 5 tests?

87 x 5 = 435

250 — 85   100

$\dfrac{82 + 70 + 98 + min + max}{5}$ = 87

⟨85⟩

6. Find the surface area of a cube with volume 27 cubic units.

V = 27
e = 3
A face = 9
SA = 9 x 6 = ⟨54⟩ sq un

Find the volume of a cube with surface area 150 square units.

SA = 150
A face = 25
e = 5
V = 5³ = ⟨125⟩ cu un

2. Plot and label the points.

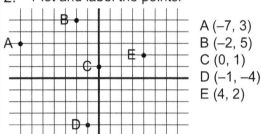

A (−7, 3)
B (−2, 5)
C (0, 1)
D (−1, −4)
E (4, 2)

7. Convert to base ten.

11011_{two}	16 + 8 + 2 + 1	27
124_{five}	1 x 25 + 2 x 5 + 4	39
221_{three}	2 x 9 + 2 x 3 + 1	25
213_{four}	2 x 16 + 1 x 4 + 3	39

3. Find the next term of the sequence. Label arithmetic (A), geometric (G), or neither (N). If A, find d. If G, find r.

G $\dfrac{1}{2}$ 1000, 500, 250,   125

N   na   0, 11, 23, 36,   50

A   5   15, 20, 25, 30, 35,   40

8. Complete the chart for a circle.

radius	diameter	area	circumference
4	**8**	16π	**8π**
3	**6**	**9π**	**6π**
5	10	**25π**	**10π**

4. Distribute.

3(x + y)   3x + 3y

2(a − b)   2a − 2b

6(2x + 4y)   12x + 24y

−5(3d + 5e)   −15d − 25e

9. Complete the table.

Original Cost	Sales Tax	Final Cost
$20.00	5%	**$21.00**
$10.00	6%	**$10.60**
$15.00	4%	**$15.60**

5. Find the probability of selecting a point from the shaded circle within the rectangle.

d = 20   r = 10

$\dfrac{A_{circ}}{A_{rec}} = \dfrac{10 \cdot 10 \cdot \pi}{30 \cdot 50}$

⟨$\dfrac{\pi}{15}$⟩

30 ... 20 ... 50

10. Evaluate.

$\sqrt{100}$ 10 $\sqrt{9}$ 3

$\sqrt{36}$ 6 $\sqrt{25}$ 5

$\sqrt{121}$ 11 $\sqrt{400}$ 20

$\sqrt{81}$ 9 $\sqrt{0}$ 0

	Level 6	Number 5

1. Find the next letter in each pattern.

skip 3

a E i M q U **y**

skip 3 backwards

Z v R n J f **B**

alternate consecutive with skip 3

B c G h L m **Q**

6.

	R	T	D
Y	15	3	45
T	20	2	40
tot	ⓘ7	5	85

Today Hal rode his bike for 2 hours at a steady rate of 20 miles per hour. Yesterday he rode his bike for 3 hours at a rate of 15 miles per hour. What was his overall rate for the two days?

2. Write the fractions in ascending order.

$\frac{26}{27}$ $\frac{28}{29}$ $\frac{25}{27}$ $\frac{31}{30}$ $\frac{25}{26}$ $\frac{29}{30}$

$\frac{25}{27}$ $\frac{25}{26}$ $\frac{26}{27}$ $\frac{28}{29}$ $\frac{29}{30}$ $\frac{31}{30}$

2 pie pieces out | One piece of each pie is missing with successively smaller pieces. | > 1

7. Answer YES or NO as to whether the numbers are valid sides of a triangle.

4, 5, 9 — N | 11, 12, 13 — Y

17, 17, 17 — Y | 10, 12, 19 — Y

15, 45, 61 — N | 8, 8, 15 — Y

3. Find the maximum area in square units of a rectangle with whole number sides and perimeter 40 units.

19 by 1 A = 19

Consider the 2 extremes: as thin as possible versus square.

10 by 10 A = ⓘ00

8. Find the median, mode, range, and outlier of the data.

Stem	Leaf
1	1 1 4 5 6 9 9
2	2 2 2 2 2
4	4 6
5	5 5 5 5
9	9

median = **22**
mode = **22**
range = 99 – 11 = **88**
outlier = **99**

4. Find the missing terms of the geometric sequences.

10, 20, 40, **80**, 160, **320**, 640

2, 8, 32, **128**, 512, **2048**, 8192

3, 15, **75**, 375, **1875**, 9375

9. Find the least common multiple.

LCM(50, 111) — 5550 — No common factors, so multiply.

LCM(100, 500) — 500 — 500 is a multiple of 100.

LCM(20, 40, 700) — 1400 — Ignore the 20 in 40. 700? No. 1400? Yes.

LCM(300, 400) — 1200 — Need factors 3, 4, 10, 10.

5. Convert measures to a decimal.

30 minutes = **0.5** hour

4 cups = **0.25** gallon

12 inches = **0.3̄** yard

1 pint = **0.125** gallon

10. Solve.

$5x + 4 = 15$ | $3x - 4 = 12$ | $7x + 3 = 20$
$5x = 11$ | $3x = 16$ | $7x = 17$
$x = \frac{11}{5}$ | $x = \frac{16}{3}$ | $x = \frac{17}{7}$

Level 6	Number 6

1. A train traveled 1200 miles in 8 hours. What was its rate in miles per hour?

$D = RT$

$\dfrac{1200}{8} = \dfrac{R \times 8}{8}$

$\boxed{150} = R$

6. A 10 by 22 rectangle has a 1 by 1 square removed from each of its corners. Find the volume in cubic units of the container formed by folding up the resulting sides.

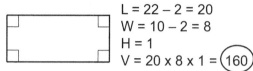

$L = 22 - 2 = 20$
$W = 10 - 2 = 8$
$H = 1$
$V = 20 \times 8 \times 1 = \boxed{160}$

2. Find the area of the irregular figure. One box equals one square unit.

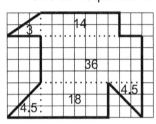

$50 + 9 + 21$

$\boxed{80}$

or use Subtraction Method

7. Five years ago Liz was 2 years older than Sam will be in 3 years. If Liz is 20 years old now, how old is Sam now?

	−5	now	+3
L	15	20	
S		⑩	13

3. Find the volume and surface area of a cylinder with diameter 20 and height 45 units.

$V = 10 \cdot 10 \cdot 45 \cdot \pi$
$V = \boxed{4500\pi}$ cu un

$SA = 100\pi + 100\pi + 20\pi \cdot 45$
$SA = 200\pi + 900\pi$
$SA = \boxed{1100\pi}$ sq un

8. Find the complement of the angle.

Comp (50°) 40°	Comp (10°) 80°
Comp (15°) 75°	Comp (75°) 15°
Comp (35°) 55°	Comp (25°) 65°
Comp (20°) 70°	Comp (30°) 60°

4. Define operation ✳ as:

X ✳ $Y = X^2 + 3XY$. For example, 3 ✳ $7 = 9 + 63 = 72$.
Find the values.

5 ✳ 4 __85__ $25 + 60$

4 ✳ 5 __76__ $16 + 60$

9. Combine like terms.

$3x + 7x - 2x$ __8x__

$9y - 3y + 8y$ __14y__

$20\pi + 5\pi - 9\pi$ __16π__

$6x - 3y + 2x - 4y$ __8x − 7y__

5. Multiply using mental math.

7.5 x 3	22.5	15.5 x 4	62
12.5 x 4	50	2.4 x 3	7.2
10.1 x 6	60.6	31.3 x 3	93.9
20.2 x 5	101	17.4 x 2	34.8

10. Of 5 test scores from 0 to 100, the mode is 65, the median is 75, the mean is 74, and the range is 20. Find the scores.

65	65	75	80	85

Place 75 1st, followed by the 65s. Next, 65 + 20 = 85. Then, 74 x 5 = 370. 370 − 290 = 80.

Level 6	Number 7

1. Three people can mow 6 lawns in 9 hours. At the same rate and for the same size lawns, how many people are needed to mow 8 lawns in 4 hours?

people	lawns	hours
3	6	9
3	2	3
3	8	12
(9)	8	4

Working in rows, keep 1, change 2 by x or ÷ (direct) or both (inverse).

2. State whether the fraction would repeat (R) or terminate (T) if converted to a decimal.

$\frac{11}{20}$ T $\frac{5}{6}$ R $\frac{7}{40}$ T

$\frac{7}{9}$ R $\frac{13}{25}$ T $\frac{5}{7}$ R

3. Find the missing sides of the similar triangles. (NTS)

5 / (7) \ 6 ; 10 / 14 \ (12) ×2

4. Write in standard form.

0.04 + 0.007 + 0.5   0.547

0.004 + 0.01 + 0.6 + 0.0009   0.6149

0.007 + 0.0008 + 0.03   0.0378

0.0002 + 0.005 + 0.8   0.8052

5. Find each central angle as a fractional part of 360°.

90° $\frac{1}{4}$ | 45° $\frac{1}{8}$ | 120° $\frac{1}{3}$

60° $\frac{1}{6}$ | 180° $\frac{1}{2}$ | 72° $\frac{1}{5}$

6. Compute using mental math.
Times 10; divide by 2 (cut in half).

465 x 5   2325 | 643 x 5   3215

348 x 5   1740 | 906 x 5   4530

246 x 5   1230 | 722 x 5   3610

387 x 5   1935 | 529 x 5   2645

7. Operate.

–1 x –9   9 | 5 x –8   –40

–7 x 6   –42 | –7 ÷ 7   –1

–8 ÷ –2   4 | –7 x –7   49

–6 ÷ 2   –3 | –9 ÷ –3   3

8. Check if the row number is divisible by the column factor.

	2	3	4	6	8	9
21,488	✓	–	✓	–	✓	–
64,872	✓	✓	✓	✓	✓	✓
45,150	✓	✓	–	✓	–	–

9. Complete.

21 ≡ 1 (mod 5) | 17 ≡ 1 (mod 2)

16 ≡ 0 (mod 4) | 30 ≡ 3 (mod 9)

19 ≡ 5 (mod 7) | 20 ≡ 4 (mod 8)

8 ≡ 2 (mod 6) | 13 ≡ 1 (mod 3)

10. 120 adults, half of whom are women, were asked to name their favorite chocolate. 30% said "milk," and half of the men said "dark." How many women said "milk"?

	milk	dark	tot
W	(6)	54	60
M	30	30	60
tot	36	84	120

An entire chart need not be completed--only enough to answer the question.

Level 6	Number 8

1. Hal bought a turkey for $53.55. The turkey weighed 12.6 pounds. Find the price per pound.

$$\frac{\text{price}}{\text{pound}} \quad \frac{53.55}{12.6}$$

$$4.25$$
$$126 \overline{)535.500}$$
$$504$$
$$315$$
$$252$$
$$630$$
$$630$$
$$0$$

($4.25)

6. Find the area of each rhombus in square units.

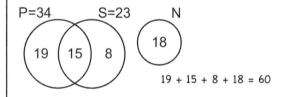

2. Rewrite the decimals in ascending order.

0.05, 5.01, 0.015, .051, .0015, 1.05

.0015 0.015 0.05 .051 1.05 5.01

7. Evaluate.

3!	3 x 2 x 1	6
0!	definition	1
5!	5 x 4 x 3 x 2 x 1	120
6!	6 x 5! = 6 x 120	720

3. Find the different ways that dimes and quarters can have a total value of $2.90. (6)

# D	x 10	# Q	x 25		# D	x 10	# Q	x 25
4	40	10	250		19	190	4	100
9	90	8	200		24	240	2	50
14	140	6	150		29	290	0	0

8. Of 60 houses, 34 need paint and 23 need shutters. If 15 need both, how many need neither?   18

P=34   S=23   N

19 15 8   18

19 + 15 + 8 + 18 = 60

4. Find the least natural number n such that kn is a perfect square.

k = 12   n = ___3___   12 = 2 x 2 x 3

k = 30   n = ___30___   30 = 2 x 3 x 5

k = 50   n = ___2___   50 = 2 x 5 x 5

An even number of each prime creates a perfect square.

9. Evaluate.

$$|2 - 6| - 6 + 9 + |-1 - 5| + 7$$
$$4 - 6 + 9 + 6 + 7$$
$$13 + 7 \qquad \underline{20}$$
$$\overline{-4 - |3 - 9| + |-5 - 4| + |1 - 8|}$$
$$-4 - 6 + 9 + 7$$
$$16 - 10 \qquad \underline{6}$$

5. Find the midpoint of the line segment with the given endpoints.

(4, 8) and (10, 8)   (7, 8)

(10, 5) and (10, 7)   (10, 6)

(1, 10) and (9, 20)   (5, 15)

(6, 3) and (12, 7)   (9, 5)

Average the xs, average the ys; or, halve the range and add to lesser.

10. 224 ounces of punch serve 35 people. How many ounces serve 50 people proportionally?

$$\frac{224}{35} = \frac{x}{50}$$

35x = 50 • 224
7x = 10 • 224
x = 10 • 32

(320)

	Level 6 Number 9

1. Translate the cipher with each letter as its own digit.

HAT	H9T	49T	**497**
+ HAT	+ H9T	+ 49T	**+ 497**
AAH	99H	994	**994**

A≠0. First look at 10s. The only way A + A = A is for the 1s to carry a 1. 2A+1 ends in A. Trying digits, A must be 9. Then H=4. T=7.

6. Find the number of 2 by 2 by 2 cubes that can completely fill a box measuring 20 by 16 by 30 of the same unit.

10 cubes by 8 cubes by 15 cubes = 1200 cubes

OR $\dfrac{\text{Volume box}}{\text{Volume 1 cube}}$ = $\dfrac{20 \times 16 \times 30}{2 \times 2 \times 2}$

2. Find the prime factorization. Work down. In the answer use exponents with primes in ascending order.

111	100	154
3 x 37	10 x 10	14 x 11
	2² x 5²	**2 x 7 x 11**

7. Write in word form: 4,065,613.007.

four million, sixty-five thousand,

six hundred thirteen and seven

thousandths

The decimal point is the "and."

3. Find the fractional part.
120 seconds out of 2 hours

$\dfrac{120}{2 \times 60 \times 60}$ $\dfrac{\cancel{120}}{2 \times \cancel{60} \times 60}$ $\dfrac{1}{60}$

8 inches out of 5 yards

$\dfrac{8}{5 \times 3 \times 12}$ $\dfrac{8^2}{5 \times 3 \times \cancel{12}_3}$ $\dfrac{2}{45}$

8. Find the number of paths from A to F, following the arrows.

7

4. Rewrite using bar notation.

6.777777777 . . . $6.\overline{7}$

8.543543543 . . . $8.\overline{543}$

2.1919191919 . . . $2.\overline{19}$

9.527892789 . . . $9.5\overline{2789}$

9. Find the perimeter in units of a square with area 49 square units.

A = 49

s = 7

P = 28

5. Multiply.

$\dfrac{3}{11}$ x 66	18	$\dfrac{8}{15}$ x 75	40
$\dfrac{7}{12}$ x 60	35	$\dfrac{3}{10}$ x 90	27
$\dfrac{4}{13}$ x 52	16	$\dfrac{7}{16}$ x 48	21

10. Evaluate for x = 2.

$x^2 - 2x + 10$	4 − 4 + 10	10
$100 - (x + 2)^2$	100 − 16	84
$x^3 - x^2 + 8x - 3$	8 − 4 + 16 − 3	17
$x^5 + 5(x + 6)$	32 + 40	72

Level 6	Number 10

1. Find the number of small squares in the 10th shape, continuing the pattern.

15

$4 + 11$

6. Write the value as an expression.

the number of cents in d dimes   10d

the number of eggs in e dozen   12e

the number of hours in d days   24d

the number of minutes in h hours   60h

2. Draw the reflection of each triangle over the x-axis. *Reflect each vertex first.*

7. Find the area of the circle inscribed in a square of area 64 square units.

$A_{sq} = 64$
$S = 8$
$D = 8$
$R = 4$
$A_{circ} = \boxed{16\pi}$

3. Divide.

$2196.5 \div 10^3$ 2.1965

$47.089 \div 10^2$ 0.47089

$3232.99 \div 10$ 323.299

$735601.5 \div 10^4$ 73.56015

8. Find the average using the rightmost digit method.

456, 459, 451, 450 $16 \div 4$ 454

295, 295, 291, 299, 295 $25 \div 5$ 295

43, 48, 47, 42, 41, 45, 49 $35 \div 7$ 45

4. Find the least value of xy.

$2 \leq x \leq 9$ $4 \leq y \leq 11$ 8

$-5 \leq x \leq 8$ $3 \leq y \leq 7$ -35

$-9 \leq x \leq -3$ $2 \leq y \leq 10$ -90

$-3 \leq x \leq 9$ $-6 \leq y \leq 6$ -54

9. Name the number 1 *before* in the given base.

456_{seven} 455_{seven} | 570_{eight} 567_{eight}

200_{three} 122_{three} | 555_{six} 554_{six}

430_{five} 424_{five} | 310_{four} 303_{four}

5. Solve.

$3x - 4 \leq 32$	$8x - 5 \geq 12$	$5x + 3 < 21$
$3x \leq 36$	$8x \geq 17$	$5x < 18$
$\boxed{x \leq 12}$	$\boxed{x \geq \dfrac{17}{8}}$	$\boxed{x < \dfrac{18}{5}}$

10. Multiply. Answer in exponential form.

$(4 \times 10^4) \times (6 \times 10^3)$ 24×10^7

$(5 \times 10^1) \times (7 \times 10^2)$ 35×10^3

$(8 \times 10^2) \times (5 \times 10^2)$ 4×10^5

$(5 \times 10^6) \times (4 \times 10^3)$ 2×10^{10}

Regroup the 10s in 40 and 20.

Level 6	Number 11

1. A school needed buses for a field trip. Would 3 buses, each holding 40 people plus the driver, be enough to transport 6 classes of 18 students each plus one teacher for each class?

bus: 40 × 3 = 120 (exact)
people: 20 × 6 = 120 (over-estimate)

(YES)

6. Solve by clearing the decimal points.

.8x = 64	.15x = .6	.07x = .49
8x = 640	15x = 60	7x = 49
x = (80)	x = (4)	x = (7)

2. Identify each as a point, line, ray, or line segment if graphed on a number line. Determine without graphing.

x = 12 ___point___

x ≥ 5 ___ray___

−3 ≤ x ≤ 19 ___line segment___

7. Remove 10 odd numbers from the first 35 whole numbers. What percent of the remaining numbers are odd?

35 whole numbers : 0 ⟶ 34
18 even, 17 odd
remove 10 odd: 25 numbers remain
18 even, 7 odd

$\dfrac{7}{25}$ = (28%)

To make the fractional part, put "is" (or "are") number over "of" number.

3. Convert to a decimal from memory.

$\frac{1}{2}$ .5	$\frac{3}{4}$ .75	$\frac{3}{8}$ .375
$\frac{3}{5}$ .6	$\frac{5}{8}$ .625	$\frac{7}{10}$ .7
$\frac{2}{3}$.$\overline{6}$	$\frac{7}{9}$.$\overline{7}$	$\frac{1}{4}$ .25

8. Find the volume and total surface area of the solid comprised of unit cubes.

V = (9) cu un

SA = (34) sq un

F = 9
B = 9
U = 5
D = 5
L = 3
R = 3

4. Find the percent mentally.

45% of 140	63	85% of 400	340
80% of 75	60	55% of 80	44
5% of 200	10	70% of 50	35
20% of 95	19	65% of 120	78

9. Find the greatest common factor.

GCF(27, 72) ___9___ 9 x 3 vs. 9 x 8

GCF(660, 880) ___220___ Ignore 10 initially. 22 x 3 vs. 22 x 4

GCF(15, 90, 120) ___15___ All are multiples of 3 and 5.

GCF(72, 225) ___9___ 9 x 8 vs. 9 x 25

5. Complete the chart of data for a circle graph.

	Number	Fraction	Percent	Degrees
Dem	120	4/8	50	180
Rep	90	3/8	37.5	135
Indep	30	1/8	12.5	45
TOTAL	240	1	100	360

10. Multiply.

```
      75,462
    x    913
     226386
      75462
   679158
   68896806
```

```
      88,235
    x  6,274
     352940
     617645
     176470
    529410
   553586390
```

Level 6	Number 12

1. Find the cost for the number specified.

7 if 3 cost $111   **$259**   111/3 = 37

9 if 4 cost $236   **$531**   236/4 = 59

11 if 5 cost $425   **$935**   425/5 = 85

8 if 6 cost $522   **$696**   522/6 = 87

6. Always draw the discards on the left.

Ninety birds were in a tree. One third flew away, and one fourth of those left were tweeting. What fraction of the original birds were quiet?

90

30   60

15   45

$\dfrac{45}{90}$ $\dfrac{1}{2}$

---

2. What time is 425 minutes after 2:15 PM?

$425 = 60 \times 7 + 5$ $2 + 7 \rightarrow 9$ $15 + 5 \rightarrow 20$

$\boxed{9\text{:}20 \text{ PM}}$

What time is 350 minutes after 10:30 PM?

$350 = 60 \times 6 - 10$ $10 + 6 \rightarrow 4$ $30 - 10 \rightarrow 20$

$\boxed{4\text{:}20 \text{ AM}}$

7. An equilateral triangle with side 10 and a square with side 10 are attached at one side. Find the perimeter of the pentagon formed.

$5 \times 10 = \boxed{50}$

---

3. Calculate the point on the number line that is:

$\dfrac{1}{2}$ of the way from −3 to 3 $6 \div 2 = 3$ 0
   $-3 + 3 = 0$

$\dfrac{1}{3}$ of the way from −4 to 5 $9 \div 3 = 3$ -1
   $-4 + 3 = -1$

$\dfrac{2}{5}$ of the way from −2 to 8 $10 \div 5 = 2$ 2
   $-2 + 4 = 2$

8. Operate and simplify.

$\dfrac{1}{2} + \dfrac{4}{7}$ $\dfrac{7 + 8}{14}$ $\dfrac{15}{14}$ $\dfrac{\cancel{15}^{5}}{\cancel{14}_{1}} \cdot \dfrac{\cancel{28}^{2}}{\cancel{3}_{1}}$

$\dfrac{6}{7} - \dfrac{3}{4}$ $\dfrac{24 - 21}{28}$ $\dfrac{3}{28}$

$\boxed{10}$

---

4. Write each fraction as a repeating decimal.

$\dfrac{7}{9}$ $.\overline{7}$	$\dfrac{3}{11}$ $.\overline{27}$	$\dfrac{7}{99}$ $.\overline{07}$
$\dfrac{13}{99}$ $.\overline{13}$	$\dfrac{2}{9}$ $.\overline{2}$	$\dfrac{22}{99}$ $.\overline{2}$

9. Label △ABC Isosceles, Equilateral, or Scalene AND Right, Acute, Obtuse, or Equiangular for 3 separate problems. CBD is also a possible answer.

m∠C = 92

AC = BC

m∠B = 90   m∠A = 40

CBD	O
I	CBD
S	R

---

5. Find the probability of selecting a point from the shaded circle within the square.

60   60

$d = 60$ $r = 30$

$\dfrac{A_{circ}}{A_{sq}} = \dfrac{30 \cdot 30 \cdot \pi}{60 \cdot 60}$

$\boxed{\dfrac{\pi}{4}}$

10. Operate.

$(9 + 6) \div 5 + 9 \times 6$

$3 + 54$ $\boxed{57}$

Do 2 groupings concurrently.

$6 \times 7 \div 3 - (-8 + 1)$

$14 - (-7)$ $\boxed{21}$

$(4 − 12 \div 3 \times 4) \times -4$

$(4 − 16) \times (−4)$

$−12 \times (−4)$ $\boxed{48}$

$30 \div (5 − 2) − 3 \times 3$

$10 − 9$ $\boxed{1}$

Simplify 6 with 3 by seeing division as a fraction.

Level 6	Number 13

1. Write the converse of the conditional statement. Are they logically equivalent? (NO)

If polygon P is a square, then polygon P is a rectangle.

If polygon P is a rectangle, then polygon P is a square.

statement TRUE
converse FALSE

2. Find the sum of the whole numbers between:

$\sqrt{11}$ and $\sqrt{99}$ ___39___ 4 + 5 + 6 + 7 + 8 + 9

$\sqrt{8}$ and $\sqrt{80}$ ___33___ 3 + 4 + 5 + 6 + 7 + 8

$\sqrt{20}$ and $\sqrt{111}$ ___45___ 5 + 6 + 7 + 8 + 9 + 10

3. Find the perimeter of the figure on the unit grid. (80)

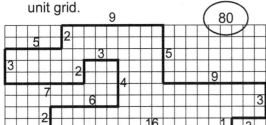

4. Operate.

$\frac{1}{5}$ (11 days   4 hours   5 minutes)

$\frac{1}{5}$ (10 d   28 h   5 m)

$\frac{1}{5}$ (10 d   25 h   185 m)

(2 d   5 h   37 m)

5. Timmy is 9 years older than Tammy who is 3 years younger than Tommy. If Tommy is 12 years old, find the sum of the three ages.

Tim 18
6
9   Tom 12   (39)
3
Tam 9

6. Operate and simplify the continued fraction. $\frac{13}{17}$

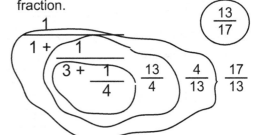

$1 + \cfrac{1}{3 + \cfrac{1}{4}}$ $\frac{13}{4}$ $\frac{4}{13}$ $\frac{17}{13}$

7. Operate.

10 − 70	−60	70 + −20	50
−35 + −35	−70	−45 + −15	−60
−20 − −40	20	40 − −20	60
−45 − 50	−95	−85 + 5	−80

8. Find the probability when tossing a die.

P(7) ___0___ none

P(number less than 10) ___1___ all

P(perfect square) $\frac{1}{3}$ 1, 4

P(perfect cube) $\frac{1}{6}$ 1

9. Add.

$\frac{9}{40}^{5} + \frac{7}{50}^{4}$ $\frac{2}{15}^{6} + \frac{7}{30}^{3} + \frac{1}{45}^{2}$

$\frac{45 + 28}{200}$ $\frac{12 + 21 + 2}{90}$

$\left(\frac{73}{200}\right)$ $\frac{35 \div 5}{90 \div 5}$ $\left(\frac{7}{18}\right)$

10. Identify membership in each set.

	N	W	Z	Q	R
100	✓	✓	✓	✓	✓
1.1	−	−	−	✓	✓
0	−	✓	✓	✓	✓
π	−	−	−	−	✓

| Level 6 | Number 14 |

1. A regular quadrilateral is called a _____ square.

Vertical, corresponding, and alternate interior angles are all _____ congruent.

Sentence is to equation as phrase is to _____ expression.

6. Two whole numbers are in the ratio 3:7. Their sum is 30. Find their product.

3:7 sum = 10
9:21 sum = 30   9 x 21 = $\boxed{189}$

Two whole numbers are in the ratio 3:8. Their sum is 44. Find their product.

3:8 sum = 11   By DPMA 320 + 64
12:32 sum = 44   32 x 12 = $\boxed{384}$

2. Name the points shown.

A (–7, –3)
B (2, –2)
C (0, 5)
D (–4, –1)
E (5, 3)

7. Six workers can do a job in 10 hours. At the same rate, how many workers are needed to do the same job in 12 hours?

W x t = w x T
6 x 10 = 5 x 12
$\boxed{5}$

Workers and time are inversely proportional-- as one goes up, the other goes down. Their product is constant.

3. Given two sides of a triangle, find the range of values for the 3rd side (use s for unknown side).

4, 20 $16 < s < 24$

8, 12 $4 < s < 20$

9, 21 $12 < s < 30$

Subtract given sides for low bound. Add for high bound. "The sum of two sides must be greater than the 3rd" to form a triangle.

8. By what number must 9! be multiplied to become 12! ?

10 x 11 x 12 = 132 x 10 = $\boxed{1320}$

By what number must 10! be divided to become 7! ?

8 x 9 x 10 = 72 x 10 = $\boxed{720}$

4. If the sum of two consecutive even numbers is 22, find their product.

10 + 12 = 22   10 x 12 = $\boxed{120}$

If the sum of two consecutive odd numbers is 24, find their product.

11 + 13 = 24   11 x 13 = $\boxed{143}$

9. Find the probability that a positive integer less than or equal to 16 is a factor of 16. $\boxed{\dfrac{5}{16}}$
1, 2, 4, 8, 16

Find the probability that a positive integer less than or equal to 30 is a factor of 30. $\dfrac{8}{30} = \boxed{\dfrac{4}{15}}$
1, 2, 3, 5, 6, 10, 15, 30

5. Operate.

$8\dfrac{3}{4} \div 4\dfrac{3}{8}$ $\dfrac{\overset{1}{35}}{4} \times \dfrac{8}{35}_1$ $\boxed{2}$

$6\dfrac{2}{9} \div 4\dfrac{2}{3}$ $\dfrac{\overset{4}{56}}{9}_3 \times \dfrac{3}{14}_1^{1}$ $\boxed{\dfrac{4}{3}}$

10. A square with area 36 square units and an equilateral triangle have equal perimeters. Find the side in units of the triangle.
A □ = 36
S □ = 6
P □ = 24
P △ = 24
S △ = $\boxed{8}$

Level 6	Number 15

1. Find the 6-digit mystery number.
Clue 1: The number is divisible by 8.
Clue 2: The leftmost 3 digits are consecutive even in ascending order.
Clue 3: The rightmost 3 digits are consecutive even in descending order.
Clue 4: The digit sum is 30.

RIGHT: 420 and 642 not div by 8
LEFT: 246 or 468; check digit sum   (246,864)

6. Given a bowl of 12 red, 15 blue, and 18 green marbles. Draw one. Find each probability.

P(R) $\dfrac{4}{15}$ $\dfrac{12 \div 3}{45 \div 3}$

P(B) $\dfrac{1}{3}$ $\dfrac{15 \div 15}{45 \div 15}$

P(G) $\dfrac{2}{5}$ $\dfrac{18 \div 9}{45 \div 9}$

2. Graph on the number line.

$-3 < x \le 4$

$x < 2$

7. Name the number of degrees, if any, of rotational symmetry for the figure.

180°   90°   180°

3. Solve by equivalent fractions.

$\dfrac{5}{x} = \dfrac{25}{30}$ (x = 6) $\dfrac{5}{9} = \dfrac{x}{36}$ (x = 20)

$\dfrac{6}{18} = \dfrac{3}{x}$ (x = 9) $\dfrac{x}{11} = \dfrac{21}{77}$ (x = 3)

8. Find the volume in cubic units of a rectangular prism with the areas of its noncongruent faces 6, 14, and 21 square units.

2 x 3 = 6   V = 2 x 3 x 7 = (42)
2 x 7 = 14
3 x 7 = 21

4. Operate on the sets.

A = {2, 4, 6, 8, 10}
B = {1, 2, 3, 4, 5, 6}
C = {4, 5, 6, 7, 8, 9}

A ∪ C =   {2, 4, 5, 6, 7, 8, 9, 10}
A ∩ C =   {4, 6, 8}
B ∪ C =   {1, 2, 3, 4, 5, 6, 7, 8, 9}

9. Compute using mental math.
Multiply by 4 mentally by doubling twice.

36 x 4   36 + 36 = 72   72 + 72   144

28 x 4   28 + 28 = 56   56 + 56   112

19 x 4   19 + 19 = 38   38 + 38   76

47 x 4   47 + 47 = 94   94 + 94   188

5. Lilah bought one book for $23.75 and another for $39.45. Find the total cost with 6% sales tax.

23.75   63.20
+ 39.45   + 3.79
63.20   ($66.99)
x .06
3.7920

Or multiply by 1.06 and omit addition.

10. Solve.

8x − 4 = 21 | 7x − 7 = −11 | 6x + 9 = −2
8x = 25 | 7x = −4 | 6x = −11
x = $\dfrac{25}{8}$ | x = $\dfrac{-4}{7}$ | x = $\dfrac{-11}{6}$

Level 6	Number 16

1. Jon scored 64, 72, and 88 out of 100 on 3 of 5 math tests. What is the minimum score he must earn on the 4th test to have an average of 80 on the 5 tests?

80 x 5 = 400   (76)
224   76   100
$$\frac{64 + 72 + 88 + \min + \max}{5} = 80$$

6. Find the surface area of a cube with volume 8 cubic units.

V = 8
e = 2
A face = 4
SA = 4 x 6 = (24) sq un

Find the volume of a cube with surface area 600 square units.

SA = 600
A face = 100
e = 10
V = 10^3 = (1000) cu un

2. A pizza shop offers 7 toppings, thick or thin crust, and red or white sauce. How many different one-topping pizzas with sauce are available?

7 x 2 x 2 = (28)

7. Convert to base ten.

1012_{three}	1 x 27 + 1 x 3 + 2	32
110101_{two}	32 + 16 + 4 + 1	53
332_{four}	3 x 16 + 3 x 4 + 2	62
343_{five}	3 x 25 + 4 x 5 + 3	98

3. Find the missing terms of the arithmetic sequences.

10, 25, 40, __55__, 70, 85, __100__

3, 12, 21, 30, __39__, 48, 57, __66__

11, 15, 19, __23__, 27, __31__, 35

8. Complete the table of values for y = –x + 3.

x	–6	–4	–3	–1	0	1	2	3	6	9
y	9	7	6	4	3	2	1	0	–3	–6

4. Reflect the point A(3, 2) over the y-axis. Then translate it four units down. What is the final point?

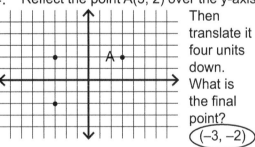

(–3, –2)

9. Find the least common multiple.

LCM(45, 900) __900__ 900 is a multiple of 45.

LCM(25, 30, 35) 1050 Need factors 5, 5, 6, 7.

LCM(22, 275) __550__ See the factor of 11 in both. Need 275 x 2.

LCM(100, 125) __500__ 25 x 4 and 25 x 5 so need 25 x 20.

5. Find the sum of each arithmetic sequence using the "First plus Last" method.

1 + 2 + 3 + 4 + 5 + 6 + 7 + 8 + 9 + 10 + 11 + 12

4 + 6 + 8 + 10 + 12 + 14 + 16 + 18 + 20 + 22

$$\frac{(1 + 12)(12)}{2} = (78)$$ $$\frac{(4 + 22)(10)}{2} = (130)$$

10. In a set of five 2-digit numbers, the mode is 75, the median is 46, the mean is 50, and the range is 61. Find the numbers.

__14__   __40__   __46__   __75__   __75__

Place 46 1st, followed by 75s. Next, 75 – 61 = 14. Then, 50 x 5 = 250. 250 – 210 = 40.

Level 6	Number 17

1. Add in the given bases.

$4_{five} + 4_{five} = \underline{13_{five}}$

$3_{six} + 5_{six} = \underline{12_{six}}$

$2_{nine} + 8_{nine} = \underline{11_{nine}}$

$3_{four} + 2_{four} = \underline{11_{four}}$

6. Find the digit d so that the 5-digit number 46,8d2 is divisible by 9. ⑦

digit sum = 20
need 27

Find the digit x so that the 4-digit number 35x4 is divisible by 36. ⑥

for div by 4, for div by 9,
x = 0, 2, 4, 6, 8   x = 6

2. Find the prime factorization. Work down. In the answer use exponents with primes in ascending order.

345	180	253
5 x 69	18 x 10	**11 x 23**
3 x 5 x 23	2 x 9 x 2 x 5	
	2^2 x 3^2 x 5	

7. Find the average using the arithmetic sequence method. The mean of an arithmetic sequence is the median.

320, 324, 328, 332 → 326

With an even number of numbers, the median is the mean of the 2 middlemost.

20, 30, 40, 50, 60, 70 → 45

60, 66, 72, 78, 84, 90 → 75

3. Find the missing sides of the similar triangles. (NTS)

11 10 ㊹ 40
⑮ 60
x4

8. Convert to standard form.

4.732×10^7	47,320,000
5.29×10^5	529,000
1.1178×10^4	11,178
7.9×10^6	7,900,000

4. Find the maximum perimeter in units of a rectangle with whole number sides and area 121 square units.

121 by 1
P = ㉔㊸ (244)

Consider the 2 extremes: as thin as possible versus square.

11 by 11
P = 44

9. Write as an algebraic expression.

twice the quantity a number plus six → $2(x + 6)$

ten minus twice a number → $10 - 2x$

one sixth a number plus four → $\frac{1}{6}x + 4$

a number increased by one, all divided by five → $\frac{x + 1}{5}$

5. Find the angle measures. (NTS)

1 2 3
4 5 6
7 8 9 10
11 12 13 14

If m∠9 = 78°, then m∠14 = 78°

If m∠6 = 43°, then m∠9 = 43°

10. Complete the table.

Decimal	Percent	Fraction
1.5	**150%**	$\frac{3}{2}$
3.25	325%	$\frac{13}{4}$
$0.\overline{6}$	$66.\overline{6}\%$	$\frac{2}{3}$

Level 6	Number 18

1. Subtract.

$$974,823,105 \qquad 629,708,544$$
$$-\,397,655,429 \qquad -\,337,839,879$$
577,167,676   **291,868,665**

6. Divide the product of the first six composite numbers by the product of the first six positive whole numbers.

$$\frac{4 \times \cancel{6} \times 8 \times 9 \times \cancel{10} \times \cancel{12}}{1 \times \cancel{2} \times \cancel{3} \times \cancel{4} \times \cancel{5} \times \cancel{6}} = 4 \times 72 = \boxed{288}$$

"Simplify first. Multiply last."

2. Find the area of the parallelograms. One box equals one square unit.

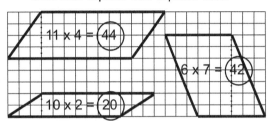

11 x 4 = (44)

6 x 7 = (42)

10 x 2 = (20)

7. The ratio of boys to girls in a school is 4:5.

Find the ratio of boys to total students.

$$\frac{B}{S} \qquad \boxed{\frac{4}{9}}$$

Find the number of students if the school has 100 boys.   x 25

$$100 + 125 = \boxed{225}$$

3. Find the simple interest on $6000 at 9% annually for 5 months.

I = PRT

$$I = \frac{\overset{5}{\cancel{6000}}}{1} \cdot \frac{9}{\cancel{100}} \cdot \frac{5}{\cancel{12}}$$

$$I = \boxed{\$225}$$

8. Complete the chart for a circle.

radius	diameter	area	circumference
6	**12**	36π	**12π**
10	**20**	**100π**	**20π**
9	18	**81π**	**18π**

4. Distribute.

2(b + 4c)   2b + 8c

4(c − 3bd)   4c − 12bd

−3(3x + 5y)   −9x − 15y

−4(2 − 4x)   −8 + 16x

9. Evaluate.

|5 − 12| − 9 + 8 + |−3 − 9| − 15

7 − 1 + 12 − 15
18 − 15   3

− 9 − |6 − 12| + |−9 − 9| + |1 − 11|

− 9 − 6 + 18 + 10
−15 + 28   13

5. Answer true (T) or false (F).

All equilateral triangles are isosceles.   T

All rhombuses are regular.   F

All rectangles are parallelograms.   T

All diameters are chords.   T

10. Calculate the area in square units of the trapezoid. (NTS)

M = (24 + 10) ÷ 2 = 17

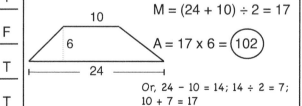

A = 17 x 6 = (102)

Or, 24 − 10 = 14; 14 ÷ 2 = 7;
10 + 7 = 17

Level 6	Number 19

1. Find the next letter in each pattern.

3 consecutive, skip 3, . . .

B	c	D	H	i	J	**N**

skip 1, 2, 3, 4, 5 backwards

Z	X	U	Q	L	**F**

skip 0, 1, 2, 3, 4, 5

A	b	D	g	K	p	**V**

6.

	R	T	D
G	7	5	35
H	4	10	40
tot	⑤	15	75

In one week, Greg biked or walked 35 miles at an average rate of 7 miles per hour, while Hank spent twice as much time yet only covered 5 miles more. What was their overall rate in mph that week?

2. Find the GCF and LCM of 135 and 165 using prime factorization.

$135 = 27 \times 5 = 3^3 \times 5$
$165 = 15 \times 11 = 3 \times 5 \times 11$

Look for one of the original numbers before multiplying.

GCF = 3×5 = ⑮

LCM = $3^3 \times 5 \times 11 = 165 \times 9$ = ⑴⑷⑻⑸ (1485)

7. Operate.

$4 \div -2$	-2	-5×5	-25
-4×-7	28	4×-3	-12
$-9 \div -9$	1	$0 \div -5$	0
-7×3	-21	-9×-9	81

3. Complete with the vocabulary word (not congruent).

α and β are __alternate exterior__

β and δ are __vertical__

8. Check if the row number is divisible by the column factor.

	2	3	6	8	9	11
37,896	✓	✓	✓	✓	–	–
63,734	✓	–	–	–	–	✓
71,487	–	✓	–	–	✓	–

4. Find the area of the triangle. One box equals one square unit.

14   6
5   35   21
                   7
140 − 76 = ⑹⑷ (64)
2   20
        20

9. Add.

	904,556,714
892,641,754	367,224,615
205,618,037	476,556,723
173,891,335	254,167,854
+ 815,773,489	+ 595,789,186
2,087,924,615	**2,598,295,092**

5. Convert measures to a decimal.

20 minutes = $0.\overline{3}$ hour

6 cups = 0.375 gallon

9 inches = 0.25 yard

2 pints = 0.25 gallon

10. Find the ratio of:

a side of a rhombus to its perimeter. $\frac{1}{4}$

a side of a square to its perimeter. $\frac{1}{4}$

a side of a regular octagon to its perimeter. $\frac{1}{8}$

Level 6	Number 20

1. A plane traveled 2450 miles at an overall rate of 350 miles per hour. How many hours did it fly?

D = RT

$\dfrac{2450}{350} = \dfrac{350 \times T}{350}$

⑦ = T

6. Find the area of each rhombus in square units.

side = 5, height = 4	A = __20__
side = 9, angles = 90°	A = __81__
side = 13, height = 12	A = __156__

2. Find the area of the irregular figure. One box equals one square unit.

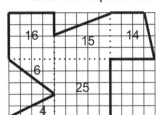

40 + 30 + 10

⑧⓪

or use Subtraction Method

7. Identify (YES/NO) whether the list of ordered pairs is a function. If NO, specify the x-value that violates the function definition.

{ (0,1), (1,2), (2,3), (3,4), (4,5) }   Y   na

{ (0,9), (1,9), (5,9), (6,9), (9,9) }   Y   na

3. Identify whether the figure is an Euler Graph--can be drawn without lifting the pencil and retracing a line.

Exactly 2 odd vertices.

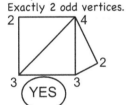 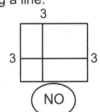

YES   NO

8. Find the complement of the angle.

Comp (70°) 20°	Comp (65°) 25°
Comp (40°) 50°	Comp (45°) 45°
Comp (80°) 10°	Comp (60°) 30°
Comp (55°) 35°	Comp (5°) 85°

4. Define operation ▲ as:

A ▲ B = A + B − AB − B³. For example, 5 ▲ 2 = 5 + 2 − 10 − 8 = −11. Find the values.

6 ▲ 3 __−36__   6 + 3 − 18 − 27

9 ▲ 1 __0__   9 + 1 − 9 − 1

9. Name the number 1 *after* in the given base.

567_{nine} 568_{nine} | 444_{five} 1000_{five}

146_{seven} 150_{seven} | 101_{two} 110_{two}

122_{three} 200_{three} | 237_{eight} 240_{eight}

5. Find the midpoint of the line segment with the given endpoints.

Average the xs, average the ys; or, halve the range and add to lesser.

(20, 8) and (30, 12) (25, 10)

(11, 11) and (11, 19) (11, 15)

(2, 13) and (12, 15) (7, 14)

(9, 7) and (19, 11) (14, 9)

10. 240 children, one third boys, named a fruit. 50% said pears, 25% said apples, and the rest said cherries. Half of the girls said pears. Half of the boys said apples. How many girls said cherries?

	P	C	A	tot
B	40	0	40	80
G	80	60	20	160
tot	120	60	60	240

| Level 6 | Number 21 |

1. Twelve cats can catch 15 mice in 2 hours. At the same rate, 20 cats can catch 50 mice in how many hours?

cats	mice	hours
12	15	2
4	5	2
20	25	2
20	50	(4)

Working in rows, keep 1, change 2 by x or ÷ (direct) or both (inverse)

6. Does the mapping represent a function?　(YES)

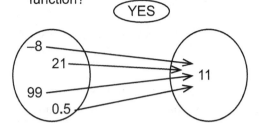

2. Given two opposite vertices of a rectangle with sides parallel to the axes. Find the other two vertices.

(11, 2) and (−7, −10)　　(11, −10)　(−7, 2)

(3, −9) and (13, −3)　　(3, −3)　(13, −9)

(−5, −8) and (0, 12)　　(−5, 12)　(0, −8)

7. Name the property exemplified.

$0 + -0 = 0$ 　　　　InPA

$0 + 0 = 0$ 　　　　IdPA

$0 \cdot 0 = 0$ 　　　　ZPM

$-k + k = 0$ 　　　　InPA

3. Find the fractional part.
12 pints out of 5 gallons

$$\frac{12}{5 \times 8} \qquad \frac{\cancel{12}^{\,3}}{5 \times \cancel{8}_{\,2}} \qquad \left(\frac{3}{10}\right)$$

8 feet out of 2 miles

$$\frac{8}{2 \times 5280} \qquad \frac{\cancel{8}^{\,1}}{2 \times \cancel{5280}_{\,660}} \qquad \left(\frac{1}{1320}\right)$$

8. Find the median, mode, range, and outlier of the data.

Stem	Leaf
1	9 9
5	2 2 3 3 3 3 4
6	6 6 6 7 7 7 8
7	2 3 5 5 6 6 7
8	1 2 3 4 5 5

median = **67**
mode = **53**
range = 85 − 19 = **66**
outlier = **19**

4. Name the solids depicted by the nets.

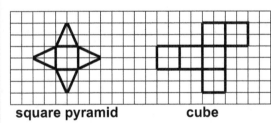

square pyramid　　　　**cube**

9. Complete.

$23 \equiv \underline{2}$ (mod 3) 　 $38 \equiv \underline{3}$ (mod 5)

$47 \equiv \underline{5}$ (mod 6) 　 $58 \equiv \underline{2}$ (mod 8)

$51 \equiv \underline{3}$ (mod 4) 　 $50 \equiv \underline{1}$ (mod 7)

$68 \equiv \underline{5}$ (mod 9) 　 $24 \equiv \underline{0}$ (mod 2)

5. Complete with CONGRUENT (C), SIMILAR (S), or NEITHER (N).

All parallelograms are 　　N

All isosceles right triangles are 　　S

All right triangles with hypotenuse of length 10 units are 　　N

10. Complete the table.

Decimal	Percent	Fraction
4.1	**410%**	$\frac{41}{10}$
0.$\overline{4}$	44.$\overline{4}$%	$\frac{4}{9}$
1.6	**160%**	$\frac{8}{5}$

Level 6	Number 22

1. A more common term for multiplicative inverse is ___reciprocal___.

The inverse operation of square rooting is ___squaring___.

Addition, subtraction, multiplication, and division are called ___operations___.

6. Find the day of the week given each separate condition.

151 days from today if today is Tuesday

151 = 147 + 4   T + 4   ___Saturday___

423 days from today if today is Friday

423 = 420 + 3   F + 3   ___Monday___

2. Answer YES or NO as to whether each solid is a polyhedron.

hexagonal pyramid   ___YES___

sphere   ___NO___

right triangular prism   ___YES___

7. Complete the truth table.

P	Q	NOT P	P AND Q	P OR Q
TRUE	TRUE	F	T	T
TRUE	FALSE	✕	F	T
FALSE	TRUE	T	F	T
FALSE	FALSE	✕	F	F

3. Find the next term of the sequence. Label arithmetic (A), geometric (G), or neither (N). If A, find d. If G, find r.

A   9   5, 14, 23, 32,   ___41___

G   3   2, 6, 18, 54,   ___162___

N   na   5, 15, 35, 65, 105,   ___155___

8. Complete the table.

Original Cost	Sales Tax	Final Cost
$35.00	4%	**$36.40**
$12.00	7%	**$12.84**
$16.00	8%	**$17.28**

4. Find the different ways that nickels and quarters can have a total value of $1.35.   (6)

# N	x 5	# Q	x 25	# N	x 5	# Q	x 25
2	10	5	125	17	85	2	50
7	35	4	100	22	110	1	25
12	60	3	75	27	135	0	0

9. Combine like terms.

$-7x + 9x - 11x$ ___$-9x$___

$10a + 12b - 5a - 14b$ ___$5a - 2b$___

$25\pi + 6\pi - 7\pi$ ___24π___

$-3y + 11x - 7y - 10x$ ___$x - 10y$___

5. Multiply using mental math.

1.3 x 4	___5.2___	4.5 x 5	___22.5___
2.5 x 6	___15___	9.1 x 9	___81.9___
16.5 x 2	___33___	19.6 x 2	___39.2___
1.7 x 3	___5.1___	3.5 x 5	___17.5___

10. Divide.

$$\begin{array}{r} 7{,}328 \\ 54\overline{)395{,}712} \\ 378 \\ \overline{177} \\ 162 \\ \overline{151} \\ 108 \\ \overline{432} \\ 432 \end{array}$$

Level 6	Number 23

1. Translate the cipher with each letter as its own digit.

MAP	MA1	M21	**721**
+ MAP	+ MA1	+ M21	**+ 721**
PEEA	1EEA	1442	**1442**

P=1, the most that can be carried from 2 addends. Then A=2, and E=4. Last, M=7.

6. Find the area in square units of a square with perimeter 36 units.

P = 36

s = 9

A = 81

---

2. Write the fractions in ascending order.

$$\frac{11}{18} \quad \frac{14}{15} \quad \frac{11}{13} \quad \frac{3}{7} \quad \frac{11}{15} \quad \frac{11}{19}$$

$$\frac{3}{7} \quad \frac{11}{19} \quad \frac{11}{18} \quad \frac{11}{15} \quad \frac{11}{13} \quad \frac{14}{15}$$

< 1/2 | the same number of successively larger pieces | 1 small piece out

7. Answer YES or NO as to whether the numbers are valid sides of a triangle.

5, 8, 12	Y	6, 6, 12	N
8, 9, 10	Y	12, 12, 12	Y
11, 19, 31	N	14, 18, 30	Y

---

3. △ABC is congruent to △DEF. Identify the six corresponding parts.

$\angle A \cong \angle D$ $\overline{AB} \cong \overline{DE}$

$\angle B \cong \angle E$ $\overline{BC} \cong \overline{EF}$

$\angle C \cong \angle F$ $\overline{AC} \cong \overline{DF}$

8. List the letters for each path or number the paths by each ray out of A. Find the number of paths from A to C, following the arrows.

6

---

4. Find the percent mentally.

80% of 60	48	95% of 80	76
30% of 90	27	45% of 120	54
40% of 80	32	25% of 160	40
75% of 200	150	35% of 40	14

9. Operate and simplify.

$17\frac{13}{15}$

$+ 18\frac{11}{15}$

$\cancel{35}\,\cancel{\frac{24}{15}}^{\,36\ \ 9}$ → $36\frac{3}{5}$

$26\frac{7}{10}\frac{21}{30}$

$- 17\frac{8}{15}\frac{16}{30}$

$9\frac{5}{30}$ → $9\frac{1}{6}$

---

5. Multiply.

$\frac{4}{9}$ x 99	44	$\frac{7}{15}$ x 45	21
$\frac{11}{15}$ x 60	44	$\frac{8}{11}$ x 88	64
$\frac{8}{13}$ x 26	16	$\frac{9}{10}$ x 80	72

10. Evaluate for x = 3.

$x^2 - 5x + 12$	$9 - 15 + 12$	6
$90 - (x + 2)^3$	$90 - 125$	−35
$x^3 - 7x - 5$	$27 - 21 - 5$	1
$x^4 + 7(x + 8)$	$81 + 77$	158

Level 6	Number 24

1. How many whole numbers are from:

497 to 726 inclusive? 726–497+1 = <u>230</u>

628 to 935 inclusive? 935–628+1 = <u>308</u>

216 to 572 inclusive? 572 –216+1 = <u>357</u>

174 to 601 inclusive? 601–174+1 = <u>428</u>

6. A 24 by 18 rectangle has a 2 by 2 square removed from each of its corners. Find the volume in cubic units of the container formed by folding up the resulting sides.

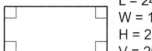

L = 24 – 4 = 20
W = 18 – 4 = 14
H = 2
V = 20 x 14 x 2 = $\boxed{560}$

2. Plot and label the points.

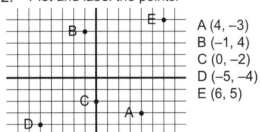

A (4, –3)
B (–1, 4)
C (0, –2)
D (–5, –4)
E (6, 5)

7. Find the average using the rightmost digit method.

987, 989, 988, 980   24 ÷ 4   <u>986</u>

517, 518, 510, 517, 518   30 ÷ 5   <u>516</u>

62, 69, 60, 65, 65, 66, 61   28 ÷ 7   <u>64</u>

3. Find the volume and surface area of a cylinder with radius 5 and height 20 units.

V = 5 • 5 • 20 • π
V = $\boxed{500\pi}$ cu un

SA = 25π + 25π + 10π • 20
SA = 50π + 200π
SA = $\boxed{250\pi}$ sq un

8. Of 50 girls, 26 are wearing coats and 21 are wearing sweaters. If 11 are wearing both, how many are wearing neither?

<u>14</u>

C=26   S=21   N

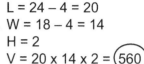

15 (11) 10   14

4. Calculate the point on the number line that is:

$\frac{1}{4}$ of the way from –4 to 4   8 ÷ 4 = 2   <u>–2</u>
    –4 + 2 = –2

$\frac{2}{3}$ of the way from –6 to 9   15 ÷ 3 = 5   <u>4</u>
    –6 + 10 = 4

$\frac{4}{5}$ of the way from –3 to 7   10 ÷ 5 = 2   <u>5</u>
    –3 + 8 = 5

9. Add.

$$\frac{7}{22}^{3} + \frac{5}{33}^{2}$$

$$\frac{21 + 10}{66}$$

$$\boxed{\frac{31}{66}}$$

$$\frac{9}{20}^{6} + \frac{7}{30}^{4} + \frac{11}{40}^{3}$$

$$\frac{54 + 28 + 33}{120}$$

$$\frac{115 ÷ 5}{120 ÷ 5} \boxed{\frac{23}{24}}$$

5. Solve.

7x + 6 > 20	9x – 3 < 17	4x + 2 ≥ 19
7x > 14	9x < 20	4x ≥ 17
$\boxed{x > 2}$	$\boxed{x < \dfrac{20}{9}}$	$\boxed{x ≥ \dfrac{17}{4}}$

10. Operate.

(18 – 3) ÷ 3 – 33 ÷ 3	(6 – 39 ÷ 3 x 2) ÷ 4
5 – 11 $\boxed{–6}$	(6 – 26) ÷ 4
Do 2 groupings concurrently.	–20 ÷ 4 $\boxed{–5}$
42 ÷ 6 x 4 – (19 – 5)	(9 + 7) ÷ 4 – 9 + 7
28 – 14 $\boxed{14}$	4 – 9 + 7 $\boxed{2}$
	Can add 4 and 7 first.

Level 6	Number 25

1. Find the total number of squares in each of the 1st through 5th figures.

1   5   14   30   55

5 = 4 + 1; 14 = 9 + 4 + 1; 30 = 16 + 9 + 4 + 1;
55 = 30 + 25 (sum of consecutive perfect squares)

6. Find the number of 3 by 3 by 3 cubes that can completely fill a box measuring 21 by 30 by 33 of the same unit.

7 cubes by 10 cubes by 11 cubes = (770) cubes

OR $\dfrac{\text{Volume box}}{\text{Volume 1 cube}} = \dfrac{21 \times 30 \times 33}{3 \times 3 \times 3}$

2. State whether the fraction would repeat (R) or terminate (T) if converted to a decimal.

$\dfrac{9}{10}$ T | $\dfrac{8}{35}$ R | $\dfrac{27}{45}$ T

$\dfrac{5}{8}$ T | $\dfrac{21}{50}$ T | $\dfrac{19}{40}$ T

7. Operate.

20 − 60   −40   −15 + −75   −90

−45 + −35   −80   −55 + 55   0

−30 − 40   −70   55 − −15   70

−25 − −35   10   −25 + −35   −60

3. Add the square matrices.

$\begin{bmatrix} 7 & 6 \\ 0 & 4 \end{bmatrix} + \begin{bmatrix} 2 & 1 \\ 9 & 4 \end{bmatrix} = \begin{bmatrix} 9 & 7 \\ 9 & 8 \end{bmatrix}$

$\begin{bmatrix} 9 & 7 \\ -2 & -4 \end{bmatrix} + \begin{bmatrix} 3 & 6 \\ -6 & -3 \end{bmatrix} = \begin{bmatrix} 12 & 13 \\ -8 & -7 \end{bmatrix}$

8. Remove 5 even numbers from the first 25 whole numbers. What percent of the remaining numbers are even?

25 whole numbers : 0 ⟶ 24
13 even, 12 odd
remove 5 even: 20 numbers remain
8 even, 12 odd

$\dfrac{8}{20}$ = (40%)

To make the fractional part, put "is" (or "are") number over "of" number.

4. Rewrite using bar notation.

7.0345714571 . . . $7.03\overline{4571}$

3.555555555 . . . $3.\overline{5}$

4.689689689 . . . $4.\overline{689}$

6.25252525 . . . $6.\overline{25}$

9. Of 5 test scores from 0 to 100, the mode is 95, the median is 90, the mean is 76, and the outlier is 20. Find the scores.

20   80   90   95   95

Place 90 1st, followed by the 95s and the 20. Then, 76 × 5 = 380. 20 + 90 + 95 + 95 = 300. 380 − 300 = 80.

5. Hal is 15 years older than Hank. Henry is 8 years younger than Hank who is 2 years older than Hale. If Hale is 13 years old, find Hal's age. (30)

Hal 30
15
Hank 15
2
Hale 13
8
6
Henr 7

10. Simplify.

$\dfrac{12!}{9!}$ 1320   10 × 11 × 12 = 132 × 10

$\dfrac{8!}{4!}$ 1680   5 × 6 × 7 × 8 = 42 × 40

Level 6	Number 26

1.

$$450$$
$$35$$
$$400$$
$$+\ 1000$$
$$\overline{\hphantom{0}1885}$$

The rounding up was insignificant compared to $150 (1885–1735).

Lynn has $1735.79 in her account. Her bills due are: rent $445.50, phone $32.86, and charge card $389.90. Does she have enough money to pay her bills and leave a required minimum balance of $1000? (NO)

6. Write the value as an expression.

the number of cents in q quarters   25q

the number of cups in g gallons   16g

the number of seconds in m minutes   60m

the number of inches in y yards   36y

2. Rewrite the decimals in ascending order.

0.87, 7.08, 8.07, .078, .0087, 7.07

.0087 0.078 0.87 7.07 7.08 8.07

7. In four years Mary will be 3 times older than Esther was 4 years ago. If Esther is 20 years old now, how old is Mary?

	−4	now	+4
M		(44)	48
E	16	20	24

3. Given two sides of a triangle, find the range of values for the 3rd side (use s for unknown side).

5, 35 $30 < s < 40$

10, 27 $17 < s < 37$

13, 30 $17 < s < 43$

Subtract given sides for low bound. Add for high bound. "The sum of two sides must be greater than the 3rd" to form a triangle.

8. Complete the table.

Principal	Rate	Interest
$40.00	3.5%	**$1.40**
$60.00	4.5%	**$2.70**
$80.00	5.5%	**$4.40**

4. Find the greatest value of x − y.

$-1 \le x \le 8$ $3 \le y \le 13$ 5

$-7 \le x \le 7$ $-4 \le y \le 6$ 11

$2 \le x \le 9$ $5 \le y \le 14$ 4

$-9 \le x \le -2$ $0 \le y \le 16$ −2

9. Label △ABC Isosceles, Equilateral, or Scalene AND Right, Acute, Obtuse, or Equiangular for 3 separate problems. CBD is also a possible answer.

m∠C = 90   CBD   R

AC = 9   AB = 9   I   CBD

m∠A = 50   m∠B = 50   I   A

5. Find the probability of selecting a point from the shaded triangle within the rectangle. Draw dotted line to see half without doing any work.

12

20

$$\frac{A_{tri}}{A_{rec}} = \frac{20 \cdot 12 \cdot .5}{20 \cdot 12}$$

$$\left(\frac{1}{2}\right)$$

10. Solve.

$|x| = -4$ x = No Sol

$|4| = -x$ x = −4

$|x| = 4$ x = ± 4

$|-4| = x$ x = 4

Level 6	Number 27

1. Find the cost for the number specified.

4 if 7 cost $231   **$132**   231/7 = 33

12 if 8 cost $456   **$684**   456 + 228

7 if 9 cost $531   **$413**   531/9 = 59

8 if 3 cost $324   **$864**   324/3 = 108

6. Solve by clearing the decimal points.

.9x = 9.9	.025x = .1	.07x = .63
9x = 99	25x = 100	7x = 63
x = (11)	x = (4)	x = (9)

2. Graph on the number line.

$-4 \le x \le 2$

$x > -2$

7. Operate and simplify the continued fraction.

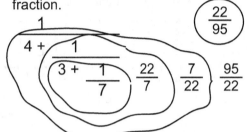

$4 + \cfrac{1}{3 + \cfrac{1}{7}}$ $\dfrac{22}{7}$ $\dfrac{7}{22}$ $\dfrac{95}{22}$ $\boxed{\dfrac{22}{95}}$

3. Find the maximum area in square units of a rectangle with whole number sides and perimeter 36 units.

17 by 1
A = 17

Consider the 2 extremes: as thin as possible versus square.

9 by 9
A = (81)

8. Find the probability when tossing a die.

P(6) $\dfrac{1}{6}$ 6

P(factor of 6) $\dfrac{2}{3}$ 1, 2, 3, 6

P(factor of 10) $\dfrac{1}{2}$ 1, 2, 5

P(less than 3) $\dfrac{1}{3}$ 1, 2

4. Name the solids depicted by the nets.

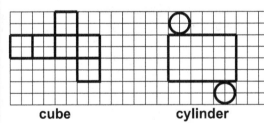

cube   **cylinder**

9. Label each polygon convex or concave.

convex   convex   concave

5. Complete the chart of data for a circle graph.

	Number	Fraction	Percent	Degrees
red	150	3/6	50	180
blue	100	2/6	33.$\overline{3}$	120
green	50	1/6	16.$\overline{6}$	60
TOTAL	300	1	100	360

10. Solve.

9x + 6 = −14	6x − 4 = 19	5x − 7 = 14
9x = −20	6x = 23	5x = 21
x = $\left(\dfrac{-20}{9}\right)$	x = $\left(\dfrac{23}{6}\right)$	x = $\left(\dfrac{21}{5}\right)$

Level 6	Number 28

1. Lara bought a cheese wheel for $39.16. The cheese weighed 5.5 pounds. Find the price per pound.

$$55 \overline{)391.600}$$
$$\quad\; 385$$
$$\quad\;\; 66$$
$$\quad\;\; 55$$
$$\quad\; 110$$
$$\quad\; 110$$
$$\qquad 0$$

$$\frac{\text{price}}{\text{pound}} \quad \frac{39.16}{5.5} \quad \boxed{\$7.12}$$

6. Always draw the discards on the left.

Sixty apples were on a tree. One third fell off, and one fifth of those left were picked. What fraction of the original apples remained?

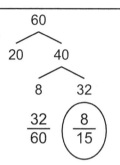

$$\frac{32}{60} \qquad \boxed{\frac{8}{15}}$$

2. Draw the reflection of each triangle over the y-axis. *Reflect each vertex first.*

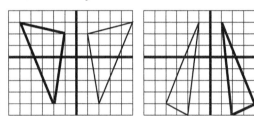

7. Write in standard form: ten billion, fifty-one million, one hundred seventy-nine thousand, five hundred one and fifty-three hundredths.

The decimal point is the "and."

10,051,179,501.53

3. Find the missing terms of the arithmetic sequences.

6, 18, 30, __42__, 54, 66, __78__

7, 15, 23, 31, __39__, 47, 55, __63__

8, 19, 30, __41__, 52, __63__, 74

8. Operate and simplify.

$$\frac{\dfrac{4}{5} - \dfrac{2}{3}}{\dfrac{4}{9} + \dfrac{2}{5}} \quad \frac{\dfrac{12-10}{15}}{\dfrac{20+18}{45}} \quad \frac{\dfrac{2}{15}}{\dfrac{38}{45}} \quad \frac{\overset{1}{\cancel{2}}}{\underset{1}{\cancel{15}}} \cdot \frac{\overset{3}{\cancel{45}}}{\underset{19}{\cancel{38}}}$$

$$\boxed{\frac{3}{19}}$$

4. Find the least natural number n such that kn is a perfect square.

k = 20 n = __5__ 20 = 2 x 2 x 5

k = 40 n = __10__ 40 = 2^3 x 5

k = 55 n = __55__ 55 = 5 x 11

An even number of each prime creates a perfect square.

9. A square with area 81 square units and an equilateral triangle have equal perimeters. Find the side in units of the triangle.

A □ = 81
S □ = 9
P □ = 36
P △ = 36
S △ = $\boxed{12}$

5. Operate.

$$6\frac{2}{3} \times 9\frac{3}{4} \qquad 6\frac{2}{5} \times 8\frac{1}{8}$$

$$\frac{\overset{5}{\cancel{20}}}{\underset{1}{\cancel{3}}} \times \frac{\overset{13}{\cancel{39}}}{\underset{1}{\cancel{4}}} \qquad \frac{\overset{4}{\cancel{32}}}{\underset{1}{\cancel{5}}} \times \frac{\overset{13}{\cancel{65}}}{\underset{1}{\cancel{8}}}$$

$$\boxed{65} \qquad \boxed{52}$$

10. Evaluate.

$\sqrt{64}$	8	$\sqrt{49}$	7
$\sqrt{16}$	4	$\sqrt{4}$	2
$\sqrt{900}$	30	$\sqrt{144}$	12
$\sqrt{1}$	1	$\sqrt{169}$	13

Level 6	Number 29

1. Find the 6-digit mystery number.
Clue 1: The number is divisible by 8.
Clue 2: The digits are in descending order.
Clue 3: The 6 digits are distinct.
Clue 4: The digit sum is 34.
Clue 5: The number contains 0 and 9.

9abcd0; a + b + c + d = 25;
7654 too small   (987,640)
cd0 div by 8, so cd as number div by 8

6. Compute using mental math.

Times 10; divide by 2 (cut in half).

284 x 5	1420	341 x 5	1705
583 x 5	2915	626 x 5	3130
745 x 5	3725	429 x 5	2145
942 x 5	4710	868 x 5	4340

2. Find the prime factorization. Work down. In the answer use exponents with primes in ascending order.

112	160	700
4 x 28	16 x 10	7 x 10 x 10
4 x 4 x 7	2^4 x 2 x 5	2^2 x 5^2 x 7
2^4 x 7	2^5 x 5	

7. Complete the table of values for $y = -3x - 1$.

x	–9	–8	–7	–5	–4	–3	–1	0	1	2
y	26	23	20	14	11	8	2	–1	–4	–7

3. Solve by equivalent fractions.

$\frac{7}{9} = \frac{63}{x}$ (x = 81) $\frac{x}{11} = \frac{32}{88}$ (x = 4)

$\frac{3}{8} = \frac{x}{56}$ (x = 21) $\frac{5}{x} = \frac{25}{35}$ (x = 7)

8. Convert to scientific notation.

3 billion	3.0×10^9	596	5.96×10^2
32,000	3.2×10^4	8150	8.15×10^3
6 million	6.0×10^6	700	7.0×10^2
210,000	2.1×10^5	9000	9.0×10^3

4. Find the area of the triangle. One box equals one square unit.

22
2 22
176 – 104 = (72)   8
64
6
18
6   16

9. Find the greatest common factor.

GCF(121, 125) 1   All 11s vs all 5s.

GCF(36, 54, 90) 18   9 in all, but also all even.

GCF(130, 150) 10   13 is prime.

GCF(15, 55, 105) 5   3 in 15 and 105 but not 55.

5. Alena bought a wallet for $19.90 and a cell phone pouch for $16.75. Find the total cost with 7% sales tax.

```
  19.90        36.65
+ 16.75      +  2.57
  36.65       ($39.22)
x   .07
  2.5655
```
Or multiply by 1.07 and omit addition.

10. Identify membership in each set.

	N	W	Z	Q	R
–7	–	–	✓	✓	✓
3/7	–	–	–	✓	✓
|–2|	✓	✓	✓	✓	✓
$0.\overline{3}$	–	–	–	✓	✓

| Level 6 | Number 30 |

---

1. Subtract.

$$801{,}567{,}955 \quad\quad 725{,}700{,}431$$
$$-\,243{,}879{,}676 \quad\quad -\,146{,}865{,}744$$
$$\mathbf{557{,}688{,}279} \quad\quad \mathbf{578{,}834{,}687}$$

6. Find the surface area of a cube with volume 1 cubic unit.

V = 1
e = 1
A face = 1
SA = 1 x 6 = ⑥ sq un

Find the volume of a cube with surface area 486 square units.

SA = 486
A face = 81
e = 9
V = 9^3 = ⑦㉙ cu un

---

2. What time is 560 minutes after 9:30 PM?
560 = 60 x 9 + 20   9 + 9 ➤ 6   30 + 20 ➤ 50
(6:50 AM)

What time is 290 minutes after 6:15 AM?
290 = 60 x 5 − 10   6 + 5 ➤ 11   15 − 10 ➤ 5
(11:05 AM)

7. Find the area of the circle inscribed in a square of area 100 square units.

A_{sq} = 100
S = 10
D = 10
R = 5
A_{circ} = (25π)

---

3. Find the simple interest on $36,000 at 3% annually for 11 months.

I = PRT

$$I = \frac{\overset{30}{\cancel{36{,}000}}}{1} \cdot \frac{3}{\cancel{100}} \cdot \frac{11}{\cancel{12}}$$

Use mental math:
3 x 3 = 9
9 x 11 = 99
99 x 10 = 990

I = ($990)

8. An equilateral triangle with side 7 is attached to a square of area 49 at each of the 4 sides. Find the perimeter in units of the figure formed.

8 x 7 = (56)

---

4. Write each fraction as a repeating decimal.

| $\frac{4}{11}$ $\overline{.36}$ | $\frac{17}{99}$ $\overline{.17}$ | $\frac{2}{33}$ $\overline{.06}$ |
| $\frac{4}{9}$ $\overline{.4}$ | $\frac{4}{99}$ $\overline{.04}$ | $\frac{35}{99}$ $\overline{.35}$ |

9. Find the least common multiple.

LCM(90, 300) 900 30 x 3, 30 x 10, so need 30 x 3 x 10.

LCM(35, 105) 105 105 is a multiple of 35.

LCM(33, 75) 825 3 x 11, 3 x 25, so need 3 x 11 x 25 = 3 x 275.

LCM(30, 40, 50) 600 Factor of 10 in all. Need 3 x 4 x 5 x 10.

---

5. Reflect the point A(4, 3) over the x-axis. Then translate it six units left. What is the final point?

(−2, −3)

10. In a set of five 2-digit numbers, the median is 60, the range is 74, the mode is 22, and the mean is 58. Find the numbers.

22   22   60   90   96

Place 60 1st, followed by the 22s. Next, 22 + 74 = 96. Then, 58 x 5 = 290. 22 + 22 + 60 + 96 = 200. 290 − 200 = 90.

Level 6	Number 31

1. Graph on the coordinate plane. One box equals one unit.

y = x   y = 2x

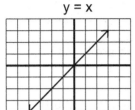 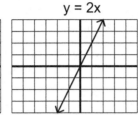

6. Given a bowl of 20 red, 24 blue, and 16 green marbles. Draw one. Find each probability.

P(R) $\frac{1}{3}$ $\frac{20 \div 20}{60 \div 20}$

P(B) $\frac{2}{5}$ $\frac{24 \div 12}{60 \div 12}$

P(G) $\frac{4}{15}$ $\frac{16 \div 4}{60 \div 4}$

2. Find the GCF and LCM of 105 and 147 using prime factorization.

$105 = 5 \times 21 = 3 \times 5 \times 7$
$147 = 3 \times 49 = 3 \times 7^2$

Look for one of the original numbers before multiplying.

GCF = 3×7 = (21)

LCM = $3 \times 5 \times 7^2 = 105 \times 7$ = (735)

7. Name the number of degrees, if any, of rotational symmetry for the figure.

none   90°   none

3. Find the missing sides of the similar triangles. (NTS)

45 55   9 (11)

(20)   4

÷5

8. Check if the row number is divisible by the column factor.

	2	3	4	6	9	10
96,150	✓	✓	–	✓	–	✓
53,936	✓	–	✓	–	–	–
81,972	✓	✓	✓	✓	✓	–

4. Operate.

$\frac{1}{4}$ (13 days   6 hours   8 minutes)

$\frac{1}{4}$ (12 d   30 h   8 m)

$\frac{1}{4}$ (12 d   28 h   128 m)

(3 d   7 h   32 m)

9. Find the perimeter in units of a square with area 16 square units.

A = 16

s = 4

(P = 16)

5. Operate on the sets.

A = {5, 10, 15, 20}
B = {1, 3, 5, 7, 9}
C = {10, 11, 12, 13, 14, 15}

A ∩ C =   {10, 15}

A ∪ C =   {5, 10, 11, 12, 13, 14, 15, 20}

B ∩ C =   { }

10. Complete the table.

Decimal	Percent	Fraction
3.2	320%	$\frac{16}{5}$
2.75	275%	$\frac{11}{4}$
$1.\overline{3}$	$100.\overline{3}\%$	$\frac{4}{3}$

Level 6	Number 32

1. A number times its reciprocal is

or digital answers ___one.___

The least probability is zero. The greatest probability is ___one.___

A percent represents parts of ___one hundred.___

6. Two whole numbers are in the ratio 2:9. Their sum is 55. Find their product.

2:9 sum = 11
10:45 sum = 55   10 x 45 = (450)

Two whole numbers are in the ratio 4:5. Their sum is 90. Find their product.

4:5 sum = 9
40:50 sum = 90   40 x 50 = (2000)

2. A buyer of a certain model car may select one of eleven colors of paint, either fabric or leather interior, a sunroof or not, and a detailing stripe or not. How many different ways is this car available?

11 x 2 x 2 X 2 = (88)

7. Six movers can do a job in 15 hours. At the same rate, how many movers are needed to do the same job in 10 hours?

w x T = W x t
6 x 15 = 9 x 10

(9)

Workers and time are inversely proportional-- as one goes up, the other goes down. Their product is constant.

3. Divide.

$6354.15 \div 10^4$ ___0.635415___

$1904.17 \div 10^3$ ___1.90417___

$29.857 \div 10$ ___2.9857___

$53467.5 \div 10^2$ ___534.675___

8. Complete the chart for a circle.

radius	diameter	area	circumference
1	2	π	2π
8	16	64π	16π
2	4	4π	4π

4. Distribute.

$-7(x - 2y)$ ___$-7x + 14y$___

$-2(-3c - d)$ ___$6c + 2d$___

$5(4a - 3b)$ ___$20a - 15b$___

$4(-2x + 7y)$ ___$-8x + 28y$___

9. Fiind the probability that a positive integer less than or equal to 100 is a factor of 100.
1, 2, 4, 5, 10, 20, 25, 50, 100 $\left(\frac{9}{100}\right)$

Fiind the probability that a positive integer less than or equal to 40 is a factor of 40.
1, 2, 4, 5, 8, 10, 20, 40 $\frac{8}{40} = \left(\frac{1}{5}\right)$

5. Find the sum of each arithmetic sequence using the "First plus Last" method.

11 + 12 + 13 + 14 + 15 + 16 + 17 + 18 + 19

3 + 6 + 9 + 12 + 15 + 18 + 21 + 24 + 27 + 30

$\frac{(11 + 19)(9)}{2}$ = (135) $\frac{(3 + 30)(10)}{2}$ = (165)

10. Calculate the area in square units of the trapezoid. (NTS)

M = (15 + 9) ÷ 2 = 12

A = 12 x 8 = (96)

9
8
15

Or, 15 − 9 = 6; 6 ÷ 2 = 3;
9 + 3 = 12

Level 6	Number 33

1. Find the area in square units of a square with perimeter 24 units.

P = 24

s = 6

(A = 36)

2. Identify each as a point, line, ray, or line segment if graphed on a number line. Determine without graphing.

x + 5 = 5 + x   line

−11 ≤ x ≤ −1   line segment

2x ≤ 12   ray

3. Find the fractional part.
15 inches out of 10 yards

$$\frac{\overset{1}{\cancel{15}}}{2\,\cancel{10} \times \cancel{36}\,_{12}}$$ ($\frac{1}{24}$)

6 cups out of 4 gallons

$$\frac{\overset{3}{\cancel{6}}}{4 \times \cancel{16}\,_{8}}$$ ($\frac{3}{32}$)

4. Find the perimeter of the figure on the unit grid.   (66)

5. Find each central angle as a fractional part of 360°.

20°	$\frac{1}{18}$	150°	$\frac{5}{12}$	240°	$\frac{2}{3}$
135°	$\frac{3}{8}$	200°	$\frac{5}{9}$	144°	$\frac{2}{5}$

6. Convert to standard form.

2.625×10^6 2,625,000

4.124356×10^5 412,435.6

3.785×10^4 37,850

1.45×10^5 145,000

7. Operate.

−2 x 9	−18	−3 x −8	24
−9 x −4	36	−8 ÷ 4	−2
−6 ÷ 6	−1	−1 x −5	5
−5 x 7	−35	0 ÷ −8	0

8. Find the average using the arithmetic sequence method. The mean of an arithmetic sequence is the median.

240, 252, 264, 276   258

10, 26, 42, 58, 74, 90   50

With an even number of numbers, the median is the mean of the 2 middlemost.

25, 35, 45, 55, 65, 75   50

9. Write as an algebraic expression.

triple the sum of two numbers   3(x + y)

triple the product of two numbers   3xy

triple the difference of two numbers   3(x − y)

triple a number plus one   3x + 1

10. Add.

690,514,582	885,043,476
463,695,743	426,886,512
554,998,264	834,516,883
+ 337,843,229	793,964,258
	+ 675,976,389
2,047,051,818	**3,616,387,518**

Level 6	Number 34

1. Dee scored 91, 77, and 87 out of 100 on 3 of 5 math tests. What is the minimum score she must earn on the 4th test to have an average of 90 on the 5 tests?

 95

90 x 5 = 450
255   95   100
$$\frac{91 + 77 + 87 + min + max}{5} = 90$$

6. Divide the product of the first five prime numbers by the product of the first four 2-digit composite numbers.

$$\frac{2 \times 3 \times 5 \times 7 \times 11}{10 \times 12 \times 14 \times 15} = \frac{11}{120}$$

2. Name the points shown.

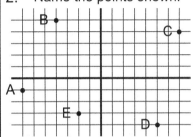

A (–7, –1)

B (–4, 5)

C (7, 4)

D (5, –4)

E (–2, –3)

7. The ratio of tulips to pansies in a garden of only those flowers is 3:8.

Find the ratio of tulips to total flowers.	If the total number of flowers is 121, find the number of pansies.
$\frac{T}{F}$ $\frac{3}{11}$	8 x 11 = 88

3. Multiply the square matrix by the constant (scalar multiplication).

$$5 \begin{bmatrix} 9 & 0 \\ 1 & 6 \end{bmatrix} = \begin{bmatrix} 45 & 0 \\ 5 & 30 \end{bmatrix}$$

$$4 \begin{bmatrix} -1 & 4 \\ 7 & -2 \end{bmatrix} = \begin{bmatrix} -4 & 16 \\ 28 & -8 \end{bmatrix}$$

8. Find the supplement of the angle.

Supp (100°)	80°	Supp (150°)	30°
Supp (20°)	160°	Supp (10°)	170°
Supp (140°)	40°	Supp (70°)	110°
Supp (50°)	130°	Supp (120°)	60°

4. If the sum of two consecutive even numbers is 38, find their product.

18 + 20 = 38   18 x 20 = 360

If the sum of two consecutive odd numbers is 44, find their product.

21 + 23 = 44   21 x 23 = 460 + 23 = 483

9. Complete the unit conversions.

2 cm	20 mm	6 L	6000 mL
4.7 m	470 cm	54 mg	.054 g
575 mL	.575 L	5 cm	.05 m

5. Answer YES or NO as to whether each solid is a polyhedron.

rectangular prism	YES
rectangular pyramid	YES
cylinder	NO

10. Divide. Answer in exponential form.

(24 x 10⁴) ÷ (4 x 10²)	6×10^2
(32 x 10⁹) ÷ (8 x 10²)	4×10^7
(36 x 10⁸) ÷ (6 x 10³)	6×10^5
(35 x 10⁷) ÷ (5 x 10³)	7×10^4

Picture the 2 groupings as parts of a fraction. Simplify.

Level 6	Number 35

1. Multiply in the given bases.

$4_{\text{five}} \times 4_{\text{five}} = \underline{31}_{\text{five}}$

$3_{\text{six}} \times 5_{\text{six}} = \underline{23}_{\text{six}}$

$4_{\text{nine}} \times 8_{\text{nine}} = \underline{35}_{\text{nine}}$

$3_{\text{four}} \times 2_{\text{four}} = \underline{12}_{\text{four}}$

6. Find the digit d such that the 5-digit number 35,7d2 is divisible by 9.

$7 + 2 = 9 \qquad 3 + 5 + 1 = 9$ ⓵

Find the digit x such that the 5-digit number 45,x18 is divisible by 11. ⑤

$12 + x - 6 = 11 \qquad x = 5$

2. Find the number of whole numbers between:

$\sqrt{27}$ and $\sqrt{83}$ $\underline{4}$ 6, 7, 8, 9

$\sqrt{6}$ and $\sqrt{39}$ $\underline{4}$ 3, 4, 5, 6

$\sqrt{35}$ and $\sqrt{125}$ $\underline{6}$ 6, 7, 8, 9, 10, 11

7. Name the property exemplified.

$a \cdot 0 = 0$ $\underline{\text{ZPM}}$

$-2(e - f) = -2e + 2f$ $\underline{\text{DPMA}}$

$-13m$ is a real number. $\underline{\text{ClPM}}$

$-54 \times \dfrac{-1}{54} = 1$ $\underline{\text{InPM}}$

3. Identify the six corresponding parts of the congruent triangles.

A B C E D F

$\underline{\angle A} \cong \underline{\angle F}$ $\qquad \overline{AB} \cong \overline{DF}$

$\underline{\angle B} \cong \underline{\angle D}$ $\qquad \overline{BC} \cong \overline{DE}$

$\underline{\angle C} \cong \underline{\angle E}$ $\qquad \overline{AC} \cong \overline{EF}$

8. Find the volume and total surface area of the solid comprised of unit cubes.

$V = \boxed{10}$ cu un

$SA = \boxed{38}$ sq un

F = 10
B = 10
U = 6
D = 6
L = 3
R = 3

4. Graph on the number line.

$0 \le x < 4$

0 4

$x < -3$

−3 0

9. Add.

$\overset{6}{\dfrac{11}{25}} + \overset{5}{\dfrac{13}{30}}$

$\dfrac{66 + 65}{150}$

$\boxed{\dfrac{131}{150}}$

$\overset{4}{\dfrac{7}{12}} + \overset{2}{\dfrac{11}{24}} + \dfrac{7}{48}$

$\dfrac{28 + 22 + 7}{48}$

$\dfrac{57 \div 3}{48 \div 3} \boxed{\dfrac{19}{16}}$

5. Convert measures to a decimal.

10 minutes $= \underline{0.1\overline{6}}$ hour

1 quart $= \underline{0.25}$ gallon

4 inches $= \underline{0.\overline{1}}$ yard

6 pints $= \underline{0.75}$ gallon

10. Find the ratio of:

a side of an equilateral triangle to its perimeter. $\boxed{\dfrac{1}{3}}$

a side of a regular pentagon to its perimeter. $\boxed{\dfrac{1}{5}}$

4 sides of a regular nonagon to its perimeter. $\boxed{\dfrac{4}{9}}$

Level 6	Number 36

1. Find the day of the week given each separate condition.

637 days from today if today is Saturday

637 is divisible by 7   _Saturday_

502 days from today if today is Thursday

502 = 497 + 5   Th + 5   _Tuesday_

---

6. Complete the divisibility rules.

If __2__ and __5__ divide a number, then 10 divides the number.

If __3__ and __7__ divide a number, then 21 divides the number.

If __4__ and __5__ divide a number, then 20 divides the number.   Not 2 and 10 (overlapping factors).

---

2. Find the area of the parallelograms. One box equals one square unit.

$9 \times 7 = \boxed{63}$

$8 \times 9 = \boxed{72}$

---

7. Convert to base ten.

10101_{two} 16 + 4 + 1   _21_

253_{six} $2 \times 36 + 5 \times 6 + 3$ _105_

1202_{three} $1 \times 27 + 2 \times 9 + 2$ _47_

165_{seven} $1 \times 49 + 6 \times 7 + 5$ _96_

---

3. Multiply by 11 mentally.
Add digits, insert ones in middle, increment leftmost.

46 x 11	506	68 x 11	748
85 x 11	935	79 x 11	869
74 x 11	814	87 x 11	957
66 x 11	726	99 x 11	1089

---

8. By what number must 7! be multiplied to become 11! ?

$8 \times 9 \times 10 \times 11 = 792 \times 10 = \boxed{7920}$

By what number must 12! be divided to become 9! ?

$10 \times 11 \times 12 = 132 \times 10 = \boxed{1320}$

---

4. Define operation ✦ as:

$W ✦ Z = 6WZ + W^4 - Z$. For example, $2 ✦ 5 = 60 + 16 - 5 = 71$.
Find the values.

$3 ✦ 0$ _81_   0 + 81 − 0

$10 ✦ 1$ _10,059_   60 + 10,000 − 1

---

9. Find the least natural number n such that kn is a perfect square.

k = 18   n = __2__ $18 = 2 \times 3 \times 3$

k = 27   n = __3__ $27 = 3 \times 3 \times 3$

k = 75   n = __3__ $75 = 3 \times 5 \times 5$

An even number of each prime creates a perfect square.

---

5. Operate and simplify.

$29 \dfrac{9}{10} \dfrac{54}{60}$

$+ 33 \dfrac{7}{12} \dfrac{35}{60}$

$62 \dfrac{89}{60} \boxed{63 \dfrac{29}{60}}$

$\overset{34}{\cancel{35}} \overset{33}{} \dfrac{3}{8} \dfrac{9}{24}$

$- 19 \dfrac{5}{6} \dfrac{20}{24}$

$\boxed{15 \dfrac{13}{24}}$

---

10. Write the converse of the conditional statement. Are they logically equivalent? ⓃⓄ

If a number is divisible by 10, then the number is not a prime.   statement TRUE

If a number is not a prime, then the number is divisible by 10.   converse FALSE   25 counterexample

Level 6	Number 37

1. Eight people can pack 72 boxes in 3 hours. For the identical boxes at the same rate, 4 people can pack how many boxes in 5 hours?

people	boxes	hours
8	72	3
8	24	1
8	120	5
4	(60)	5

Working in rows, keep 1, change 2 by x or ÷ (direct) or both (inverse).

6.

	R	T	D
to	20	6	120
fr	30	4	120
tot	(24)	10	240

Mia traveled 120 miles at 20 miles per hour. She returned along the same road at 30 miles per hour. What was her overall rate in miles per hour for the round trip?

2. Convert to a decimal from memory.

$\frac{2}{5}$ .4 $\frac{9}{10}$ .9 $\frac{1}{8}$ .125

$\frac{7}{8}$ .875 $\frac{1}{6}$ .1$\overline{6}$ $\frac{4}{5}$ .8

$\frac{5}{9}$.$\overline{5}$ $\frac{1}{3}$.$\overline{3}$ $\frac{5}{6}$ .8$\overline{3}$

7. Operate.

−80 − 15 _−95_   15 − −80 _95_

−80 + −15 _−95_   −15 + 80 _65_

−80 − −15 _−65_   80 − 15 _65_

15 − 80 _−65_   80 − −15 _95_

3. Find the maximum perimeter in units of a rectangle with whole number sides and area 900 square units.

[rectangle] 900 by 1   P = (1802)

[square]   Consider the 2 extremes: as thin as possible versus square.   30 by 30   P = 120

8. Find the volume in cubic units of a rectangular prism with the areas of its noncongruent faces 15, 21, and 35 square units.

3 x 5 = 15   V = 3 x 5 x 7 = (105)
3 x 7 = 21
5 x 7 = 35

4. Find the percent mentally.

15% of 300 _45_   60% of 45 _27_

85% of 200 _170_   30% of 70 _21_

20% of 75 _15_   55% of 20 _11_

45% of 160 _72_   40% of 95 _38_

9. Compute using mental math.

Multiply by 4 mentally by doubling twice.

39 x 4   39 + 39 = 78   78 + 78   _156_

27 x 4   27 + 27 = 54   54 + 54   _108_

18 x 4   18 + 18 = 36   36 + 36   _72_

44 x 4   44 + 44 = 88   88 + 88   _176_

5. Find the angle measures. (NTS)

If m∠ 8 = 54°, then m∠ 11 = _54°_

If m∠ 8 = 38°, then m∠ 4 = _38°_

10. Multiply.

```
      59,324
   x     685
     296620
     474592
     355944
    40636940
```

```
      46,715
   x   8,723
      140145
       93430
      327005
      373720
    407494945
```

Level 6	Number 38
1. A bus traveled 792 miles in 11 hours. What was its rate in miles per hour? D = RT $\frac{792}{11} = \frac{R \times 11}{11}$ (72) = R	6. A 18 by 26 rectangle has a 3 by 3 square removed from each of its corners. Find the volume in cubic units of the container formed by folding up the resulting sides. L = 26 – 6 = 20 W = 18 – 6 = 12 H = 3 V = 20 x 12 x 3 = (720)
2. Identify whether the figure is an Euler Graph--can be drawn without lifting the pencil and retracing a line. **NO** 2 ⎴ 2 3 3 ⎴ 3 3 ⎴ 3 3 ⎴ 3 2 **YES** 2 An Euler Graph must have exactly 0 or 2 vertices with an odd number of impinging segments (one to start, one to stop).	7. Identify (YES/NO) whether the set of ordered pairs is a function. If NO, specify the x-value that violates the function rule. { (0,7), (5,4), (6,4), (5,3), (1,9) } <u>N</u> <u>5</u> { (1,2), (9,2), (0,2), (4,5), (0,1) } <u>N</u> <u>0</u>
3. Find the missing terms of the arithmetic sequences. 9, 16, 23, <u>30,</u> 37, 44, <u>51</u> 2, 15, 28, 41, <u>54,</u> 67, 80, <u>93</u> 6, 12, 18, <u>24,</u> 30, <u>36,</u> 42	8. Of 70 students, 55 study Spanish and 28 study French. If 19 study both, how many study neither? <u>6</u> S=55 F=28 N 36 (19) 9 (6)
4. Answer true (T) or false (F). The shortest chord is a radius. <u>F</u> A trapezoid cannot have exactly one right angle. <u>T</u> Regular polygons are similar. <u>F</u> The longest chord is a diameter. <u>T</u>	9. Combine like terms. –3y + 6x – 6y <u>6x – 9y</u> –15π + 6π + 9π <u>0</u> 11w – 15x + 7x – 3w <u>8w – 8x</u> 5a – 7b + 4a – 2b <u>9a – 9b</u>

5. Find the probability of selecting a point from the shaded area within the rectangle.
 	
$\frac{A_{sh}}{A_{rec}} = \frac{800}{6300}$
 	
70 ⎿ 20 ⎿ 40 ⎿ 90
 	
($\frac{8}{63}$)
	10. Divide. **7,985**
	45) 359,325
	     315
	     443
	     405
	     382
	     360
	     225
	     225

Level 6	Number 39

1. Translate the cipher with each letter as its own digit.

TAT	TAT	5A5	**565**
+ TAB	+ TA1	+ 5A1	**+ 561**
BBUA	11UA	11UA	**1126**

B=1, the most that can be carried from 2 addends. In 100s, T+T=2T cannot be odd unless 1 carried, so T=5. In 1s, A=6. Then U=2.

6. Does the mapping represent a function?

NO

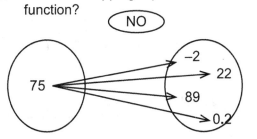

2. Find the prime factorization. Work down. In the answer use exponents with primes in ascending order.

220	165	187
22 x 10	15 x 11	**11 x 17**
2 x 11 x 2 x 5	**3 x 5 x 11**	
2² x 5 x 11		

7. Answer YES or NO as to whether the numbers are valid sides of a triangle.

5, 9, 13	Y	6, 15, 25	N
12, 24, 36	N	20, 30, 40	Y
15, 20, 40	N	23, 23, 23	Y

3. Complete with the vocabulary word (not congruent).

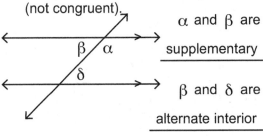

α and β are ___supplementary___

β and δ are ___alternate interior___

8. Find the probability if tossing 2 dice.

P(sum of 2)	$\frac{1}{36}$	1&1
P(sum of 3)	$\frac{1}{18}$	1&2 or 2&1
P(sum of 2 OR 3)	$\frac{1}{12}$	the above 3 options
P(sum of 4)	$\frac{1}{12}$	1&3 or 3&1 or 2&2

4. Name the solids depicted by the nets.

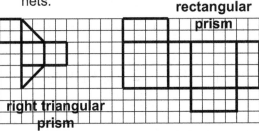

rectangular prism

right triangular prism

9. Complete.

97 ≡ _7_ (mod 9)	45 ≡ _1_ (mod 4)
19 ≡ _4_ (mod 5)	44 ≡ _2_ (mod 7)
40 ≡ _4_ (mod 6)	38 ≡ _2_ (mod 3)
16 ≡ _0_ (mod 2)	71 ≡ _7_ (mod 8)

5. Complete with CONGRUENT (C), SIMILAR (S), or NEITHER (N).

All rectangles with width 5 cm are	N
All hexagons are	N
All isosceles right triangles with legs 5 units are	C

10. Find the missing terms of the geometric sequences.

5, 15, _45_, 135, _405_, 1215

7, 70, _700_, _7000_, 70,000

5, 30, _180_, 1080, _6480_, 38,880

Level 6	Number 40

1. The intersection of the axes is called the _____ origin.

The axes separate the coordinate plane into four _____ quadrants.

The greatest chord of a circle is called a _____ diameter.

6. Draw the discards on the left. 100

One hundred cookies were on a plate. One fifth were broken. Of those whole, one fifth were later eaten. The whole cookies that remained were what fraction of the original cookies?

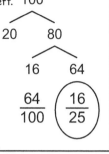

$\dfrac{64}{100}$ $\left(\dfrac{16}{25}\right)$

2. Draw the translation of each triangle 5 units left and 3 units up. Move vertices first.

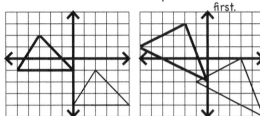

7. Find the average using the rightmost digit method.

88, 89, 83, 87, 81, 82 $30 \div 6$ 85

738, 734, 736, 739, 733 $30 \div 5$ 736

26, 29, 23, 25, 24, 28, 27 $42 \div 7$ 26

3. Calculate the point on the number line that is:

$\dfrac{3}{4}$ of the way from –8 to 8 $16 \div 4 = 4$ 4
      $-8 + 12 = 4$ ——

$\dfrac{3}{5}$ of the way from –9 to 6 $15 \div 5 = 3$ 0
      $-9 + 9 = 0$ ——

$\dfrac{5}{6}$ of the way from –9 to 9 $18 \div 6 = 3$ 6
      $9 - 3 = 6$ ——

8. Complete the table of values for $y = 4x - 3$.

x	–2	–1	0	3	4	5	7	8	10	12
y	–11	–7	–3	9	13	17	25	29	37	45

4. Solve.

$2x - 7 \le 19$	$5x - 9 \ge 14$	$3x + 7 > 18$
$2x \le 26$	$5x \ge 23$	$3x > 11$
$\boxed{x \le 13}$	$x \ge \dfrac{23}{5}$	$x > \dfrac{11}{3}$

9. Evaluate.

$|3 - 16| - 5 - 15 + |-6 - 5| + 20$
$13 - 20 + 11 + 20$
$13 + 11$ 24

$- 9 - |1 - 17| - |-2 - 2| + |10 - 1| - 11$
$- 9 - 16 - 4 + 9 - 11$
$-20 - 11$ –31

5. Multiply using mental math.

18.7 x 2	37.4	9.2 x 6	55.2
2.7 x 3	8.1	6.4 x 5	32
1.6 x 4	6.4	11.5 x 4	46
8.5 x 5	42.5	5.9 x 3	17.7

10. 300 children, one third girls, named a color. 40% said blue, 25% said red, and the rest said green. Half of the girls said red. Half of the boys said blue. How many boys said green?

	Bl	Re	Gr	tot
B	100	25	75	200
G	20	50	30	100
tot	120	75	105	300

Level 6	Number 41

1. Graph on the coordinate plane. One box equals one unit.

y = 3x   y = –x

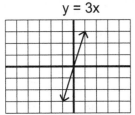 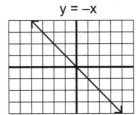

6. Find the area in square units of a square with perimeter 40 units.

P = 40

s = 10

$\left(A = 100\right)$

2. Graph on the number line.

x < 4

1 < x < 5

7. Complete the unit conversions.

89 cm	.89 m	36 m	360 dm
2.9 cm	29 mm	0.9 L	900 mL
.71 km	710 m	70 mg	.07 g

3. Given two opposite vertices of a rectangle with sides parallel to the axes. Find the other two vertices.

(7, –4) and (–9, 5)   (7, 5) (–9, –4)

(–8, 10) and (5, 7)   (–8, 7) (5, 10)

(11, 18) and (9, 8)   (11, 8) (9, 18)

8. Convert to standard form.

9.11673295×10^7 91,167,329.5

35.353×10^3 35,353

6.7×10^9 6,700,000,000

8.819×10^2 881.9

4. Rewrite using bar notation.

2.1999999999 . . . $2.1\overline{9}$

7.843843843 . . . $7.\overline{843}$

4.0075757575 . . . $4.00\overline{75}$

12.31093109 . . . $12.\overline{3109}$

9. Complete the table.

Principal	Rate	Interest
$50.00	2.5%	**$1.25**
$90.00	1.5%	**$1.35**
$20.00	6.5%	**$1.30**

5. Find the midpoint of the line segment with the given endpoints.

(22, 2) and (44, 8)   (33, 5)

(17, 17) and (23, 19)   (20, 18)

(27, 4) and (27, 6)   (27, 5)

(34, 34) and (24, 34)   (29, 34)

Average the xs, average the ys; or, halve the range and add to lesser.

10. Solve.

8x + 3 = 30	3x – 2 = –9	7x + 5 = 30
8x = 27	3x = –7	7x = 25
$x = \left(\dfrac{27}{8}\right)$	$x = \left(\dfrac{-7}{3}\right)$	$x = \left(\dfrac{25}{7}\right)$

Level 6	Number 42

1. Find the two missing numbers of the pattern.

6, 24, 60, 120, 210, <u>336</u>, <u>504</u>

1 x 2 x 3
2 x 3 x 4
3 x 4 x 5   6 x 7 x 8   7 x 8 x 9
4 x 5 x 6
5 x 6 x 7

2. Find the area of the irregular figure. One box equals one square unit.

6   12
52
3   8

60 + 12 + 9

(81)

or use Subtraction Method

3. Given two sides of a triangle, find the range of values for the 3rd side (use s for unknown side).

11, 21   <u>10 < s < 32</u>
15, 25   <u>10 < s < 40</u>
19, 40   <u>21 < s < 59</u>

Subtract given sides for low bound. Add for high bound. "The sum of two sides must be greater than the 3rd" to form a triangle.

4. Find the greatest value of xy.

$5 \le x \le 11$ $4 \le y \le 9$ <u>99</u>
$-9 \le x \le -1$ $-8 \le y \le -2$ <u>72</u>
$-8 \le x \le -3$ $0 \le y \le 15$ <u>0</u>
$4 \le x \le 12$ $-5 \le y \le 10$ <u>120</u>

5. Operate.

$5\dfrac{5}{6} \div 3\dfrac{4}{7}$

$\dfrac{\overset{7}{\cancel{35}}}{6} \times \dfrac{7}{\underset{5}{\cancel{25}}}$

$\boxed{\dfrac{49}{30}}$

$7\dfrac{3}{5} \div 9\dfrac{1}{2}$

$\dfrac{\overset{2}{\cancel{38}}}{5} \times \dfrac{2}{\underset{1}{\cancel{19}}}$

$\boxed{\dfrac{4}{5}}$

6. Write the value as an expression.

the number of years in d decades   <u>10d</u>

the number of cents in d dollars   <u>100d</u>

the number of feet in y yards   <u>3y</u>

the number of cups in q quarts   <u>4q</u>

7. Randomly select one letter from each word. Find the probability that the letter is a vowel.

CALENDARS $\dfrac{3}{9} = \dfrac{1}{3}$

TELESCOPES $\dfrac{4}{10} = \dfrac{2}{5}$

TELEVISION $\dfrac{5}{10} = \dfrac{1}{2}$

8. Complete the table.

Original Cost	Sales Tax	Final Cost
$42.00	3%	**$43.26**
$14.00	5%	**$14.70**
$23.00	6%	**$24.38**

9. Name the number 1 *before* in the given base.

1110_{two} <u>1101_{two}</u> 200_{five} <u>144_{five}</u>

212_{three} <u>211_{three}</u> 450_{six} <u>445_{six}</u>

230_{four} <u>223_{four}</u> 436_{seven} <u>435_{seven}</u>

10. Of 5 test scores from 0 to 100, the mode is 90, the median is 85, the mean is 72, and the outlier is 15. Find the scores.

<u>15</u>   <u>80</u>   <u>85</u>   <u>90</u>   <u>90</u>

Place 85 1st, followed by the 90s and the 15. Then, 72 x 5 = 360. 15 + 85 + 90 + 90 = 280. 360 − 280 = 80.

| Level 6 | Number 43 |

1. If a store needs 9 registers open for 150 customers, how many registers are needed for 250 customers?

$$\frac{9}{150} = \frac{x}{250}$$

Or,

$$150x = 9 \cdot 250$$
$$6x = 9 \cdot 10$$
$$x = \boxed{15}$$

9 R   150 C
3 R   50 C
15 R   250 C

6. Operate and simplify the continued fraction.

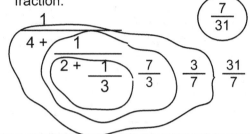
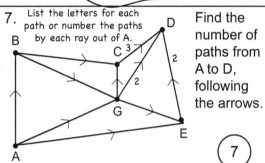

$\boxed{\dfrac{7}{31}}$

1 , $4 +$, $2 +$, $\dfrac{1}{3}$, $\dfrac{7}{3}$, $\dfrac{3}{7}$, $\dfrac{31}{7}$

2. Write the fractions in ascending order.

$$\frac{17}{18} \quad \frac{15}{19} \quad \frac{13}{22} \quad \frac{16}{19} \quad \frac{15}{22} \quad \frac{7}{19}$$

$$\mathbf{\frac{7}{19}} \quad \mathbf{\frac{13}{22}} \quad \mathbf{\frac{15}{22}} \quad \mathbf{\frac{15}{19}} \quad \mathbf{\frac{16}{19}} \quad \mathbf{\frac{17}{18}}$$

< 1/2   compare left & right pairs; then compare middle pair   almost 1

7. List the letters for each path or number the paths by each ray out of A. Find the number of paths from A to D, following the arrows.

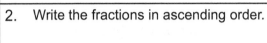

$\boxed{7}$

3. Solve by equivalent fractions.

$$\frac{1}{4} = \frac{15}{x} \quad \boxed{x = 60} \qquad \frac{x}{7} = \frac{27}{63} \quad \boxed{x = 3}$$

$$\frac{7}{8} = \frac{x}{48} \quad \boxed{x = 42} \qquad \frac{4}{x} = \frac{16}{52} \quad \boxed{x = 13}$$

8. Remove 7 odd numbers from the first 32 whole numbers. What percent of the remaining numbers are odd?

32 whole numbers: 0 ⟶ 31
16 even, 16 odd
remove 7 odd: 25 numbers remain
16 even, 9 odd

$$\frac{9}{25} = \boxed{36\%}$$

To make the fractional part, put "is" (or "are") number over "of" number.

4. Find the area of the triangle. One box equals one square unit.

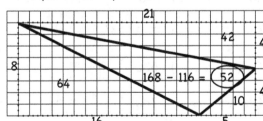

21
42   4
168 − 116 = $\boxed{52}$
64
4
10
16   5
8

9. Define operation ★ as:

$A \star B = |B - A| - 5AB$. For example, $12 \star 2 = 10 - 120 = -110$.

Find the values.

$10 \star 3$ $\underline{-143}$ 7 − 150

$5 \star 20$ $\underline{-485}$ 15 − 500

5. Sam is 20 years younger than Cam. Nan is 6 years older than Cam and 8 years younger than Lam who is 50 years old. Find the sum of the four ages.

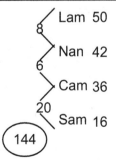

Lam 50
8
Nan 42
6
Cam 36
20
Sam 16
$\boxed{144}$

10. Identify membership in each set.

	N	W	Z	Q	R
$\sqrt{36}$	✓	✓	✓	✓	✓
3.6	–	–	–	✓	✓
−36	–	–	✓	✓	✓
6/3	✓	✓	✓	✓	✓

Level 6	Number 44

1. Draw one of each kind of trapezoid.

isosceles	right	scalene

Answers may vary.

6. Two whole numbers are in the ratio 2:3. Their sum is 50. Find their product.

2:3 sum = 5
20:30 sum = 50   20 x 30 = (600)

Two whole numbers are in the ratio 5:6. Their sum is 55. Find their product.

5:6 sum = 11
25:30 sum = 55   25 x 30 = (750)

2. Operate and simplify.

$$24 \frac{7}{8} \frac{35}{40}$$
$$+ 47 \frac{9}{20} \frac{18}{40}$$
$$71 \frac{53}{40} \quad \left(72 \frac{13}{40}\right)$$

$$\overset{32}{\cancel{33}} \frac{5}{12} \frac{\cancel{45}}{36} ^{51}$$
$$- 16 \frac{17}{18} \frac{34}{36}$$
$$\left(16 \frac{17}{36}\right)$$

7. Find the area of the circle inscribed in a square of area 16 square units.

$A_{sq} = 16$
$S = 4$
$D = 4$
$R = 2$
$A_{circ} = $ (4π)

3. Divide.

$435.096 \div 10^2$ ___4.35096___

$4.17 \div 10^3$ ___0.00417___

$5281.9 \div 10$ ___528.19___

$56297.5 \div 10^4$ ___5.62975___

8. Label ΔABC Isosceles, Equilateral, or Scalene AND Right, Acute, Obtuse, or Equiangular for 3 separate problems. CBD is also a possible answer.

m∠B = 60	CBD CBD
AC = BC = AB	E, I E, A
m∠B = 90 AB = 5	CBD R

4. Find the different ways that pennies and dimes can have a total value of $1.11. (12)

# P	x 1	# D	x 10		# P	x 1	# D	x 10
1	1	11	110		D = 5,	4,	3,	2
11	11	10	100		101	101	1	10
D = 9,	8,	7,	6		110	110	0	0

9. Find the least common multiple.

LCM(10, 11, 13) ___1430___ No common factors, so multiply. Use 11 rule.

LCM(27, 36, 45) ___540___ 9x3, 9x4, 9x5. Need 9 x 60.

LCM(30, 70, 90) ___630___ Ignore the 30 (in 90). Need 7 x 9 x 10.

LCM(260, 390) ___780___ 13 x 2 x 10, 13 x 3 x 10. Need 13 x 6 x 10.

5. Multiply the square matrix by the constant (scalar multiplication).

$$7 \begin{bmatrix} 9 & 11 & -5 \\ -7 & 0 & 8 \\ -1 & 10 & 6 \end{bmatrix} = \begin{bmatrix} 63 & 77 & -35 \\ -49 & 0 & 56 \\ -7 & 70 & 42 \end{bmatrix}$$

10. Simplify.

$$\frac{10!}{7!} \quad \underline{720} \quad 8 \times 9 \times 10$$

$$\frac{7!}{4!} \quad \underline{210} \quad 5 \times 6 \times 7$$

| Level 6 | Number 45 |

1. Find the 6-digit mystery number.
Clue 1: The number is divisible by 9.
Clue 2: The digits are consecutive in descending order.

543210 DS = 15 NO
654321 DS = 21 NO   (765,432)
765432 DS = 27 YES
876543 DS = 33 NO
987654 DS = 39 NO

6. Solve by clearing the decimal points.

.12x = .6	.05x = .9	.11x = .88
12x = 60	5x = 90	11x = 88
x = (5)	x = (18)	x = (8)

2. Identify each as a point, line, ray, or line segment if graphed on a number line. Determine without graphing.

$3 \le x \le 23$ — line segment

$x = -13$ — point

$x \le -15$ — ray

7. Find the average using the arithmetic sequence method.

The mean of an arithmetic sequence is the median. With an even number of numbers, the median is the mean of the 2 middlemost.

510, 540, 570, 600 — 555

11, 33, 55, 77, 99, 121 — 66

56, 64, 72, 80, 88, 96 — 76

3. Operate on the sets.

A = {3, 6, 9, 12, 15}
B = {6, 8, 10, 12, 14}
C = {6, 12, 18, 24}

B ∩ C = {6, 12}
A ∩ C = {6, 12}
B ∪ C = {6, 8, 10, 12, 14, 18, 24}

8. Check if the row number is divisible by the column factor.

	3	4	8	9	10	11
73,568	–	✓	✓	–	–	✓
42,900	✓	✓	–	–	✓	✓
81,664	–	✓	✓	–	–	✓

4. Write in standard form.

0.005 + 0.02 + 0.1 + 0.0007 — 0.1257

0.0001 + 0.08 + 0.5 — 0.5801

0.02 + 0.008 + 0.6 — 0.628

0.003 + 0.0007 + 0.09 — 0.0937

9. Find the probability if drawing from a deck of cards.

P(jack OR king) — $\frac{2}{13}$ — 8 out of 52

P(jack of club) — $\frac{1}{52}$ — 1 card

P(club) — $\frac{1}{4}$ — 1 of 4 suits or 13 of 52 cards

5. Cameron bought a device for $45.25 and another for $38.99. Find the total cost with 8% sales tax.

```
  45.25        84.24
+ 38.99      +  6.74
  84.24      ($90.98)
x   .08
  6.7392
```
Or multiply by 1.08 and omit addition.

10. Evaluate for x = 4.

$15 + 6x + x^2$ — 15 + 24 + 16 — 55

$80 - (x + 3)^2$ — 80 − 49 — 31

$x^3 - x^2 + 7x - 7$ — 64 − 16 + 28 − 7 — 69

$(x - 6) + 4(x + 6)$ — −2 + 40 — 38

Level 6	Number 46

1. Find the cost for the number specified.

9 if 4 cost $132   **$297**   132 + 132 + 33

7 if 6 cost $612   **$714**   612/6 = 102

12 if 5 cost $235   **$564**   235/5 = 47

7 if 10 cost $490   **$343**   490/10 = 49

6. Find the surface area of a cube with volume 216 cubic units.

V = 216
e = 6
A face = 36
SA = 36 x 6 = 216 sq un

Find the volume of a cube with surface area 96 square units.

SA = 96
A face = 16
e = 4
V = 4^3 = 64 cu un

2. Find the fractional part.
150 seconds out of 5 hours

$$\frac{\overset{}{\cancel{150}}}{\underset{2}{5 \times \cancel{60} \times 60}} \quad \left(\frac{1}{120}\right)$$

35 days out of 30 weeks

$$\frac{\cancel{35}}{7 \times \cancel{30}\,_{6}} \quad \left(\frac{1}{6}\right)$$

7. Six years ago Ari was 7 years older than Amy will be in 2 years. If Ari is 30 years old now, how old is Amy now?

	–6	now	+2
Ari	24	30	32
Amy		15	17

3. Find the next term of the sequence. Label arithmetic (A), geometric (G), or neither (N). If A, find d. If G, find r.

G $\frac{1}{4}$ 2560, 640, 160,   40

N   na   5, 6, 5, 6, 5,   6

N   na   10, 20, 11, 21, 12,   22

8. Operate and simplify.

$$\frac{\frac{8}{9} + \frac{2}{7}}{\frac{6}{7} - \frac{2}{3}} \quad \frac{\frac{56 + 18}{63}}{\frac{18 - 14}{21}} \quad \frac{\frac{74}{63}}{\frac{4}{21}} \quad \frac{\overset{37}{\cancel{74}}}{\underset{3}{\cancel{63}}} \cdot \frac{\overset{1}{\cancel{21}}}{\underset{2}{\cancel{4}}}$$

$$\left(\frac{37}{6}\right)$$

4. Find the least natural number n such that kn is a perfect square.

k = 24   n =   6   24 = 2^3 x 3

k = 44   n =   11   44 = 2 x 2 x 11

k = 70   n =   70   70 = 2 x 5 x 7

An even number of each prime creates a perfect square.

9. A square with area 9 square units and an equilateral triangle have equal perimeters. Find the side in units of the triangle.

A □ = 9
S □ = 3
P □ = 12
P △ = 12
S △ = 4

5. Reflect the point A(5, 1) over the y-axis. Then translate it 3 units up. What is the final point?

$(-5, 4)$

10. Solve.

$|-1| = x$
x =   1

$|1| = -x$
x =   –1

$|x| = 1$
x =   ± 1

$|x| = -1$
x =   No Sol

Level 6	Number 47

1. Write the converse of the conditional statement. Are they logically equivalent? (NO)

If a shape is a polygon, then it has sides.

statement TRUE
converse FALSE

If a shape has sides, then it is a polygon.

The sides could be open or crossing.

6. Find the number of 4 by 4 by 4 cubes that can completely fill a box measuring 24 by 32 by 44 of the same unit.

6 cubes by 8 cubes by 11 cubes = (528) cubes

OR $\dfrac{\text{Volume box}}{\text{Volume 1 cube}} = \dfrac{24 \times 32 \times 44}{4 \times 4 \times 4}$

2. State whether the fraction would repeat (R) or terminate (T) if converted to a decimal.

$\dfrac{26}{25}$ T $\dfrac{13}{26}$ T $\dfrac{8}{11}$ R

$\dfrac{8}{9}$ R $\dfrac{13}{60}$ R $\dfrac{6}{75}$ T

7. Operate.

$12 \div -4$	-3	-7×-9	63
$-36 \div -6$	6	8×-2	-16
$-48 \div 6$	-8	$32 \div -8$	-4
-9×5	-45	$-14 \div -7$	2

3. Find the missing sides of the similar triangles. (NTS)

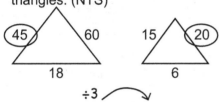

÷3

8. Find the greatest common factor.

GCF(40, 48, 120)   8   Ignore 120 (multiple of 40). 8 x 5, 8 x 6

GCF(39, 65, 260)   13   13 x 3, 13 x 5, 13 x 20

GCF(77, 91, 700)   7   91 = 7 x 13

GCF(41, 83, 101)   1   All prime

4. Operate.

$\dfrac{1}{6}$ (27 days   10 hours   12 minutes)

24 d	82 h	12 m
24 d	78 h	252 m
4 d	13 h	42 m

9. Add.

$\dfrac{4}{15}^{7} + \dfrac{9}{35}^{3}$ $\dfrac{3}{11}^{10} + \dfrac{3}{22}^{5} + \dfrac{7}{55}^{2}$

$\dfrac{28 + 27}{105}$ $\dfrac{30 + 15 + 14}{110}$

$\dfrac{55 \div 5}{105 \div 5}$ $\left(\dfrac{11}{21}\right)$ $\left(\dfrac{59}{110}\right)$

5. Multiply.

$\dfrac{9}{22} \times 88$ 36 $\dfrac{5}{13} \times 65$ 25

$\dfrac{7}{17} \times 51$ 21 $\dfrac{9}{25} \times 75$ 27

$\dfrac{8}{19} \times 38$ 16 $\dfrac{9}{16} \times 80$ 45

10. Complete the table.

Decimal	Percent	Fraction
2.8	**280%**	$\dfrac{14}{5}$
1.4	140%	$\dfrac{7}{5}$
3.3	**330%**	$\dfrac{33}{10}$

Level 6	Number 48

1. A cafeteria offers 10 main dishes, 6 vegetables, 5 starches, 4 beverages, and 3 desserts. How many different meals are available with one main dish, one vegetable, one starch, one beverage, and one dessert?

10 x 6 x 5 x 4 x 3 = (3600)

6. Divide the product of the first five 2-digit positive integers by the product of the first five 2-digit composite integers greater than 20.

$$\frac{\overset{2}{\cancel{10}} \times \cancel{11} \times \overset{2}{\cancel{12}} \times \cancel{13} \times \cancel{14}}{\underset{3}{\cancel{21}} \times \underset{2}{\cancel{22}} \times \underset{2}{\cancel{24}} \times \underset{5}{\cancel{25}} \times \underset{2}{\cancel{26}}} = \left(\frac{1}{30}\right)$$

2. Rewrite the decimals in ascending order.

1.34, 4.13, .143, .134, .1434, 3.14

.134 .143 .1434 1.34 3.14 4.13

7. Write in standard form: five billion, fourteen million, two hundred eighty-nine thousand, seven hundred eleven and sixteen thousandths.

The decimal point is the "and."

5,014,289,711.016

3. Find the simple interest on $9000 at 8% annually for 10 months.

I = PRT

$$I = \frac{\overset{90 \quad 30 \quad 2}{\cancel{9000}}}{1} \cdot \frac{\cancel{8}}{\cancel{100}} \cdot \frac{\cancel{10}}{\underset{3}{\cancel{12}}}$$

I = ($600)

8. Find the supplement of the angle.

Supp (30°)	150°	Supp (90°)	90°
Supp (110°)	70°	Supp (130°)	50°
Supp (160°)	20°	Supp (40°)	140°
Supp (60°)	120°	Supp (170°)	10°

4. Distribute.

6(8x – 4y)	48x – 24y
9(3a + 7c)	27a + 63c
–5(5x – 7y)	–25x + 35y
–7(–6b + 8d)	42b – 56d

9. Calculate the area in square units of the trapezoid. (NTS)

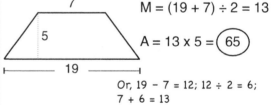

M = (19 + 7) ÷ 2 = 13

A = 13 x 5 = (65)

Or, 19 – 7 = 12; 12 ÷ 2 = 6; 7 + 6 = 13

5. Find the sum of each arithmetic sequence using the "First plus Last" method.

5 + 10 + 15 + 20 + 25 + 30 + 35 + 40 + 45 + 50

4 + 8 + 12 + 16 + 20 + 24 + 28 + 32 + 36

$$\frac{(5 + 50)(10)}{2} = \left(275\right) \quad \frac{(4 + 36)(9)}{2} = \left(180\right)$$

10. Evaluate.

$\sqrt{1600}$	40	$\sqrt{10{,}000}$	100
$\sqrt{8100}$	90	$\sqrt{2500}$	50
$\sqrt{625}$	25	$\sqrt{225}$	15
$\sqrt{490{,}000}$	700	$\sqrt{3600}$	60

Level 6	Number 49

1. Given two opposite vertices of a rectangle with sides parallel to the axes. Find the other two vertices.

(11, 3) and (−3, −6)   (11, −6)   (−3, 3)

(−8, 8) and (7, −7)   (−8, −7)   (7, 8)

(12, −6) and (14, 7)   (12, 7)   (14, −6)

6. Find the digit d such that the 4-digit number 58d2 is divisible by 12.

for div by 4, d = 1,   for div by 3,   ③
   3, 5, 7   5 + 8 + 2 = 15

Find the digit x such that the 4-digit number 59x2 is divisible by 11.   ⑥

9 + 2 = 11   5 + 6 = 11   11 − 11 = 0

2. Find the GCF and LCM of 150 and 525 using prime factorization.

150 = 25 x 6 = 2 x 3 x 5^2
525 = 25 x 21 = 3 x 5^2 x 7
GCF = 3 x 5^2 = 3 x 25 = ⟨75⟩

LCM = 2 x 3 x 5^2 x 7 = 2 x 525 = ⟨1050⟩

Look for one of the original numbers before multiplying.

7. Name the property exemplified.

k + 0 = k   IdPA

a(b + c) = ab + ac   DPMA

(x + y)z = xz + yz   DPMA

$w + \dfrac{-1}{w} = 0$ InPA

3. Find the maximum area in square units of a rectangle with whole number sides and perimeter 80 units.

39 by 1
A = 39

Consider the 2 extremes: as thin as possible versus square.

20 by 20
A = ⟨400⟩

8. Find the volume in cubic units of a rectangular prism with the areas of its noncongruent faces 14, 22, and 77 square units.

2 x 7 = 14   V = 2 x 7 x 11 = ⟨154⟩
2 x 11 = 22
7 x 11 = 77

4. Solve.

$8x - 4 \le 11$	$6x - 5 \ge 31$	$9x + 5 < 23$
$8x \le 15$	$6x \ge 36$	$9x < 18$
$x \le \dfrac{15}{8}$	$x \ge 6$	$x < 2$

9. Write as an algebraic expression.

six more than six times a number   6x + 6

triple the quantity of a number minus four   3(x − 4)

half the difference of two numbers $\dfrac{x - y}{2}$

four times the sum of a number and three   4(x + 3)

5. Convert measures to a decimal.

12 minutes   =   0.2   hour

8 cups   =   0.5   gallon

8 inches   = $0.\overline{2}$ yard

3 hours   =   0.125   day

10. Compute using mental math.

Divide by 4 mentally by halving twice.

500 ÷ 4	250	125	636 ÷ 4	318	159
340 ÷ 4	170	85	560 ÷ 4	280	140
920 ÷ 4	460	230	300 ÷ 4	150	75
760 ÷ 4	380	190	912 ÷ 4	456	228

Level 6	Number 50

1. A polygon may be concave or ___convex___.

The number belonging to the set of whole numbers but not counting numbers is ___zero___.

The number added to each successive term of an arithmetic sequence is the common ___difference___.

2. What time is 495 minutes after 7:40 PM?
495 = 60 x 8 + 15   7 + 8 ➔ 3   40 + 15 ➔ 55

(3:55 AM)

What time is 355 minutes after 4:35 AM?
355 = 60 x 6 − 5   4 + 6 ➔ 10   35 − 5 ➔ 30

(10:30 AM)

3. Find the area of the parallelograms. One box equals one square unit.

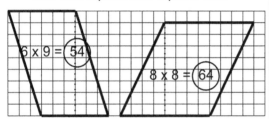

6 x 9 = (54)
8 x 8 = (64)

4. Define operation ❤ as:

A ❤ B = $|A^2 - B^2|$ − A + B. For example,
4 ❤ 5 = $|16 - 25|$ − 4 + 5 = 10.
Find the values.

2 ❤ 4 ___14___   12 − 2 + 4

3 ❤ 7 ___44___   40 − 3 + 7

5. Find the missing terms of the geometric sequences.

13, 26, ___52___, 104, 208, 416, ___832___

2, 10, ___50___, ___250___ 1250, 6250

5, 20, ___80___, ___320___, ___1280___, 5120

6. Identify (YES/NO) whether each set of ordered pairs is a function. If NO, specify the x-value that violates the function rule.

{ (1,3), (2,3), (6,7), (7,6), (0,9) }   Y   na

{ (9,9), (8,9), (8,8), (6,5), (5,6) }   N   8

7. Convert to base ten.

2431_{five} 2 x 125 + 4 x 25 + 3 x 5 + 1   366

257_{nine} 2 x 81 + 5 x 9 + 7   214

437_{eight} 4 x 64 + 3 x 8 + 7   287

1203_{four} 1 x 64 + 2 x 16 + 3   99

8. Complete the chart for a circle.

radius	diameter	area	circumference
7	14	49π	14π
11	22	121π	22π
20	40	400π	40π

9. Complete the chart.

Original Cost	Discount	New Cost
$65.00	3%	**$63.05**
$90.00	8%	**$82.80**
$60.00	7%	**$55.80**

10. Multiply. Answer in exponential form.

(5 x 10^5) x (6 x 10^4) 3×10^{10}

(3 x 10^7) x (4 x 10^1) 12×10^8

(5 x 10^3) x (12 x 10^3) 6×10^7

(3 x 10^4) x (5 x 10^4) 15×10^8

Regroup the 10s in 30 and 60.

Level 6	Number 51

1. Eight chefs can prepare 21 dinners in 3 hours. At the same rate for the same size meals, how many chefs can prepare 14 dinners in 4 hours?

chefs	dinners	hours
8	21	3
8	7	1
8	28	4
⨀4	14	4

Working in rows, keep 1, change 2 by x or ÷ (direct) or both (inverse).

6.

	R	T	U	
AM	12	5	60	
PM	6	10	60	
tot	⨀8	15	120	

In the AM Lily used her exercise machine for 5 minutes at a steady rate of 12 output units per minute. In the PM she did the same amount of units but at half the rate. Find her overall rate for the day.

2. Find the prime factorization. Work down. In the answer use exponents with primes in ascending order.

135	190	315
5 x 27	19 x 10	5 x 63
3³ x 5	**2 x 5 x 19**	5 x 7 x 9
		3² x 5 x 7

7. Find the probability if tossing 2 dice.

P(sum of 12) $\frac{1}{36}$ 6&6

P(sum of 11) $\frac{1}{18}$ 6&5 or 5&6

P(sum of 11 OR 12) $\frac{1}{12}$ all 3 above

P(sum of 10) $\frac{1}{12}$ 6&4 or 4&6 or 5&5

3. Subtract the square matrices.

$$\begin{bmatrix} 11 & 16 & 9 \\ 10 & 0 & 7 \\ 5 & 3 & 13 \end{bmatrix} - \begin{bmatrix} 7 & 9 & 2 \\ 0 & 4 & 8 \\ 8 & 5 & 1 \end{bmatrix} = $$

$$\begin{bmatrix} \mathbf{4} & \mathbf{7} & \mathbf{7} \\ \mathbf{10} & \mathbf{-4} & \mathbf{-1} \\ \mathbf{-3} & \mathbf{-2} & \mathbf{12} \end{bmatrix}$$

8. Find the mean, median, mode, and range of the data.

Stem	Leaf
1	0 0 5 5 5
2	5 5
3	0 0 5 5 7
6	3
7	5 5

mean = **33**
median = **30**
mode = **15**
range = 75 – 10 = **65**

20 + 45 + 50 + 60 + 70 + 100 + 150 = 495
495/15 = 33

4. Name the solids depicted by the nets.

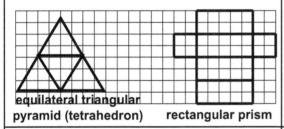

equilateral triangular pyramid (tetrahedron)   **rectangular prism**

9. Complete the table of values for y = –2x + 1.

x	–9	–8	–6	–3	–1	0	2	4	5	6
y	19	**17**	**13**	7	3	1	**–3**	**–7**	**–9**	–11

5. Complete the chart of data for a circle graph.

	Number	Fraction	Percent	Degrees
vanilla	175	**7/10**	**70**	**252**
choco	50	**2/10**	**20**	**72**
strawb	**25**	**1/10**	**10**	**36**
TOTAL	250	**1**	**100**	**360**

10. Solve.

4x – 4 = 17
4x = 21
x = ⨀$\frac{21}{4}$

6x – 8 = 11
6x = 19
x = ⨀$\frac{19}{6}$

9x + 15 = 5
9x = –10
x = ⨀$\frac{-10}{9}$

Level 6	Number 52

1. Cam scored 75, 83, and 84 out of 100 on 3 of 5 math tests. What is the minimum score she must earn on the 4th test to have an average of 85 on the 5 tests?

$85 \times 5 = 425$ (83)

242   83   100

$$\frac{75 + 83 + 84 + min + max}{5} = 85$$

6. Find the day of the week given each separate condition.

355 days from today if today is Monday

$355 = 350 + 5$ M + 5   Saturday

282 days from today if today is Sunday

$282 = 280 + 2$ Su + 2   Tuesday

2. Find the next term of the sequence. Label arithmetic (A), geometric (G), or neither (N). If A, find d. If G, find r.

G   4   5, 20, 80, 320,   1280

N   na   36, 17, 34, 19, 32,   21

A   7   8, 15, 22, 29, 36,   43

7. An equilateral triangle with side 9 has a congruent triangle attached at one side and a square with side 9 attached at another. Find the perimeter of the figure formed.

$9 \times 6 =$ (54)

3. Find the volume and surface area of a cylinder with diameter 6 and height 12 units.

$V = 3 \cdot 3 \cdot 12 \cdot \pi$
$V =$ (108π) cu un

$SA = 9\pi + 9\pi + 6\pi \cdot 12$
$SA = 18\pi + 72\pi$
$SA =$ (90π) sq un

8. Of 90 cell phones, 65 take pictures and 42 offer internet. If 26 do both, how many do neither?   9

P=65   I=42   N

39   26   16   9

4. Write each fraction as a repeating decimal.

$\frac{63}{99}$ $.\overline{63}$ $\frac{8}{33}$ $.\overline{24}$ $\frac{8}{99}$ $.\overline{08}$

$\frac{8}{11}$ $.\overline{72}$ $\frac{8}{9}$ $.\overline{8}$ $\frac{48}{99}$ $.\overline{48}$

9. Combine like terms.

$-13w + 7s + 7w - 11s$ $-6w - 4s$

$-6\pi + 7\pi - 12\pi + 14\pi$ 3π

$-6x + 6\pi - 8x + 8\pi$ $14\pi - 14x$

$7a - 9b + 7a - 9b$ $14a - 18b$

5. Operate.

$7\frac{7}{9} \times 5\frac{1}{7}$

$\frac{\overset{10}{\cancel{70}}}{\underset{1}{9}} \times \frac{\overset{4}{\cancel{36}}}{\underset{1}{7}}$

(40)

$2\frac{5}{8} \times 6\frac{2}{3}$

$\frac{\overset{7}{\cancel{21}}}{\underset{2}{8}} \times \frac{\overset{5}{\cancel{20}}}{\underset{1}{3}}$

$\frac{35}{2}$

10. Operate.

$(7 + 7) \div 7 + 7 \times 7$
$2 + 49$ (51)
Do 2 groupings concurrently.

$5 \times (5 - 5 \div 5 \times 5)$
$5 \times (5 - 5)$ (0)
5×0

$6 \times 6 \div 6 - (6 - 6)$
$6 - 0$ (6)

$(9 + 9) + 9 - 9 \div 9$
$18 + 9 - 1$ (26)

Level 6	Number 53

1. An ice cream shop offers 1-scoop cones--either a sugar or waffle cone; with one of 6 ice creams or 5 sherbets; and nuts, sprinkles, both, or no topping. How many different cones could be made?

$2 \times 11 \times 4 =$ (88)

6. Compute using mental math.

Times 10; divide by 2 (cut in half).

845 x 5	4225	569 x 5	2845
706 x 5	3530	363 x 5	1815
484 x 5	2420	422 x 5	2110
667 x 5	3335	928 x 5	4640

---

2. Graph on the number line.

$x > 1$

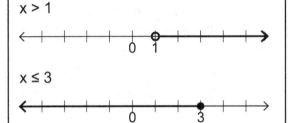

$x \le 3$

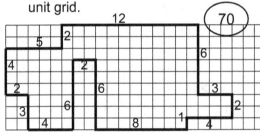

7. Answer YES or NO as to whether the numbers are valid sides of a triangle.

7, 10, 15	Y	8, 8, 18	N
42, 43, 44	Y	5, 15, 25	N
14, 14, 14	Y	12, 21, 32	Y

---

3. Find the perimeter of the figure on the unit grid.

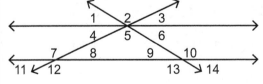

(70)

8. Name the number 1 *after* in the given base.

102_{three}	110_{three}	525_{six}	530_{six}
600_{nine}	601_{nine}	233_{four}	300_{four}
377_{eight}	400_{eight}	1011_{two}	1100_{two}

---

4. Operate and simplify.

$$38 \frac{23}{24}$$
$$+ 49 \frac{11}{24}$$
$$\overline{} \begin{array}{c} 88 \ 10 \\ 87 \ \frac{34}{24} \end{array}$$ (88 $\frac{5}{12}$)

$$\overset{44}{\cancel{45}} \frac{2}{15} \frac{10}{75}$$
$$- 28 \frac{16}{25} \frac{48}{75}$$
$$\overline{}$$ (16 $\frac{37}{75}$)

9. Find the perimeter in units of a square with area 400 square units.

A = 400

s = 20

(P = 80)

---

5. Find the angle measures. (NTS)

[figure of intersecting lines with angles 1–14]

If m∠6 = 55°, then m∠1 = __55°__

If m∠6 = 86°, then m∠14 = __86°__

10. Evaluate for x = 5.

$x^2 - 5x + 13$	$25 - 25 + 13$	13
$25 + (2x + 5)^2$	$25 + 225$	250
$x^3 - x - 3x$	$125 - 5 - 15$	105
$2(x + 2) + 3(x + 3)$	$14 + 24$	38

Level 6	Number 54

1. A car traveled 340 miles at an overall rate of 68 miles per hour. How many hours did this trip take?

D = RT

$\dfrac{340}{68} = \dfrac{68 \times T}{68}$

$\boxed{5} = T$

6. A 20 by 34 rectangle has a 2 by 2 square removed from each of its corners. Find the volume in cubic units of the container formed by folding up the resulting sides.

L = 34 − 4 = 30
W = 20 − 4 = 16
H = 2
V = 30 x 16 x 2 = $\boxed{960}$

---

2. Name the points shown.

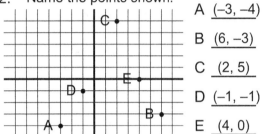

A (−3, −4)
B (6, −3)
C (2, 5)
D (−1, −1)
E (4, 0)

7. Nine girls can mow the neighborhood lawns in 80 minutes. At the same rate, 6 girls could mow those lawns in how many hours? Watch for the unit conversion.

W x t = w x T
9 x 80 = 6 x 120

$\boxed{2}$

Workers and time are inversely proportional-- as one goes up, the other goes down. Their product is constant.

---

3. Multiply by 11 mentally.
Add digits, insert ones in middle, increment leftmost.

84 x 11	924	67 x 11	737
77 x 11	847	49 x 11	539
95 x 11	1045	38 x 11	418
58 x 11	638	88 x 11	968

8. By what number must 6! be multiplied to become 10! ?

7 x 8 x 9 x 10 = 72 x 7 x 10 = $\boxed{5040}$

By what number must 11! be divided to become 8! ?

11 x 10 x 9 = 99 x 10 = $\boxed{990}$

---

4. Answer about cubes.

Cubes are rectangular solids. (T or F) T

Number of faces 6

Number of edges 12

Number of vertices 8

9. Add.

$\dfrac{11}{40}^{3} + \dfrac{7}{60}^{2}$

$\dfrac{33 + 14}{120}$

$\boxed{\dfrac{47}{120}}$

$\dfrac{2}{25}^{6} + \dfrac{3}{50}^{3} + \dfrac{7}{75}^{2}$

$\dfrac{12 + 9 + 14}{150}$

$\dfrac{35 \div 5}{150 \div 5}\boxed{\dfrac{7}{30}}$

---

5. Find the probability of selecting a point from the shaded area within the rectangle.

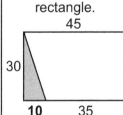

$\dfrac{A_{sh}}{A_{rec}} = \dfrac{5 \cdot 30}{45 \cdot 30}$

$\boxed{\dfrac{1}{9}}$

Not multiplying allows easy simplifying.

10. Rewrite using bar notation.

15.21452145 . . . $15.\overline{2145}$

14.945454545 . . . $14.9\overline{45}$

9.0333333 . . . $9.0\overline{3}$

6.305305305 . . . $6.\overline{305}$

MAVA Math: Middle Reviews Solutions Copyright © 2013 Marla Weiss

Level 6	Number 55

1. Numbered pegs are spaced equi-distant and consecutively around a circle. Number 55 is directly opposite number 89. How many pegs are around the circle?

33 numbers

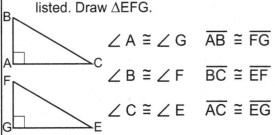

89 ( ) 55

Make a smaller problem.

33 numbers

2. Write the fractions in ascending order.

$$\frac{5}{34} \quad \frac{7}{36} \quad \frac{5}{35} \quad \frac{1}{8} \quad \frac{7}{35} \quad \frac{1}{6}$$

$$\frac{1}{8} \quad \frac{5}{35} \quad \frac{5}{34} \quad \frac{1}{6} \quad \frac{7}{36} \quad \frac{7}{35}$$

Order 1/8, 5/35 = 1/7, 1/6, and 7/35 = 1/5. Insert remaining two by pairwise comparisons.

3. Given △ABC and the congruence as listed. Draw △EFG.

∠A ≅ ∠G $\overline{AB} ≅ \overline{FG}$

∠B ≅ ∠F $\overline{BC} ≅ \overline{EF}$

∠C ≅ ∠E $\overline{AC} ≅ \overline{EG}$

4. Complete the chart.

Original Cost	Discount	New Cost
$45.00	2%	**$44.10**
$70.00	5%	**$66.50**
$80.00	6%	**$75.20**

5. Find each central angle as a fractional part of 360º.

10º	$\frac{1}{36}$	105º	$\frac{7}{24}$

320º $\frac{8}{9}$

40º $\frac{1}{9}$ 300º $\frac{5}{6}$ 30º $\frac{1}{12}$

6. Complete the unit conversions.

2.6 L	2600 mL	9 kg	9000 g
3.8 dm	0.38 m	3 cm	.00003 km
3.1 km	31,000 dm	41 cm	4.1 dm

7. Complete the table.

Principal	Rate	Interest
$30.00	8.5%	**$2.55**
$70.00	7.5%	**$5.25**
$10.00	9.5%	**$0.95**

8. Find the probability when drawing from a deck of cards.

P(king) $\frac{1}{13}$ 4 out of 52

P(red card) $\frac{1}{2}$ 26 of 52 cards or 2 of 4 suits

P(red king) $\frac{1}{26}$ 2 out of 52

9. Label each polygon convex or concave.

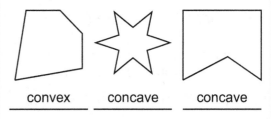

convex   concave   concave

10. 400 adults, one fourth men, named a drink. 30% said milk, 45% said juice, and the rest said coffee. Half of the women said juice. Half of the men said milk. How many women said coffee?

	M	J	C	tot
M	50	30	20	100
W	70	150	80	300
tot	120	180	100	400

| Level 6 | Number 56 |

1. Operate in the given bases.

$6_{eight} + 5_{eight}$ = $\underline{13_{eight}}$

$3_{five} \times 4_{five}$ = $\underline{22_{five}}$

$6_{nine} \times 7_{nine}$ = $\underline{46_{nine}}$

$6_{seven} + 6_{seven}$ = $\underline{15_{seven}}$

6. Draw the discards always on the left.

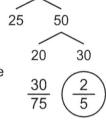

Seventy-five berries were on a bush. One third were spoiled. Of those good, two fifths were picked. The good, unpicked berries were what fraction of the original berries?

$\dfrac{30}{75}$ $\left(\dfrac{2}{5}\right)$

2. Draw the translation of each triangle 6 units right and 6 units down. Move vertices first.

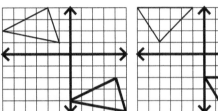

7. The ratio of wins to losses in a series of games with no ties is 4:9.

Find the ratio of wins to total games.

$\dfrac{W}{G}$ $\left(\dfrac{4}{13}\right)$

Find the number of games if the series had 117 losses. x 13

$52 + 117 = \boxed{169}$

3. Calculate the point on the number line that is:

$\dfrac{1}{7}$ of the way from −7 to 7 $\quad 14 \div 7 = 2 \quad$ −5
$\qquad\qquad\qquad\qquad -7 + 2 = -5$ ——

$\dfrac{4}{9}$ of the way from −2 to 7 $\quad 9 \div 9 = 1 \quad$ 2
$\qquad\qquad\qquad\qquad -2 + 4 = 2$ ——

$\dfrac{3}{8}$ of the way from −7 to 9 $\quad 16 \div 8 = 2 \quad$ −1
$\qquad\qquad\qquad\qquad -7 + 6 = -1$ ——

8. Complete the table of values for $y = 5x + 2$.

x	−2	−1	0	2	5	7	10	11	12	15
y	−8	−3	2	12	27	37	52	57	62	77

4. If the sum of two consecutive even numbers is 62, find their product.

$30 + 32 = 62 \qquad 30 \times 32 = \boxed{960}$

If the sum of two consecutive odd numbers is 40, find their product.

$19 + 21 = 40 \quad 21 \times 19 = 380 + 19 = \boxed{399}$

9. Label △ABC Isosceles, Equilateral, or Scalene AND Right, Acute, Obtuse, or Equiangular for 3 separate problems. CBD is also a possible answer.

m∠A = 100 $\qquad\qquad \dfrac{CBD}{}$ $\dfrac{O}{}$

AB = 10 $\qquad\qquad \dfrac{CBD}{E, I}$ $\dfrac{CBD}{E, A}$

m∠A = 60 $\quad$ m∠C = 60

5. Find the midpoint of the line segment with the given endpoints.

Average the xs, average the ys; or, halve the range and add to lesser.

(37, 18) and (13, 20) $\quad \underline{(25, 19)}$

(44, 35) and (16, 15) $\quad \underline{(30, 25)}$

(27, 1) and (25, 13) $\quad \underline{(26, 7)}$

(60, 15) and (30, 21) $\quad \underline{(45, 18)}$

10. Of 5 test scores from 0 to 100, the mode is 23, the median is 64, the range is 60, and the mean is 54. Find the scores.

23 $\quad$ 23 $\quad$ 64 $\quad$ 77 $\quad$ 83

Place 64 1st, followed by the 23s. Next, 23 + 60 = 83. Then, 54 x 5 = 270. 23 + 23 + 64 + 83 = 193. 270 − 193 = 77.

Level 6	Number 57

1. Find the GCF mentally.

GCF(121, 143)   11   Middle digit is sum of left and right.

GCF(175, 225)   25   Think of quarters. 25x7, 25x9.

GCF(48, 80, 160)   16   16 in all. 3 vs. 5, so no other factors.

GCF(360, 480)   120   Think of multiples of 12. See 10.

6. Given a bowl of 24 red, 16 blue, and 40 green marbles. Draw one. Find each probability.

P(R) $\dfrac{3}{10}$ $\dfrac{24}{80}$

P(B) $\dfrac{1}{5}$ $\dfrac{16}{80}$

P(G) $\dfrac{1}{2}$ $\dfrac{40}{80}$

2. Find the GCF and LCM of 242 and 275 using prime factorization.

$242 = 2 \times 121 = 2 \times 11^2$
$275 = 25 \times 11 = 5^2 \times 11$
GCF = $\boxed{11}$

$LCM = 2 \times 5^2 \times 11^2 = 2 \times 25 \times 121 =$
$50 \times 121 = \boxed{6050}$

7. Operate and simplify the continued fraction.

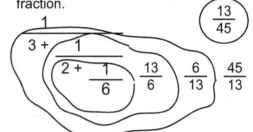

$\dfrac{13}{45}$

$3 + \cfrac{1}{2 + \cfrac{1}{6}}$ $\dfrac{13}{6}$ $\dfrac{6}{13}$ $\dfrac{45}{13}$

3. Identify each as a point, line, ray, or line segment if graphed on a number line. Determine without graphing.

$0 \le x \le 17$ line segment

$3(x + 1) = 3x + 3$ line

$x = 14$ point

8. Convert to scientific notation.

53,000	5.3×10^4	340	3.4×10^2
745,000	7.45×10^5	6600	6.6×10^3
4 million	4.0×10^6	2,540	2.54×10^3
5 billion	5.0×10^9	952	9.52×10^2

4. Find the area of the triangle. One box equals one square unit.

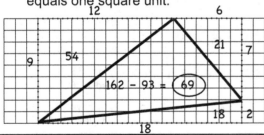

9. Add.

765,279,614	790,631,555
953,678,276	248,775,629
299,446,772	841,665,884
+ 410,678,876	608,288,556
	+ 552,899,383
2,429,083,538	**3,042,261,007**

5. Operate on the sets.

A = {4, 9, 13, 14, 19}
B = {2, 3, 5, 7, 11, 13}
C = {13, 14, 15, 16, 17, 18, 19}

A ∪ B =   {2, 3, 4, 5, 7, 9, 11, 13, 14, 19}

A ∩ C =   {13, 14, 19}

A ∩ B =   {13}

10. Complete the table.

Decimal	Percent	Fraction
2.7	270%	$\dfrac{27}{10}$
$0.\overline{5}$	$55.\overline{5}\%$	$\dfrac{5}{9}$
$1.\overline{1}$	$111.\overline{1}\%$	$\dfrac{10}{9}$

Level 6	Number 58

1. Mirena bought a side of beef for $54.12. The beef weighed 8.8 pounds. Find the price per pound.

$$\frac{price}{pound} \quad \frac{54.12}{8.8} \quad \boxed{\$6.15}$$

$$\begin{array}{r} 6.15 \\ 88\overline{)541.20} \\ \underline{528} \\ 132 \\ \underline{88} \\ 440 \\ \underline{440} \\ 0 \end{array}$$

6. Two whole numbers are in the ratio 5:8. Their sum is 65. Find their product.

5:8 sum = 13
25:40 sum = 65   25 x 40 = $\boxed{1000}$

Two whole numbers are in the ratio 4:7. Their sum is 33. Find their product.

4:7 sum = 11   By DPMA 240 + 12
12:21 sum = 33   12 x 21 = $\boxed{252}$

2. Identify whether the figure is an Euler Graph--can be drawn without lifting the pencil and retracing a line.

3 3 3

other vertices are 2 or 4

NO

3

3

YES

An Euler Graph must have exactly 0 or 2 vertices with an odd number of impinging segments (one to start, one to stop).

7. Find the area of the circle inscribed in a square of area 36 square units.

A_{sq} = 36
S = 6
D = 6
R = 3
A_{circ} = $\boxed{9\pi}$

3. Find the missing terms of the arithmetic sequences.

11, 16, 21, __26__, __31__, 36, __41__

3, 17, 31, 45, __59__, 73, 87, __101__

1, 21, 41, __61__, __81__, __101__, 121

8. Find the average using the rightmost digit method.

671, 677, 673, 671   12 ÷ 4   __673__

481, 482, 488, 486, 483   20 ÷ 5   __484__

95, 94, 99, 98, 98, 96, 99   49 ÷ 7   __97__

4. Solve.

$7x - 4 > 24$	$2x + 11 \geq 9$	$5x + 9 < 5$
$7x > 28$	$2x \geq -2$	$5x < -4$
$\boxed{x > 4}$	$\boxed{x \geq -1}$	$\boxed{x < \dfrac{-4}{5}}$

9. Find the least common multiple.

LCM(12, 72, 90)   360   Ignore the 12 (in 72). 18x4, 18x5, need 18x20.

LCM(125, 175)   875   25x5, 25x7, need 25x35. 750+125.

LCM(33, 66, 77)   462   Ignore 33 (in 66). Need 11x6x7 = 42x11.

LCM(140, 210)   420   Both have 10. 7x2, 7x3, need 7x6x10.

5. Operate.

$$9\frac{4}{5} \times 1\frac{3}{7}$$

$$\overset{7}{\underset{1}{\frac{49}{5}}} \times \overset{2}{\underset{1}{\frac{10}{7}}}$$

$$\boxed{14}$$

$$4\frac{1}{6} \times 3\frac{1}{5}$$

$$\overset{5}{\underset{3}{\frac{25}{6}}} \times \overset{8}{\underset{1}{\frac{16}{5}}}$$

$$\boxed{\frac{40}{3}}$$

10. Divide.

$$\begin{array}{r} 6{,}457 \\ 66\overline{)426{,}162} \\ \underline{396} \\ 301 \\ \underline{264} \\ 376 \\ \underline{330} \\ 462 \\ \underline{462} \end{array}$$

Level 6	Number 59

1. How many whole numbers are from:

137 to 345 inclusive? 345–137+1 = <u>209</u>

248 to 593 inclusive? 593–248+1 = <u>346</u>

604 to 752 inclusive? 752–604+1 = <u>149</u>

491 to 885 inclusive? 885–491+1 = <u>395</u>

6. Answer YES or NO as to whether the numbers are valid sides of a triangle.

7, 11, 18	<u>N</u>	10, 10, 10	<u>Y</u>
3, 4, 5	<u>Y</u>	7, 8, 15	<u>N</u>
8, 9, 19	<u>N</u>	3, 20, 21	<u>Y</u>

2. Find the sum of the whole numbers between:

$\sqrt{15}$ and $\sqrt{75}$ <u>30</u> $4+5+6+7+8$

$\sqrt{5}$ and $\sqrt{35}$ <u>12</u> $3+4+5$

$\sqrt{18}$ and $\sqrt{128}$ <u>56</u> $5+6+7+8+9+$ $10+11$

7. Name the number of degrees, if any, of rotational symmetry for the figure.

90°   180°   180°

3.

	R	T	D
#1	60	4	240
#2	40	6	240
tot	(48)	10	480

Last week Sara drove to her sister's home 240 miles away in 4 hours. This week due to road construction and traffic the same trip took 6 hours. What was her overall rate for the two trips?

8. Check if the row number is divisible by the column factor.

	1	4	5	8	9	10
62,028	✓	✓	–	–	✓	–
55,860	✓	✓	✓	–	–	✓
73,737	✓	–	–	–	✓	–

4. Solve by equivalent fractions.

$\dfrac{4}{15} = \dfrac{12}{x}$ (x = 45) $\dfrac{x}{13} = \dfrac{25}{65}$ (x = 5)

$\dfrac{7}{12} = \dfrac{x}{48}$ (x = 28) $\dfrac{9}{x} = \dfrac{18}{28}$ (x = 14)

9. Add.

$\dfrac{2}{15}^{7} + \dfrac{2}{21}^{5}$ $\dfrac{1}{10}^{7} + \dfrac{3}{14}^{5} + \dfrac{6}{35}^{2}$

$\dfrac{14+10}{3 \times 5 \times 7}$ $\dfrac{7+15+12}{2 \times 5 \times 7}$

$\dfrac{24}{3 \times 5 \times 7} \dfrac{\div 3}{\div 3} \left(\dfrac{8}{35}\right)$ $\dfrac{34}{2 \times 5 \times 7} \dfrac{\div 2}{\div 2} \left(\dfrac{17}{35}\right)$

5. Complete with CONGRUENT (C), SIMILAR (S), or NEITHER (N).

All squares are   <u>S</u>

All rectangles are   <u>N</u>

All regular pentagons with sides 12 inches are   <u>C</u>

10. Find the perimeter in units of a square with area 9 square units.

A = 9

s = 3

(P = 12)

Level 6	Number 60

1. Of 85 coffees, 43 are decaffeinated and 35 are flavored. If 17 are both, how many are neither? 24

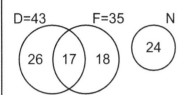

D=43   F=35   N

26 17 18   24

6. Complete the divisibility rules.

If __2__ and __7__ divide a number, then 14 divides the number.

If __2__ and __11__ divide a number, then 22 divides the number.

If __3__ and __8__ divide a number, then 24 divides the number.   Not 2 & 12 or 4 & 6 (overlapping factors).

2. Find the area of the irregular figure. One box equals one square unit.

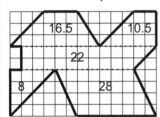

16.5   10.5
22
8   28

50 + 8 + 27

85

or use Subtraction Method

7. Evaluate.

4!	4 x 3 x 2 x 1	24
2!	2 x 1	2
1!		1
7!	7 x 6 x 5 x 4 x 3 x 2 x 1	5040

3. Find the different ways that dimes and half-dollars can have a total value of $2.20.   5

# D	x 10	# H	x 50	# D	x 10	# H	x 50
2	20	4	200	17	170	1	50
7	70	3	150	22	220	0	0
12	120	2	100				

8. Complete the table.

Original Cost	Sales Tax	Final Cost
$30.00	8%	**$32.40**
$25.00	9%	**$27.25**
$45.00	6%	**$47.70**

4. Distribute.

$-9(2x - 3y)$ $-18x + 27y$

$8(a + 4b)$ $8a + 32b$

$7(-7x - 5y)$ $-49x - 35y$

$-6(-6d + 5e)$ $36d - 30e$

9. Fiind the probability that a positive integer less than or equal to 36 is a factor of 36. $\frac{9}{36} = \left(\frac{1}{4}\right)$
1, 2, 3, 4, 6, 9, 12, 18, 36

Fiind the probability that a positive integer less than or equal to 50 is a factor of 50. $\frac{6}{50} = \left(\frac{3}{25}\right)$
1, 2, 5, 10, 25, 50

5. Reflect the point A(6, 4) over the x-axis. Then translate it 3 units up. What is the final point?

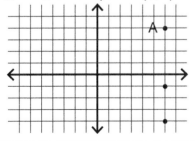

A

(6, −1)

10. Solve.

$\lvert x \rvert = 23$ $x = \pm 23$	$\lvert -2 \rvert = x$ $x = 2$
$\lvert x \rvert = -11$ $x =$ No Sol	$\lvert 5 \rvert = -x$ $x = -5$

Level 6	Number 61

1. A 19 by 32 rectangle has a 1 by 1 square removed from each of its corners. Find the volume in cubic units of the container formed by folding up the resulting sides.

L = 32 – 2 = 30
W = 19 – 2 = 17
H = 1
V = 30 x 17 x 1 = (510)

6. Solve by clearing the decimal points.

.008x = 5.6	.005x = .8	.04x = .28
8x = 5600	5x = 800	4x = 28
x = (700)	x = (160)	x = (7)

2. Find the prime factorization. Work down. In the answer use exponents with primes in ascending order.

120	125	200
12 x 10	5 x 25	8 x 25
4 x 3 x 5 x 2	5^3	2^3 x 5^2
2^3 x 3 x 5		

7. Operate.

–45 ÷ –5	9	–4 x –7	28
16 ÷ –4	–4	9 x –6	–54
–25 ÷ 5	–5	18 ÷ –6	–3
–6 x 5	–30	–20 ÷ –5	4

3. Find the missing sides of the similar triangles. (NTS)

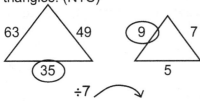

÷7

8. Find the average using the arithmetic sequence method. The mean of an arithmetic sequence is the median. With an even number of numbers, the median is the mean of the 2 middlemost.

500, 550, 600, 650	575
11, 29, 47, 65, 83, 101	56
10, 36, 62, 88, 114	62

4. Find the percent mentally.

85% of 300	255	15% of 240	36
45% of 180	81	65% of 80	52
80% of 90	72	70% of 110	77
35% of 60	21	5% of 700	35

9. Compute using mental math.

Divide by 4 mentally by halving twice.

700 ÷ 4	350	175	660 ÷ 4	330	165
900 ÷ 4	450	225	720 ÷ 4	360	180
640 ÷ 4	320	160	624 ÷ 4	312	156
560 ÷ 4	280	140	600 ÷ 4	300	150

5. Terry bought a textbook for $57.89 and another for $36.75. Find the total cost with 6% sales tax.

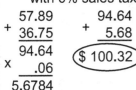

$ 100.32

Or multiply by 1.06 and omit addition.

10. Identify membership in each set.

	N	W	Z	Q	R
$\sqrt{-9}$	–	–	–	–	–
1.1	–	–	–	✓	✓
2.$\overline{6}$	–	–	–	✓	✓
4/9	–	–	–	✓	✓

Level 6	Number 62

1. The number raised to an exponent is called a _____ base.

The formula for the area of a rectangle is the height times the _____ base.

The decimal number system with ten digits uses ten as the _____ base.

6. Divide the product of the first four odd composite numbers greater than 27 by the first six prime numbers.

$$\frac{\overset{3}{\cancel{33}} \times \cancel{35} \times \cancel{39} \times 45}{2 \times 3 \times 5 \times \cancel{7} \times \cancel{11} \times \cancel{13}} = \boxed{\frac{135}{2}}$$

Simplify the 3 and 13 with 39, the 5 and 7 with 35, and 11 with 33.

2. Operate.

−90 − −25	−65	90 − −25	115
−90 + −25	−115	−25 + 90	65
−90 − 25	−115	25 − 90	−65
90 − 25	65	25 − −90	115

7. Four chickens can lay a fixed amount of eggs in 20 minutes. How many chickens would be needed to lay the same number of eggs in 16 minutes?

w x T = W x t
4 x 20 = 5 x 16   Workers and time are inversely proportional-- as one goes up, the other goes down. Their product is constant.

⑤

3. Divide.

$9831.5 \div 10^3$	9.8315
$80.721 \div 10^2$	0.80721
$7843.74 \div 10$	784.374
$825493.5 \div 10^4$	82.54935

8. Find the volume in cubic units of a rectangular prism with the areas of its noncongruent faces 10, 26, and 65 square units.

2 x 5 = 10   V = 2 x 5 x 13 = ⑬⓪
2 x 13 = 26
5 x 13 = 65

4. Find the least value of xy.

3 ≤ x ≤ 11	−2 ≤ y ≤ 8	−22
−4 ≤ x ≤ 5	1 ≤ y ≤ 6	−24
−8 ≤ x ≤ −2	3 ≤ y ≤ 10	−80
−7 ≤ x ≤ 5	−7 ≤ y ≤ 5	−35

9. Evaluate.

−7 + |9 − 17| − 17 + 9 + |−9 − 7| + 7
8 − 17 + 9 + 16   combine −7 and 7
17 − 17 + 16   16

− 10 − |2 − 9| + |−6 − 7| + |0 − 9| − 12
− 10 − 7 + 13 + 9 − 12
−17 + 22 − 12   −7

5. Answer YES or NO as to whether each solid is a polyhedron.

cone	NO
cube	YES
pentagonal pyramid	YES

10. Simplify.

$\frac{10!}{8!}$ 90   9 x 10

$\frac{11!}{9!}$ 110   10 x 11

Level 6	Number 63

1. Find the 6-digit mystery number.
Clue 1: Of the 6 distinct digits, the only odd digit is 3 in the ones place.
Clue 2: The number is less than 270,000.
Clue 3: The 3 rightmost digits form a number divisible by 3.
Clue 4: The number is divisible by 11.
Must start with 2. R must be 603, 063, 483, or 843. 260,843

6. Find the number of 5 by 5 by 5 cubes that can completely fill a box measuring 25 by 40 by 60 of the same unit.

5 cubes by 8 cubes by 12 cubes = 480 cubes

OR $\dfrac{\text{Volume box}}{\text{Volume 1 cube}} = \dfrac{25 \times 40 \times 60}{5 \times 5 \times 5}$

---

2. State whether the fraction would repeat (R) or terminate (T) if converted to a decimal.

$\dfrac{29}{40}$ T | $\dfrac{7}{8}$ T | $\dfrac{21}{60}$ T

$\dfrac{5}{11}$ R | $\dfrac{19}{20}$ T | $\dfrac{8}{15}$ R

7. Write in word form: 12,103,746.02.

twelve million, one hundred three

thousand, seven hundred forty-six

and two hundredths

The decimal point is the "and."

---

3. Complete with the vocabulary word (not congruent).

α and δ are

vertical

β and δ are

corresponding

8. Remove 9 even numbers from the first 49 whole numbers. What percent of the remaining numbers are even?
49 whole numbers: 0 ⟶ 48
25 even, 24 odd
remove 9 even: 40 numbers remain
16 even, 24 odd
$\dfrac{16}{40} = \dfrac{8}{20}$ 40%

To make the fractional part, put "is" (or "are") number over "of" number.

---

4. Find the maximum perimeter in units of a rectangle with whole number sides and area 64 square units.

64 by 1
P = 130

Consider the 2 extremes: as thin as possible versus square.
8 by 8
P = 32

9. Complete.

31 ≡ 1 (mod 2)	27 ≡ 3 (mod 8)
26 ≡ 5 (mod 7)	56 ≡ 2 (mod 6)
49 ≡ 4 (mod 5)	34 ≡ 2 (mod 4)
62 ≡ 2 (mod 3)	50 ≡ 5 (mod 9)

---

5. Bo is 25 years older than Joe who is 4 years older than Jo. Mo, age 20, is 8 years older than Jo. How old is Bo?

Bo 41
25 — Mo 20
4
Joe 16
4
Jo 12

41

10. Operate.

$(8 + 8) \div 8 + 8 \times 8$
2 + 64   66
Do 2 groupings concurrently.

$(6 + 12 \div 6 \times 2) \div 5$
$(6 + 4) \div 5$ 2
The ÷ and × have equal weight so occur L to R.

$9 \times 9 \div 3 - 20 - 5$
27 − 25   2
Never need to see 81.

$(4 + 12) \div 4 - 7 + 2$
4 − 7 + 2   −1
Can add 4 and 2.

MAVA Math: Middle Reviews Solutions Copyright © 2013 Marla Weiss

Level 6	Number 64

1. Joe scored 69, 78, and 81 out of 100 on 3 of 5 math tests. What is the minimum score he must earn on the 4th test to have an average of 82 on the 5 tests?

82 x 5 = 410 $\boxed{82}$

228 — 82 100

$$\frac{69 + 78 + 81 + min + max}{5} = 82$$

6. Randomly select one letter from each word. Find the probability that the letter is a vowel.

ARITHMETIC $\frac{4}{10} = \frac{2}{5}$

NOTEBOOK $\frac{4}{8} = \frac{1}{2}$

SHOPPING $\frac{2}{8} = \frac{1}{4}$

2. Plot and label the points.

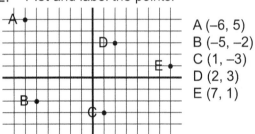

A (–6, 5)
B (–5, –2)
C (1, –3)
D (2, 3)
E (7, 1)

7. Find the volume and surface area of a cylinder with radius 4 and height 11 units.

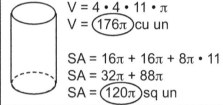

$V = 4 \cdot 4 \cdot 11 \cdot \pi$
$V = \boxed{176\pi}$cu un

$SA = 16\pi + 16\pi + 8\pi \cdot 11$
$SA = 32\pi + 88\pi$
$SA = \boxed{120\pi}$sq un

3. Find the simple interest on $15,000 at 6% annually for 9 months.

I = PRT

$$I = \frac{\overset{75}{\cancel{15,000}}}{1} \cdot \frac{\overset{1}{\cancel{6}}}{\cancel{100}} \cdot \frac{9}{\underset{2}{\cancel{12}}}$$

I = $\boxed{\$675}$

8. Operate and simplify.

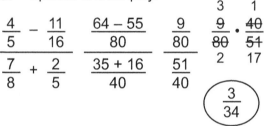

$$\frac{\frac{4}{5} - \frac{11}{16}}{\frac{7}{8} + \frac{2}{5}} \quad \frac{64 - 55}{35 + 16} \quad \frac{9}{80} \quad \overset{3}{\cancel{9}} \cdot \overset{1}{\cancel{40}}$$

$$\frac{80}{40} \quad \frac{51}{40} \quad \frac{\cancel{80}}{2} \cdot \frac{\cancel{54}}{17}$$

$\boxed{\frac{3}{34}}$

4. Given two sides of a triangle, find the range of values for the 3rd side (use s for unknown side).

7, 34 $\underline{27 < s < 41}$
12, 28 $\underline{16 < s < 40}$
18, 39 $\underline{21 < s < 57}$

Subtract given sides for low bound. Add for high bound. "The sum of two sides must be greater than the 3rd" for a triangle to form.

9. Convert to base ten.

111001_{two} 32 + 16 + 8 + 1 $\underline{57}$

3043_{five} 3 x 125 + 4 x 5 + 3 $\underline{398}$

11112_{three} 81 + 27 + 9 + 3 + 2 $\underline{122}$

2121_{four} 2 x 64 + 1 x 16 + 2 x 4 + 1 $\underline{153}$

5. Find the sum of each arithmetic sequence using the "First plus Last" method.

6 + 7 + 8 + 9 + 10 + 11 + 12 + 13 + 14 + 15 + 16

10 + 17 + 24 + 31 + 38 + 45 + 52 + 59

$$\frac{(6 + 16)(11)}{2} = \boxed{121} \qquad \frac{(10 + 59)(8)}{2} = \boxed{276}$$

10. A square with area 144 square units and an equilateral triangle have equal perimeters. Find the side in units of the triangle.

A □ = 144
S □ = 12
P □ = 48
P △ = 48
S △ = $\boxed{16}$

Level 6	Number 65

1. 300 bees can make 15 pounds of honey in 8 weeks. At the same rate, how many pounds of honey will 400 bees make in 12 weeks?

bees	pounds	weeks
300	15	8
1200	15	2
200	15	12
400	(30)	12

Working in rows, keep 1, change 2 by x or ÷ (direct) or both (inverse).

6. Operate and simplify the continued fraction.

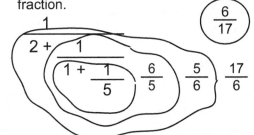

$\left(\dfrac{6}{17}\right)$

$$2 + \cfrac{1}{1 + \cfrac{1}{5}} \qquad \cfrac{6}{5} \qquad \dfrac{5}{6} \qquad \dfrac{17}{6}$$

2. Find the GCF and LCM of 120 and 440 using prime factorization.

$120 = 12 \times 10 = 8 \times 3 \times 5 = 2^3 \times 3 \times 5$
$440 = 4 \times 11 \times 10 = 2^3 \times 5 \times 11$

Look for one of the original numbers before multiplying.

$GCF = 2^3 \times 5 = \boxed{40}$

$LCM = 2^3 \times 3 \times 5 \times 11 = 3 \times 440 = \boxed{1320}$

7. Name the property exemplified.

$5 + 6$ is a real number	C*l*PA
$7(a + b) = 7(b + a)$	CPA
$7(x + y) = (x + y) \cdot 7$	CPM
(a) $\dfrac{1}{a} = 1$	InPM

3. Find the perimeter of the figure on the unit grid.

13   (86)   6

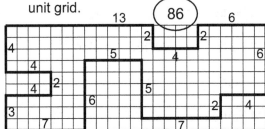

8. Find the volume and total surface area of the solid comprised of unit cubes.

$V = \boxed{9}$ cu un
$SA = \boxed{34}$ sq un

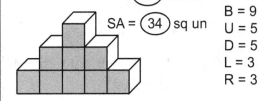

F = 9
B = 9
U = 5
D = 5
L = 3
R = 3

4. Operate.

$\dfrac{1}{3}$ (35 days   7 hours   6 minutes)

$\dfrac{1}{3}$ (33 d   55 h   6 m)

$\dfrac{1}{3}$ (33 d   54 h   66 m)

$\left(11\ d \qquad 18\ h \qquad 22\ m\right)$

9. Write as an algebraic expression.

the quotient of two numbers plus one	$\dfrac{x}{y} + 1$
the sum of double a number and triple a number	$2x + 3y$
half the product of two numbers	$\dfrac{1}{2}(xy)$
five times a number minus two	$5x - 2$

5. Multiply.

$\dfrac{11}{35} \times 70$ 22 $\dfrac{5}{12} \times 72$ 30

$\dfrac{19}{30} \times 90$ 57 $\dfrac{3}{8} \times 96$ 36

$\dfrac{8}{13} \times 39$ 24 $\dfrac{7}{18} \times 36$ 14

10. Multiply.

```
    35,924
  x    852
    71848
  179620
  287392
  30607248
```

```
     21,657
  x   3,428
    173256
     43314
     86628
     64971
   74240196
```

Level 6	Number 66

1. Draw the discards on the left.

Seventy-seven marbles were in a jar. One seventh, not solid-colored, were removed. Of those left, one third were red. What fraction of the original marbles were solid-colored and not red?

77

11 — 66

22 — 44

$\frac{44}{77}$ $\left(\frac{4}{7}\right)$

6. Write the value as an expression.

the number of years in c centuries — 100c

the number of pints in g gallons — 8g

the number of seconds in h hours — 3600h

the number of days in y leap years — 366y

2. Rewrite the decimals in ascending order.

6.25, .652, 6.125, .265, 6.5, 2.56

.265   .652   2.56   6.125   6.25   6.5

7. Seven years ago Kal's age was twice Pam's age in eight years. If Pam is 15 years old now, how old is Kal now?

	–7	now	+8
K	46	(53)	
P	8	15	23

3. What time is 515 minutes after 8:55 AM?

$515 = 60 \times 9 - 25$ $8 + 9 \to 5$ $55 - 25 \to 30$

(5:30 PM)

What time is 265 minutes after 5:20 AM?

$265 = 60 \times 4 + 25$ $5 + 4 \to 9$ $20 + 25 \to 45$

(9:45 AM)

8. Convert measures to a decimal.

8 hours = $0.\overline{3}$ day

2 cups = 0.125 gallon

2 feet = $0.\overline{6}$ yard

4 hours = $0.1\overline{6}$ day

4. Write each fraction as a repeating decimal.

$\frac{20}{44}$ $.4\overline{5}$ | $\frac{10}{33}$ $.\overline{30}$ | $\frac{1}{9}$ $.\overline{1}$

$\frac{6}{11}$ $.\overline{54}$ | $\frac{29}{99}$ $.\overline{29}$ | $\frac{36}{66}$ $.5\overline{4}$

Simplify 20/44 and 36/66 first.

9. Calculate the area in square units of the trapezoid. (NTS)

8

7

20

$M = (20 + 8) \div 2 = 14$

$A = 14 \times 7 = (98)$

Or, 20 – 8 = 12; 12 ÷ 2 = 6;
8 + 6 = 14

5. Multiply using mental math.

8.2 x 8	65.6	15.6 x 5	78
11.3 x 7	79.1	4.9 x 4	19.6
5.5 x 9	49.5	8.9 x 3	26.7
6.4 x 6	38.4	36.5 x 2	73

10. Divide. Answer in exponential form.

$(63 \times 10^6) \div (7 \times 10^2)$ 9×10^4

$(45 \times 10^7) \div (9 \times 10^5)$ 5×10^2

$(64 \times 10^8) \div (8 \times 10^4)$ 8×10^4

$(42 \times 10^9) \div (6 \times 10^3)$ 7×10^6

Picture the 2 groupings as parts of a fraction. Simplify.

| Level 7 | Number 1 |

1. Primes such as 5 and 7 that are two apart are called _____ twin.

A number equivalent to being "pulled out of a hat" is called _____ random.

The numbers 1, 4, 9, 16, 25, and 36 are _____ perfect squares.

6. Find the area and perimeter of the rectangle with vertices:

$(-5, 2)$, $(3, 2)$, $(3, 8)$, and $(-5, 8)$ | $(3, 3)$, $(3, 10)$, $(9, 10)$, and $(9, 3)$

$b = 8$ ⟨A = 48⟩ $b = 6$ ⟨A = 42⟩
$h = 6$ ⟨P = 28⟩ $h = 7$ ⟨P = 26⟩

2. Find the shaded area in square units.

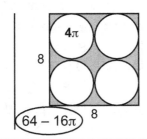

12, 12 — $(144 - 36\pi)$

8, 8, 4π — $(64 - 16\pi)$

7. Operate.

5^{-3}	$\dfrac{1}{125}$	4^4	256	6^{-2}	$\dfrac{1}{36}$
7^{-2}	$\dfrac{1}{49}$	17^0	1	10^{-3}	$\dfrac{1}{1000}$
		2^{10}	1024		
19^{-1}	$\dfrac{1}{19}$	3^5	243	11^{-2}	$\dfrac{1}{121}$

3. Add the rectangular matrices.

$$\begin{bmatrix} 11 & 8 \end{bmatrix} + \begin{bmatrix} 9 & 7 \end{bmatrix} = \begin{bmatrix} 20 & 15 \end{bmatrix}$$

$$\begin{bmatrix} 0 & 16 \\ 6 & 5 \\ -3 & 13 \end{bmatrix} + \begin{bmatrix} 4 & 12 \\ -1 & -8 \\ -6 & -4 \end{bmatrix} = \begin{bmatrix} 4 & 28 \\ 5 & -3 \\ -9 & 9 \end{bmatrix}$$

8. Answer YES or NO as to whether the 3 numbers form sides of a right triangle.

3, 4, 5 — Y | 9, 16, 25 — N

5, 6, 7 — N | 6, 6, 6 — N

6, 8, 10 — Y | 7, 24, 25 — Y
3, 4, 5 x 2

4. Find the area of the rectangle bounded by the lines $y = -3$, $x = -1$, $x = 9$, and $y = -11$.

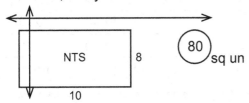

NTS 8 ⟨80⟩ sq un

10

9. Find the average using the arithmetic sequence method.

5.9, 7.9, 9.9, 11.9, 13.9, 15.9 — 10.9

10.2, 14.3, 18.4, 22.5, 26.6, 30.7 — 20.45

24.25, 26.35, 28.45, 30.55 — 27.4

5. Find the percent increase. $\dfrac{change}{original}$

From 25 to 50
$\dfrac{25}{25}$ = ⟨100%⟩

From 25 to 40
$\dfrac{15}{25}$ = ⟨60%⟩

From 25 to 30
$\dfrac{5}{25}$ = ⟨20%⟩

From 25 to 45
$\dfrac{20}{25}$ = ⟨80%⟩

10. Multiply and simplify by regrouping.

7 • (3 YD 2 FT 8 IN)

21 YD 14 FT 56 IN

21 YD 18 FT 8 IN

⟨27 YD 0 FT 8 IN⟩

Level 7	Number 2

1. Find the average rate in miles per hour of a plane that flies 90 miles in 18 minutes.

D = RT

$90 = R \times \dfrac{18}{60}$

$R = 90 \times \dfrac{60}{18} = 5 \times 60$

$R = \boxed{300}$

2. What time is 548 minutes before 4:30 AM?

$548 = 60 \times 9 + 8 \quad 4 - 9 \to 7 \quad 30 - 8 \to 22$

$\boxed{\text{7:22 PM}}$

What time is 475 minutes before 6:10 PM?

$475 = 60 \times 8 - 5 \quad 6 - 8 \to 10 \quad 10 + 5 \to 15$

$\boxed{\text{10:15 AM}}$

3. Given two sides of a triangle, find the range of values for the perimeter (s for unknown side, p for perimeter).

2, 22	$20 < s < 24$	$44 < p < 48$	Add 2 sides to each of LO & HI of 3rd side.
7, 13	$6 < s < 20$	$26 < p < 40$	
8, 24	$16 < s < 32$	$48 < p < 64$	

4. Define operation ★ as:

A ★ B = | 2A − B | + 3AB. For example, 3 ★ 8 = 2 + 72 = 74.

Find the values.

2 ★ 4.5 $\underline{27.5}$ 0.5 + 27

2.5 ★ 5 $\underline{37.5}$ 0 + 37.5

5. Find the midpoint of the line segment with the given endpoints.

(7.5, −2) and (10.5, 8)	$\underline{(9, 3)}$	Average the xs, average the ys.
(−6, 5.1) and (−8, 7.9)	$\underline{(-7, 6.5)}$	
(5.4, −5) and (9.2, 19)	$\underline{(7.3, 7)}$	
(−4, 3) and (12, −7)	$\underline{(4, -2)}$	

6. Find the area of an isosceles triangle with sides 10, 13, 13.

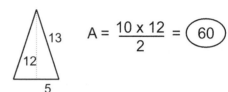

$A = \dfrac{10 \times 12}{2} = \boxed{60}$

7. Find the area of the circle inscribed in a square of area 81 square units.

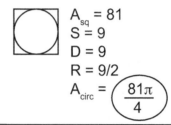

$A_{sq} = 81$

$S = 9$

$D = 9$

$R = 9/2$

$A_{circ} = \boxed{\dfrac{81\pi}{4}}$

8. Find the complement of the angle.

Comp (52°)	38°	Comp (44°)	46°
Comp (29°)	61°	Comp (16°)	74°
Comp (37°)	53°	Comp (63°)	27°
Comp (78°)	12°	Comp (81°)	9°

9. Find the slope and intercepts of the lines.

line	slope	y-intercept	x-intercept
y = 4x + 8	4	(0, 8)	(−2, 0)
2x + y = 4	−2	(0, 4)	(2, 0)
y = 6	0	(0, 6)	none

Picture 0 inserted to find intercepts mentally.

10. A square with area 100 square units and an equilateral triangle have equal perimeters. Find the side in units of the triangle.

A □ = 100

S □ = 10

P □ = 40

P △ = 40

S △ = $\boxed{13.\overline{3}}$

Level 7	Number 3

1. Write the inverse of the conditional statement. Are they logically equivalent? (NO)

If polygon P is a square, then polygon P is a rectangle.

statement TRUE
inverse FALSE

If polygon P is a not a square, then _____ polygon P is a not a rectangle.

P could be a rectangle.

6.

	R	T	D
B	5	2	10
S	3.5	4	14
tot	(4)	6	24

Bo walked for 2 hours at a rate of 5 miles per hour. Sam walked for 4 hours at a rate of 3.5 miles per hour. What was their average rate?

2. Find the prime factorization. Work down. In the answer use exponents with primes in ascending order.

1000	1250	4900
10^3	125×10	49×10^2
$\mathbf{2^3 \times 5^3}$	$5^3 \times 2 \times 5$	$7^2 \times 2^2 \times 5^2$
	$\mathbf{2 \times 5^4}$	$\mathbf{2^2 \times 5^2 \times 7^2}$

7. Answer YES or NO as to whether the numbers are valid sides of a triangle.

6.1, 6.3, 12.3 — Y   7.2, 6.4, 13.9 — N

4.6, 7.2, 11.8 — N   7.1, 9.2, 15.8 — Y

3.3, 3.3, 3.3 — Y   3.4, 5.6, 7.8 — Y

3. Find the area of the rhombus in square units given both diagonals.

D = 22   d = 15

$A = \dfrac{22 \times 15}{2}$

$= 11 \times 15 = (165)$

D = 37   d = 20

$A = \dfrac{37 \times 20}{2}$

$= 37 \times 10 = (370)$

8. Check if the row number is divisible by the column factor.

	2	3	5	9	15	18
24,930	✓	✓	✓	✓	✓	✓
46,455	–	✓	✓	–	✓	–
93,636	✓	✓	–	✓	–	✓

4. Graph on the coordinate plane. One box equals one unit.

y = x + 1   y = –x + 2

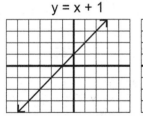 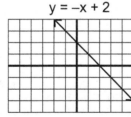

9. Find the reciprocal in the given mod.

reciprocal of 6 mod 7   6   *6x6=36, 36−35=1*

reciprocal of 3 mod 5   2   *3x2=6, 6−5=1*

reciprocal of 2 mod 3   2   *2x2=4, 4−3=1*

reciprocal of 4 mod 6   none

5. Find the face diagonal of a cube with

edge = 6 $6\sqrt{2}$ e = 6

area of face = 81 $9\sqrt{2}$ e = 9

volume = 8 $2\sqrt{2}$ e = 2

edge = 7 $7\sqrt{2}$ e = 7

10. Find the sum of the angles of each polygon.

pentagon   540   *180 x 3*

decagon   1440   *180 x 8 = 800 + 640*

22-gon   3600   *180 x 20*

Level 7	Number 4

1. The angles of a quadrilateral are in the ratio 1:2:3:4. Find the angles.

sum angles = 360

1 + 2 + 3 + 4 = 10

360 ÷ 10 = 36

angles = ⟨ 36, 72, 108, 144 ⟩

6. Mentally calculate the midline of each trapezoid.

base	BASE	midline
12.4	35.6	**24**
11.1	23.2	**17.15**
30.4	40.4	**35.4**

2. Find the equal number of nickels, quarters, and half dollars that together make $10.40.

$5x + 25x + 50x = 1040$

$80x = 1040$

$8x = 104$

$x = ⟨ 13 ⟩$

7. Convert to base ten.

$T91_{eleven}$	10 x 121 + 9 x 11 + 1	1310
287_{twenty}	2 x 400 + 8 x 20 + 7	967
$T2E_{twelve}$	10 x 144 + 2 x 12 + 11	1475
$365_{fifteen}$	3 x 225 + 6 x 15 + 5	770

3. Multiply by 11 mentally.
Leave left and right digits. Insert pairwise digit sums.

123 x 11	1353	352 x 11	3872
422 x 11	4642	804 x 11	8844
516 x 11	5676	253 x 11	2783
724 x 11	7964	816 x 11	8976

8. Complete the chart for a circle.

radius	diameter	area	circumference
$\frac{9}{2}$	9	$\frac{81\pi}{4}$	9π
$\frac{7}{2}$	7	$\frac{49\pi}{4}$	7π
1.1	2.2	1.21π	2.2π

4. Find x in degrees. (NTS)

$x°$ ⟨ 110 ⟩ by supplement

70 by triangle

75°   35°

9. Of 7 2-digit multiples of 10, the mode is 30, the median is 40, the mean is 50, and the range is 80. Find the numbers.

10   30   30   40   70   80   90

Place 10, 90, 40, and the 30s first. The sum = 50 x 7 = 350. Then 350 − 200 = 150. Only 70 and 80 work next or else 90 would be a mode.

5. Solve.

$-3x - 4 \leq 29$	$8x - 5 \geq 13$	$9 - 2x < 22$
$-3x \leq 33$	$8x \geq 18$	$-2x < 13$
⟨ $x \geq -11$ ⟩	$x \geq \frac{18}{8}$	⟨ $x > \frac{-13}{2}$ ⟩
	$x \geq \frac{9}{4}$	

10. Find the volume in cubic units of a square pyramid with edge 12 and altitude 8 units.

$$\frac{12 \times 12 \times 8}{3} = 4 \times 12 \times 8 = 48 \times 8$$

$$= 320 + 64 = ⟨ 384 ⟩$$

	Level 7	Number 5

1. Four girls can mow 30 lawns in 5 days. Eight girls can mow how many lawns in a day and a half?

girls	lawns	days
4	30	5
8	30	$\frac{5}{2}$
8	6	$\frac{1}{2}$
8	(18)	$\frac{3}{2}$

6. Find the area of parallelogram ABCD with vertices A(–3, 7), B(6, 7), C(–2, –5), and D(–11, –5).

B = 9, H = 12

A = 9 x 12 = (108)

2. Graph on the number line.

x > 1 AND x > 0

x ≤ 3 OR x > 4

7. Operate.

38 ÷ –2	–19	–15 x 5	–75
–3 x –16	48	6 x –12	–72
–99 ÷ –9	11	0 ÷ –21	0
–15 x 4	–60	–11 x 34	–374

3. List all subsets of {A, B}.

{ }

{A}

{B}

{A, B}

8. Find the volume in cubic units of a rectangular prism with the areas of its noncongruent faces 30, 35, and 42 square units.

30 = 2 x 3 x 5 = 5 x 6
35 = 5 x 7
42 = 2 x 3 x 7 = 6 x 7

V = 5 x 6 x 7 = 30 x 7 = (210)

4. Find the percent mentally.

2% of 150	3	12% of 700	84
4% of 950	38	4% of 325	13
6% of 800	48	11% of 900	99
8% of 50	4	18% of 200	36

9. Operate in the given mod.

3 + 4 + 5	≡	0	(mod 6)
7 x 4 x 8	≡	8	(mod 9)
(5 x 4) + (5 x 3)	≡	0	(mod 7)
7 + (3 x 6)	≡	1	(mod 8)

Can simplify value partway rather than at end.

5. Find the perimeter of a rectangle with area 143 square units and length 13 units.

A = 143
L = 13
W = 11
SP = 24
(P = 48)

10. Find the area of the parallelogram.

8, 15, 17 right Δ

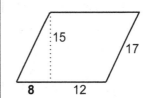

15, 17, 8, 12

A = 20 x 15

A = (300)

Level 7	Number 6

1. Find the day of the week given each separate condition.

219 days ago if today is Monday

−217 = Mon   <u>Saturday</u>

354 days ago if today is Tuesday

−357 = Tues   <u>Friday</u>

6. Operate and simplify.

$$5\frac{4}{5} + 6\frac{5}{8} + 3\frac{13}{20}$$

$$5\frac{32}{40} + 6\frac{25}{40} + 3\frac{26}{40}$$

$$14\frac{83}{40} \qquad \boxed{16\frac{3}{40}}$$

2. Operate.

−78 − −12	<u>−66</u>	51 − −18	<u>69</u>
−28 + −33	<u>−61</u>	−37 + 89	<u>52</u>
−73 − 37	<u>−110</u>	32 − 88	<u>−56</u>
53 − 99	<u>−46</u>	14 − −76	<u>90</u>

7. For a trapezoid:

A = 230, b = 16, h = 10. Find B.

M = 230 / 10 = 23

16, 23, ⓷⓪

A = 400, B = 65, h = 10. Find b.

M = 400 / 10 = 40

⒂ , 40, 65

Calculate by arithmetic sequence.

3. Find the volume of the cone in cubic units.

radius = 9
height = 11

$$V = \frac{9 \cdot 9 \cdot \pi \cdot 11}{3}$$

$$= 27 \cdot 11 \cdot \pi$$

$$= \boxed{297\pi}$$

8. Complete the table of values for 3x + 5y = 30

x	−5	−2	−1	0	2	5	10	15	20	25
y	9	$\frac{36}{5}$	$\frac{33}{5}$	6	$\frac{24}{5}$	3	0	−3	−6	−9

4. Solve by cross multiplication.

$$\frac{12}{13} = \frac{5}{x} \qquad \frac{2}{5} = \frac{x}{6} \qquad \frac{14}{x} = \frac{3}{5}$$

12x = 65   5x = 12   3x = 70

$$\boxed{x = \frac{65}{12}} \quad \boxed{x = \frac{12}{5}} \quad \boxed{x = \frac{70}{3}}$$

9. Find the supplement of the angle.

Supp (63°)	<u>117°</u>	Supp (119°)	<u>61°</u>
Supp (102°)	<u>78°</u>	Supp (166°)	<u>14°</u>
Supp (48°)	<u>132°</u>	Supp (131°)	<u>49°</u>
Supp (156°)	<u>24°</u>	Supp (57°)	<u>123°</u>

5. Find the least natural number n such that kn is a perfect cube.

k = 6   n = <u>36</u>   6=2x3, 4x9=36

k = 10   n = <u>100</u>   10=2x5, 4x25=100

k = 15   n = <u>225</u>   15=3x5, 9x25=225

Three of each prime create a perfect cube.

10. Solve.

$$|x + 5| = 15 \qquad |x − 2| = 19$$

x = <u>10, −20</u>   x = <u>21, −17</u>

$$|x − 3| = 12 \qquad |x + 6| = 13$$

x = <u>15, −9</u>   x = <u>7, −19</u>

Level 7	Number 7

1. Answer as indicated.

16 is what % of 20?	16 is 20% of what number?	What is 16% of 20?
$\dfrac{16}{20} = \dfrac{80}{100}$	$16 = \dfrac{1}{5}(x)$	10% is 2 5% is 1 1% is .2
(80%)	(80)	(3.2)

2. Answer as indicated.

20 is what % of 16?	20 is 16% of what number?	What is 20% of 16?
$\dfrac{20}{16} = \dfrac{5}{4}$	$20 = \dfrac{4}{25}(x)$	10% is 1.6
(125%)	(125)	(3.2)

3. Find the volume and surface area of a sphere with radius 3 units.

$$V = \frac{4 \cdot \pi \cdot 3 \cdot 3 \cdot 3}{3} = \boxed{(36\pi)}\text{cu un}$$

$$SA = 4 \cdot \pi \cdot 3 \cdot 3 = \boxed{(36\pi)}\text{sq un}$$

4. Identify as a cube or not a cube.

YES NO
NO
NO

5. Operate on the sets.

A = {4, 5, 6, 7, 9}
B = {1, 2, 3, 5, 7, 9}
U = {1, 2, 3, 4, 5, 6, 7, 8, 9, 10}

A′	=	{1, 2, 3, 8, 10}
B′	=	{4, 6, 8, 10}
(A ∩ B)′	=	{1, 2, 3, 4, 6, 8, 10}

6. Multiply by 25 mentally.

Divide by 4 (cut in half twice). Tack on two 0s.

4872 x 25	121,800	840 x 25	21,000
1660 x 25	41,500	560 x 25	14,000
3244 x 25	81,100	524 x 25	13,100
3684 x 25	92,100	380 x 25	9500

7. Name the property exemplified.

If x = y, then y = x.	SPE
xy = yx	CPM
126 = 126	RPE
If x = y and y = z, then x = z.	TPE

8. Convert to scientific notation.

0.56	5.6×10^{-1}
0.01439	1.439×10^{-2}
0.0091	9.1×10^{-3}
0.0287	2.87×10^{-2}

9. Operate.

$1 + 7 \div 5 \times 10 + 3 \div 6$	$5 \div 2 \times 6 \div 5 - 1 \div 5$
$1 + 14 + .5$ (15.5)	$15 \div 5 - .2$ (2.8)
$9 + 5 \div 3 \times 9 - 4 \div 10$	$(4 + 3) \div 4 \times 8 - .9$
$9 + 15 - .4$ (23.6)	$14 - .9$ (13.1)

10. A 3 by 3 by 3 cube is painted on the outside and then cut into 27 1 by 1 by 1 cubes. How many of the little cubes have each number of faces painted?

3 faces?	8	1 face?	6
2 faces?	12	0 faces?	1

Level 7	Number 8

1. The average of 10 numbers is 19. Subtract 8 from four of these numbers. Find the new average.

sum = 10 x 19 = 190
new sum = 190 − 32 = 158
new average = 158/10 = (15.8)

6. Five years ago Jay was the same age as Matt now. Now Paul is twice Matt's age. If Jay will be 18 in 3 years, how old was Paul 5 years ago?

	−5	now	+3
J		10	15	18
M			10	
P		(15)	20	

2. Picking 4 bushels of peaches in 25 minutes, how many bushels can be picked in 5 hours at the same rate?

$\frac{4}{25} = \frac{x}{5 \cdot 60}$ $x = 4 \cdot 12$
$x = $ (48)

$x = \frac{4 \cdot 5 \cdot 60}{25}$

7. Find the volume of a wedge cut at a 60° central angle from a cylinder with radius 10 and height 6.

$\frac{60}{360} = \frac{1}{6}$ $V = \frac{10 \cdot 10 \cdot 6 \cdot \pi}{6}$

(100π)

3. Find the simple interest on $12,000 at 5.5% annually for 20 months.

I = PRT

$I = \frac{12,000}{1} \cdot \frac{5.5}{100} \cdot \frac{20}{12}$

I = 55 x 20 = ($1100)

8. Multiply by 11 mentally.

123 x 11   1353
2345 x 11   25,795
452 x 11   4972
5436 x 11   59,796

Write from R to L. For 2345: Write the 5. Then 5+4 = 9; 4+3 = 7; 3+2=5. Write the 2.

4. 10! contains how many factors of 12?

10 · 9 · 8 · 7 · 6 · 5 · 4 · 3 · 2
2· 3·3·4·2

(4)

9. Convert 582 in base ten to base five.

582
− 500
82
− 75
7
− 5
2

4	3	1	2
125	25	5	1

(4312 five)

5. Find the probability of selecting a point from the shaded area within the rectangle.

$\frac{A_{sh}}{A_{rec}} = \frac{240 - 16}{240}$
$= \frac{224}{240} = \frac{56}{60}$

($\frac{14}{15}$)

10. Evaluate.

$\sqrt{1.21}$ 1.1 $\sqrt{.0144}$ .12
$\sqrt{.541^2}$ .541 $\sqrt{6.25}$ 2.5
$\sqrt{2.56}$ 1.6 $\sqrt{.97^2}$ .97
$\sqrt{.0016}$ .04 $\sqrt{2.25}$ 1.5

| Level 7 | Number 9 |

1. Find the number of ways to park 5 different colored cars in a row with the blue one in the middle.

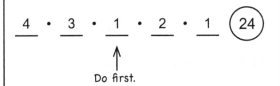

$\underline{4} \cdot \underline{3} \cdot \underline{1} \cdot \underline{2} \cdot \underline{1}$ (24)

Do first.

6. Count the number of paths from A to B moving only right and/or up.

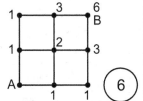

(6)

Rather than tracing paths with a pencil or listing them with labeled vertices, mark initial vertices with the # of paths to the next point; sum paths for successive vertices.

2. Count the total number of triangles in the picture.

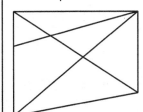

1: 5
2: 6
3: 2
4: 1
5: 0
6: 0

(14)

7. Give the simplified fractional part that the first decimal is of the second.

| $\frac{1.04}{4.16}$ | $\frac{104}{416}$ $\left(\frac{1}{4}\right)$ | $\frac{5.2}{21.6}$ | $\frac{52}{216}$ $\left(\frac{13}{54}\right)$ |

| $\frac{20.1}{21.3}$ | $\frac{201}{213}$ $\left(\frac{67}{71}\right)$ | $\frac{3.12}{5.16}$ | $\frac{312}{516}$ $\left(\frac{26}{43}\right)$ |

3. Find the perimeter of the figure on the unit grid.

(62)

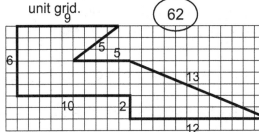

8. Write as an equation.

The quotient of two numbers minus two is ten. $\frac{x}{y} - 2 = 10$

Twice the quantity six plus a number is nine. $2(6 + x) = 9$

Ten less than a number is twice the number. $x - 10 = 2x$

The positive difference of a number and five is one. $|x - 5| = 1$

4. Without graphing on the number line, identify each as 2 points, 2 rays, line, open ray, or open/half-open segment.

$-9 < x < 17$ open segment

$x = 6$ OR $x = 18$ 2 points

$x > 17$ open ray

9. Find the point equidistant from points $(-1, 4)$, $(7, 4)$, $(7, -2)$, and $(-1, -2)$.

Points are vertices of a rectangle.
b = 8; 8/2 = 4; 7 – 4 = 3
h = 6; 6/2 = 3; 4 – 3 = 1

(3, 1)

5. Operate using mental math.

$8 - 2.17$	5.83	$4 \div .25$	16
$11 - 5.6$	5.4	$8 \div .125$	64
$15 - 3.3$	11.7	$9 \div .3$	30
$20 - .19$	19.81	$18 \div .5$	36

10. A football team won 6 games and lost 30. What fraction of its games did the team win?

$\frac{W}{G} = \frac{6}{36} = \left(\frac{1}{6}\right)$

In a class of 40 students, 24 are girls. What fraction of the students are boys?

$\frac{B}{S} = \frac{16}{40} = \left(\frac{2}{5}\right)$

Level 7	Number 10

1. Arrange the digits 2, 3, 5, 6, 7, and 9 into two 3-digit numbers with the:

greatest positive difference	least positive difference
976 Max L to R. **− 235** Min L to R. ―――― **741**	**623** Min L to R. **− 597** Max L to R. ―――― First, 100s 1 **26** apart. 723 − 695 = 28 356 − 297 = 59

6. Find each term for the repeating sequence EFGHEFGH... .

cycle length 4

1600th	H	4 divides 1600; R0
590th	F	4 divides 588; R2
233rd	E	4 divides 232; R1

2. Of 250 children:
28 like juice, milk, & soda;
50 like juice & milk;
60 like milk & soda;
30 like juice & soda;
150 like milk;
65 like juice;   250 − 205
102 like soda.
How many like none?   45

J=65   M=150
13   22   68
28
2   32
40
S=102

7. Find the average using the rightmost digit methods.

50, 50, 50, 50, 52 keep 50	2 ÷ 5	50.4
120, 121, 124, 127, 129 keep 120	21 ÷ 5	124.2
111, 116, 116, 118 keep 110	21 ÷ 4	115.25

3. Write each fraction as a repeating decimal.

$\dfrac{4}{45}$ $.0\overline{8}$	$\dfrac{1}{300}$ $.00\overline{3}$	$\dfrac{1}{900}$ $.00\overline{1}$
$\dfrac{7}{90}$ $.0\overline{7}$	$\dfrac{7}{330}$ $.0\overline{21}$	$\dfrac{17}{990}$ $.0\overline{17}$

Convert 45 to 90, 300 to 900, 330 to 990.

8. Answer as a simplified fraction.

The two 100s in each % simplify.

15% is what fractional part of 35%? $\dfrac{15}{35}$ $\dfrac{3}{7}$

12% is what fractional part of 66%? $\dfrac{12}{66}$ $\dfrac{2}{11}$

20% is what fractional part of 85%? $\dfrac{20}{85}$ $\dfrac{4}{17}$

4. Calculate the point on the number line that is:

$\dfrac{3}{5}$ of the way from −1 to 3   4/5 = .8   **1.4**
 −1 + 2.4

$\dfrac{1}{2}$ of the way from −2 to 7   9/2 = 4.5   **2.5**
 −2 + 4.5

$\dfrac{1}{4}$ of the way from 9 to −5   14/4 = 3.5   **5.5**
 9 − 3.5

9. Find the remainder without dividing.

567,934 ÷ 10 567,930 div by 10	R 4
253,819 ÷ 11 253,814 div by 11	R 5
140,143 ÷ 14 140,140 div by 2, 7	R 3
936,616 ÷ 15 936,615 div by 3, 5	R 1

5. Operate.

$\begin{array}{r} 4238_{nine} \\ + 3756_{nine} \\ \hline 8105_{nine} \end{array}$ $\begin{array}{r} 3011_{four} \\ - 1223_{four} \\ \hline 1122_{four} \end{array}$ $\begin{array}{r} 1243_{five} \\ + 2234_{five} \\ \hline 4032_{five} \end{array}$

10. A rectangular prism has sides 3, 4, and 5. Find the 3 distinct face diagonals.

3, 4,   (5)   Pythagorean primitive

3, 5,   (√34)   9 + 25 = 34

4, 5,   (√41)   16 + 25 = 41

Level 7	Number 11

1. Draw the discards on the left. 48

Of 48 people, 1/8 ate salad, 1/4 ate pizza, and the rest ate soup. Of the soup eaters, 1/3 had chicken, 1/6 had tomato, and the rest had beef. The beef soup eaters are what fractional part of the total group?

6 12 30

10 5 15

$\frac{15}{48}$ $\left(\frac{5}{16}\right)$

6. Translate the cipher with each letter as its own digit.

$$\begin{array}{r} \text{SPAP} \\ +\text{ARRT} \\ \hline \text{SSSSP} \end{array} \quad \begin{array}{r} \text{1P9P} \\ +9220 \\ \hline 11,11\text{P} \end{array} \quad \begin{array}{r} \mathbf{1898} \\ +\mathbf{9220} \\ \hline \mathbf{11118} \end{array}$$

S=1, the most that can be carried from 2 addends. In 1s, T=0. In 1000s, A=9. In 10s, R=2. In 100s, P=8.

2. Find the missing sides of the similar triangles. (NTS)

3.5   5.6   10.5   (16.8)

(4.4)   13.2

x3

7. Evaluate for x = 2.

$(x-3)^3 - x^2 + 5x^0$	$-1 - 4 + 5$	0
$(2x)^{-1} + 5x^2 - 8x$	$.25 + 20 - 16$	4.25
$x^{-1} + (x-2)^7 - 10$	$.5 + 0 - 10$	-9.5
$2x^{-1} + (5x-8)^0$	$1 + 1$	2

3. Estimate the fraction to the nearest whole number.

0.25   ~~500~~ 20   ~~9~~

$$\frac{0.248 \times 502.06 \times 9.017}{8.88 \times 25.015}$$

~~9~~   ~~25~~

(5)

1/4 of 20 is 5.

8. By what fractional part of:

$\frac{3}{4}$ does $\frac{2}{4}$ exceed $\frac{1}{4}$?	$\left(\frac{1}{3}\right)$
$\frac{4}{5}$ does $\frac{3}{5}$ exceed $\frac{2}{5}$?	$\left(\frac{1}{4}\right)$
$\frac{5}{6}$ does $\frac{5}{6}$ exceed $\frac{1}{6}$?	$\left(\frac{4}{5}\right)$

4. Name the specified lattice point in the picture. One box = one unit.

twice as far from A as from B (3, 4)

twice as far from B as from A (0, 1)

(0,0)

9. Complete the unit conversions.

2 cm^2	200	mm^2	1.3 cm^2	130	mm^2
3 cm^3	3000	mm^3	47 mm^3	.047	cm^3
.34 cm^3	340	mm^3	(2 cm)^3	8000	mm^3

5. Find the height of an isosceles right triangle with each given area.

A = 32
$s^2 = 64$
s = h = (8)

A = 50
$s^2 = 100$
s = h = (10)

An isosceles right triangle is half of a square.

10. Solve.

$5(2x - 7) - 4 = 16$
$10x - 35 = 20$
$10x = 55$
$x = \left(\dfrac{11}{2}\right)$

$8(3x - 7) + 2 = 9$
$24x - 56 = 7$
$24x = 63$
$x = \left(\dfrac{21}{8}\right)$

Level 7	Number 12

1. Guy bought 11 pounds of beef for $90.75. What was the cost per pound?

$$\frac{\text{cost}}{\text{pound}} = \frac{90.75}{11} = \boxed{\$8.25}$$

9075 div by 11

The value to the left of "per" is the numerator.
The value to the right of "per" is the denominator.

6. Solve.

$5\sqrt{x} + 9 = 134$	$4\sqrt{x} - 7 = 193$
$5\sqrt{x} = 125$	$4\sqrt{x} = 200$
$\sqrt{x} = 25$	$\sqrt{x} = 50$
$x = \boxed{625}$	$x = \boxed{2500}$

2. If 16 is written as the product of 3 positive integers, each greater than 1, how many of them are even?

2 x 2 x 4 $\boxed{3}$

7. Find AC if BC = 18, CD = 14, and AD = 40.

3, 4, 5 x8
24, 32, 40

3, 4, 5 x6

18, 24, $\boxed{30}$

3. Multiply the rectangular matrix by the constant (scalar multiplication).

$$2\begin{bmatrix} 28 & .5 & -4 \\ 0 & 17 & 39 \end{bmatrix} = \begin{bmatrix} 56 & 1 & -8 \\ 0 & 34 & 78 \end{bmatrix}$$

$$9\begin{bmatrix} 9 & 2 & 11 & 0 \\ 7 & 5 & -1 & 10 \end{bmatrix} = \begin{bmatrix} 81 & 18 & 99 & 0 \\ 63 & 45 & -9 & 90 \end{bmatrix}$$

8. Find the statistics for the first 10 prime numbers.

2, 3, 5, 7, 11, 13, 17, 19, 23, 29

n	10	median	12
sum	129	mode	none
mean	12.9	range	27

4. Find the sum of all 2-digit factors of 140.

1, 2, 4, 5, 7, 10

14, 20, 28, 35, 70, 140

10 + 14 + 20 + 28 + 35 + 70 = $\boxed{177}$

9. Distribute.

$-5a(b - 3c + 8)$	$-5ab + 15ac - 40a$
$7c(a + 4b - c)$	$7ac + 28bc - 7c^2$
$9x(2a + 4b + c)$	$18ax + 36bx + 9cx$
$-3b(-6 + 5b - 3c)$	$18b - 15b^2 + 9bc$

5. Find the least value of x ÷ y.

$10 \le x \le 20$	$4 \le y \le 5$	$10 \div 5$	2
$-4 \le x \le 24$	$2 \le y \le 8$	$-4 \div 2$	-2
$-8 \le x \le -2$	$4 \le y \le 10$	$-8 \div 4$	-2
$-8 \le x \le 16$	$-2 \le y \le 8$	$16 \div -2$	-8

10. A book was on sale for $23.66. That was 9% off the regular price. What was the regular price?

.91P = 23.66

$$\frac{2366}{91} = \frac{338}{13} = \boxed{\$26.00}$$

Level 7	Number 13

1. Find the 6-digit mystery number.
Clue 1: The number is divisible by 12.
Clue 2: The number is a palindrome.
Clue 3: The number has 3 distinct digits.
Clue 4: The digits are even and positive.
Clue 5: The number is less than 430,000.

Ends in 24, 28, 48, 64, 68, or 84
0 not positive   426,624

6. Find the area and perimeter of the triangle with vertices:

(0, 0), (4, 0), and (0, −3).	(0, −15), (8, −15), and (8, 0)
legs = 3 & 4	legs = 8 & 15
hypotenuse = 5	hypotenuse = 17
A = 6	A = 60
P = 12	P = 40

2. Find the shaded area in square units.

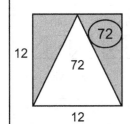 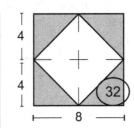

7. Combine like terms.

$5(x + 2) - 3(4 - x) + 8x - 1$	$16x - 3$
$3(x + y) - 2x - 4(x + 1) - 4y$	$-3x - y - 4$
$-5(x + 3) - 2(2 - x) + 3x - 7$	-26
$10 + 4(x - 3) - 6x - 2(x + 5)$	$-4x - 12$

3. Answer ALWAYS, SOMETIMES, or NEVER.

A central angle is acute.	S
Two right angles are supplementary.	A
Vertical angles are right.	S
A straight angle is obtuse.	N
Vertical angles are complementary.	S

8. Find the angle in degrees formed by clock hands at 9:30.

360°/12 = 30° between neighboring numbers

30 x 3.5 = 105

Hour hand is halfway between 9 and 10.

4. Write each repeating decimal as a simplified fraction.

$.\overline{18}$ $\dfrac{18}{99}$ $\dfrac{2}{11}$ $.\overline{75}$ $\dfrac{75}{99}$ $\dfrac{25}{33}$

$.0\overline{18}$ $\dfrac{2}{110}$ $\dfrac{1}{55}$ $.0\overline{75}$ $\dfrac{25}{330}$ $\dfrac{5}{66}$

9. Find the opposite in the given mod.

opposite of 4 mod 7   3   4+3=7, 7−7=0

opposite of 3 mod 5   2   3+2=5, 5−5=0

opposite of 2 mod 3   1   2+1=3, 3−3=0

opposite of 1 mod 6   5   1+5=6, 6−6=0

5. Find the percent increase. $\dfrac{change}{original}$

From 50 to 55	From 24 to 36
$\dfrac{5}{50}$ = 10% $\dfrac{10}{100}$	$\dfrac{12}{24}$ = 50% $\dfrac{1}{2}$
From 20 to 80	From 12 to 21
$\dfrac{60}{20}$ = 300% $\dfrac{3}{1}$	$\dfrac{9}{12}$ = 75% $\dfrac{3}{4}$

10. How many perfect squares are between 400 and 3600?

$20^2 = 400$
$60^2 = 3600$
21 to 59;   39   numbers

Do not list the perfect squares but rather the numbers that generate them. Subtract and add 1 to get the number of numbers.

| Level 7 | Number 14 |

1. One and one half hours are what fraction of time between noon on Monday and midnight of Friday of the same week?

$$\frac{1.5}{4.5 \times 24} = \frac{1}{3 \times 24} = \boxed{\frac{1}{72}}$$

6. Find the area of an isosceles triangle with sides 10, 10, 16.

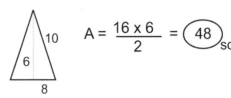

$$A = \frac{16 \times 6}{2} = \boxed{48} \text{ sq un}$$

2. Reflect each point over the line y = x. Then translate as specified.

(1, 4)	(4, 1)	up 3	(4, 4)
(−2, −5)	(−5, −2)	down 2	(−5, −4)
(6, −3)	(−3, 6)	left 4	(−7, 6)

7. Find the perimeter of the 9th shape, continuing the pattern.

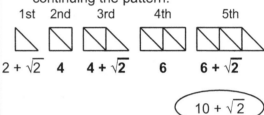

1st 2nd 3rd 4th 5th

$2 + \sqrt{2}$ 4 $4 + \sqrt{2}$ 6 $6 + \sqrt{2}$

$$\boxed{10 + \sqrt{2}}$$

3. Find the missing terms of the arithmetic sequences.

100, _85_, 70, 55, _40_, 25, 10

80, 76, _72_, 68, 64, 60, _56_, 52

45, 46.5, _48_, 49.5, _51_, 52.5

8. Evaluate.

$$-9 - |15 \div 3| - 4 \times 9 \div 3 + |-3 - 6| \times 8$$
$-9 - 5 - 12 + 72$
$-14 + 60$ ___46___

$$- 10 \times |3 - 8| \div |0 - 2| + |7 - 9| - 12$$
$- 10 \times 5 \div 2 + 2 - 12$
$-25 - 10$ ___−35___

4. Find x in degrees. (NTS)

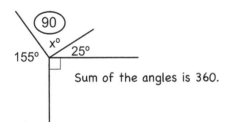

Sum of the angles is 360.

9. Find the slope and intercepts of the lines.

line	slope	y-intercept	x-intercept
y = 3x + 9	3	(0, 9)	(−3, 0)
3x − y = 12	3	(0, −12)	(4, 0)
x = 5	undef	none	(5, 0)

Picture 0 inserted to find intercepts mentally.

5. Find the midpoint of the line segment with the given endpoints.

(8.5, 16) and (6.5, −1)	(7.5, 7.5)
(−4, 4.2) and (−14, 8.8)	(−9, 6.5)
(2.3, −6) and (6.3, −7)	(4.3, −6.5)
(−6, 1) and (20, −9)	(7, −4)

Average the xs, average the ys.

10. Copy the prime factorizations from page 29. Then find the GCF and LCM.

| 112 | 160 | 700 |
| $2^4 \times 7$ | $2^5 \times 5$ | $2^2 \times 5^2 \times 7$ |

GCF (112, 160, 700) = $2^2 \times 5$ = **20**

LCM (112, 160, 700) = $2^5 \times 5^2 \times 7$ = **5600**

Level 7	Number 15

1. The time on a 12-hour circular clock is 10:00 PM. What will be the time after the minute hand goes around 5 times?

(3:00 AM)

A full circle of the minute hand is one hour.

6. Operate and simplify the continued fraction.

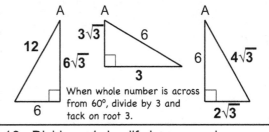

$\left(2\,\dfrac{17}{28}\right)$

$3 - \cfrac{1}{2 + \cfrac{2}{3 + \cfrac{2}{3}}}$ $\dfrac{11}{28}$ $\dfrac{11}{3}$ $\dfrac{6}{11}$ $\dfrac{28}{11}$

2. Find the probability that after spinning each wheel, the sum of the numbers at the arrows is even.

$\left(\dfrac{1}{2}\right)$

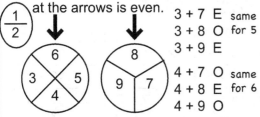

3 + 7 E same
3 + 8 O for 5
3 + 9 E
4 + 7 O same
4 + 8 E for 6
4 + 9 O

7. Operate.

37^{-1}	$\dfrac{1}{37}$	9^3	729	20^{-2}	$\dfrac{1}{400}$
		17^2	289		
4^{-3}	$\dfrac{1}{64}$	2^6	64	6^{-3}	$\dfrac{1}{216}$
25^{-2}	$\dfrac{1}{625}$	14^0	1	11^{-3}	$\dfrac{1}{1331}$

3. Find the area of an equilateral triangle with edge 20.

$$\dfrac{s^2 \cdot \sqrt{3}}{4} = \dfrac{20 \cdot 20 \cdot \sqrt{3}}{4} = \left(100\sqrt{3}\right) \text{ sq un}$$

8. Divide by 5 mentally.

Double; move decimal point 1 place to left.

$4514 \div 5$	902.8	$274 \div 5$	54.8
$2637 \div 5$	527.4	$383 \div 5$	76.6
$1729 \div 5$	345.8	$492 \div 5$	98.4
$3216 \div 5$	643.2	$181 \div 5$	36.2

4. Find the area of the triangle bounded by both axes and the line $y = -x + 4$.

(8) sq un

9. Find the missing sides of the triangles with m∠A = 30°. (NTS)

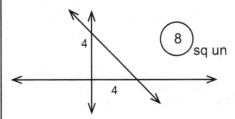

When whole number is across from 60°, divide by 3 and tack on root 3.

5. Find the area in square units of a rectangle with perimeter 54 and length 15 units.

P = 54
SP = 27
L = 15
W = 12
(A = 180)

10. Divide and simplify by regrouping.

(8 DAYS 6 HR 30 MIN 20 SEC) ÷ 4

(8 DAYS 4 HR 150 MIN 20 SEC) ÷ 4

(8 DAYS 4 HR 148 MIN 140 SEC) ÷ 4

(2 DAYS 1 HR 37 MIN 35 SEC)

Level 7	Number 16

1. Write the digit 3 to the right of a 2-digit number. The new 3-digit number is 372 more than the original 2-digit number. Find the original number.

original XY   XY3
new XY3   − 372   ⟨41⟩
            XY
                Y = 1
                X = 4

6. Find the probability of tossing a sum of 5 with 3 standard dice.

3, 1, 1
1, 3, 1
1, 1, 3 $\dfrac{6}{6 \times 6 \times 6}$ = ⟨$\dfrac{1}{36}$⟩
1, 2, 2
2, 1, 2
2, 2, 1

2. What time is 425 minutes before 4:20 AM?
$425 = 60 \times 7 + 5$ $4 - 7 \rightarrow 9$ $20 - 5 \rightarrow 15$
⟨9:15 PM⟩

What time is 350 minutes before 1:15 PM?
$360 = 60 \times 6 - 10$ $1 - 8 \rightarrow 7$ $15 + 10 \rightarrow 25$
⟨7:25 AM⟩

7. Show Goldbach's Conjecture: Even integers > 2 are the sum of 2 primes. Answers may vary.

	17 + 19		**11 + 11**		**3 + 29**
36	**7 + 29**	22	**5 + 17**	32	**13 + 19**
6	**3 + 3**	12	**5 + 7**	40	**17 + 23**
8	**3 + 5**	4	**2 + 2**	44	**13 + 31**
10	**5 + 5**	14	**7 + 7**	56	**19 + 37**
	3 + 7		**3 + 11**		**13 + 43**

3. Draw the reflection of each triangle over the line x = −1.

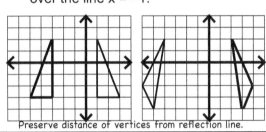

Preserve distance of vertices from reflection line.

8. Operate and simplify.

$$\dfrac{\dfrac{12}{11} \div \dfrac{36}{55}}{\dfrac{7}{9} \times \dfrac{45}{21}} \quad \dfrac{\dfrac{12}{11} \times \dfrac{55}{36}}{\dfrac{7}{9} \times \dfrac{45}{21}} \quad \dfrac{\dfrac{5}{3}}{\dfrac{5}{3}} \quad ⟨1⟩$$

4. Define operation ♥ as:

A ♥ B = |B³ − A²| − B + A. For example,
4 ♥ 2 = |8 − 16| − 2 + 4 = 10.
Find the values.

0.5 ♥ 2   6.25   7.75 − 2 + 0.5

1.1 ♥ 1   0.31   0.21 − 1 + 1.1

9. Find the principal for a time of 1 year.

principal	rate	interest	Get 1%. Then × 100.
$2000	9%	$180	9% is 180
$7000	6%	$420	1% is 20
$500	7%	$35	100% is 2000
$1600	4%	$64	

5. Harry can make a cigar out of 5 cigar butts. He finds 25 cigar butts. How many total cigars would he be able to smoke?   ⟨6⟩

After he makes and smokes 5 cigars, he will have 5 new butts to make one more cigar.

10. A rectangular prism has sides 6, 7, and 8. Find the 3 distinct face diagonals.

6, 7, ⟨$\sqrt{85}$⟩   36 + 49 = 85

6, 8, ⟨10⟩   3, 4, 5 ×2

7, 8, ⟨$\sqrt{113}$⟩   49 + 64 = 113

Level 7 Number 17

1. Interest that is earned on interest is called _compound_.

In a parallelogram, adjacent angles are _supplementary_.

A synonym for a member of a set is _element_.

6. Find the sum in degrees of the supplement of $(3x)°$ and the complement of $(5x)°$.

$180 - 3x$
$90 - 5x$
$\boxed{270 - 8x}$

2. Find the prime factorization. Work down. In the answer use exponents with primes in ascending order.

1210	1430	1024
$11^2 \times 2 \times 5$	143×10	2^{10}
$\mathbf{2 \times 5 \times 11^2}$	$11 \times 13 \times 2 \times 5$	Memorize
	$\mathbf{2 \times 5 \times 11 \times 13}$	powers of 2.

7. Find the next letter in each pattern.

Skip 0, 2, 4, 6, 8
a B e J q **Z**

Skip 1, 3, 5, 7 backwards
Z x T n **F**

Skip 1, 6, 1, 6, 1
a C j L s **U**

3. Find the probability when drawing once from a standard deck of cards.

P(heart OR red) $\dfrac{1}{2}$ red

P(heart AND red) $\dfrac{1}{4}$ heart only

P(1-digit prime OR club) $\dfrac{25}{52}$ 2,3,5,7 3 suits + 13 clubs = 25

8. Answer YES or NO as to whether the 3 numbers form sides of a right triangle.

Equilateral is not right.
12, 12, 12	N	10, 24, 26	Y
		5, 12, 13 x 2	
11, 12, 13	N	13, 24, 29	N

Ones digits alone can rule out a triplet.
| 5, 12, 13 | Y | 15, 36, 39 | Y |
| | | 5, 12, 13 x 3 | |

4. Find the percent.

14% of 100	14	12% of 100	12
14% of 150	21	12% of 300	36
14% of 200	28	12% of 450	54
14% of 250	35	12% of 225	27

9. Find the mean, median, mode, and range of the data in the frequency table. n = 10 sum = 890

number	frequency
100	3
90	4
80	2
70	1

mean = 89
median = 90
mode = 90
range = 30

5. Label the angles given the 2 marked. (NTS)

33° **89** 58°
58 89 33
122 58 33 147
58 122 147 33

10. Answer as indicated.

16 is what % of 40?	15% of what number is 13.5?	260 is 40% of what number?
$\dfrac{16}{40} = \dfrac{4}{10}$	$\dfrac{15}{100}(x) = \dfrac{27}{2}$	40% is 260 20% is 130 100% is 650
$\boxed{40\%}$	$\dfrac{27}{2} \ \dfrac{100}{15} \ \boxed{90}$	$\boxed{650}$

MAVA Math: Middle Reviews Solutions Copyright © 2013 Marla Weiss

Level 7	Number 18

1. A plane flies 300 miles at 450 miles per hour in how many minutes?

$D = RT$
$300 = 450 \times T$

$T = \dfrac{300}{450} = \dfrac{2}{3}$ ⟨**40**⟩

6. Find the statistics for the factors of 60.

1, 2, 3, 4, 5, 6, 10, 12, 15, 20, 30, 60

n	12	median	8
sum	168	mode	none
mean	14	range	59

2. Find the next term in the sequence by successive differences.

0   12   25   41   62   90 ⟨**127**⟩
  12 **13** **16** **21** **28** **37**
    1   **3**   **5**   **7**   **9**

Compute the differences moving down in rows until a pattern emerges. Then build back up.

7. If Jim types 3 pages a minute and Joe types 5 pages a minute, after how many hours will Joe be 40 pages ahead of Jim?

2 pgs   1min ahead
40 pgs 20 min ahead ⟨$\dfrac{1}{3}$⟩

3. Given two sides of a triangle, find the range of values for the perimeter (s for unknown side, p for perimeter).

10, 31	$21 < s < 41$	$62 < p < 82$
15, 45	$30 < s < 60$	$90 < p < 120$
13, 17	$4 < s < 30$	$34 < p < 60$

Add 2 sides to each of LO & HI of 3rd side.

8. Five and one fourth hours are what fraction of the time between 12:00 PM on Tuesday and 12:00 AM on Saturday of the same week?

$\dfrac{5\ 1/4\ \text{HR}}{3\ 1/2\ \text{DY}} = \dfrac{\dfrac{21}{4}}{\dfrac{7}{2} \times 24} = \dfrac{21}{4} \cdot \dfrac{1}{84}$ ⟨$\dfrac{1}{16}$⟩

4. Solve by cross multiplication.

$\dfrac{6}{x} = \dfrac{7}{11}$ | $\dfrac{x}{7} = \dfrac{4}{9}$ | $\dfrac{8}{11} = \dfrac{5}{x}$

$7x = 66$ | $9x = 28$ | $8x = 55$

$x = \left\langle\dfrac{66}{7}\right\rangle$ | $x = \left\langle\dfrac{28}{9}\right\rangle$ | $x = \left\langle\dfrac{55}{8}\right\rangle$

9. Solve.

$4(x - 3) + 5x - 10 = 3(x + 4) - 5 - 3x$
$4x - 12 + 5x - 10 = 3x + 12 - 5 - 3x$
$9x - 22 = 7$
$9x = 29$

$x = \left\langle\dfrac{29}{9}\right\rangle$

5. If 9 pears cost the same as 8 apples, find the ratio of the cost of 18 pears to 18 apples.

$9P = 8A$

$\dfrac{9P}{9A} = \dfrac{8A}{9A}$

$\dfrac{P}{A} = \left\langle\dfrac{8}{9}\right\rangle$

The ratio would remain constant blown up by any factor.

10. A square with area 121 square units and an equilateral triangle have equal perimeters. Find the side in units of the triangle.

A □ = 121
S □ = 11
P □ = 44
P △ = 44
S △ = ⟨$14.\overline{6}$⟩

| Level 7 | Number 19 |

1. Ten chefs can prepare 45 meals in 4 hours. Five chefs can prepare 27 meals in how many hours?

chefs	meals	hours
10	45	4
10	9	$\frac{4}{5}$
10	27	$\frac{12}{5}$
5	27	$\boxed{\frac{24}{5}}$

or 4 hr & 48 min

6. Mia, who lives in NYC with heavy traffic, went 5 miles from her home to a store in a taxi in 1/2 hour. She returned home on a bus that took 45 minutes. What was her overall rate in mph for the round trip?

	R	T	D
To	not needed	$\frac{1}{2}$	5
Fr		$\frac{3}{4}$	5
tot	⑧	$\frac{5}{4}$	10

2. Graph on the number line.

x = –4 OR x = 0 OR x = 5

x > 4 AND x < 2   impossible--no such x, so no graph

7. Estimate the fraction to the nearest whole number.

$$\frac{\cancel{7}}{\cancel{350}} \quad \cancel{0.5} \quad \cancel{35}$$

$$\frac{351.41 \times 0.4929 \times 34.79}{50.012 \times 2.453}$$

$$\frac{\cancel{50} \quad \cancel{2.5}}{5} \qquad ㊾$$

3. Find the area of the rhombus in square units given both diagonals.

D = 40 d = 36

$A = \dfrac{40 \times 36}{2}$

$= 20 \times 36 = \boxed{720}$

D = 46 d = 11

$A = \dfrac{46 \times 11}{2}$

$= 23 \times 11 = \boxed{253}$

8. An area 96 by 135 feet is equivalent in size to an area 72 yards by how many yards?

96/3 = 32
135/3 = 45
32 • 45 = 72 • x
4 • 45 = 9 • x
x = 4 • 5   ⑳

4. Find the volume and surface area of a sphere with radius 6 units.

$V = \dfrac{4 \cdot \pi \cdot 6 \cdot 6 \cdot 6}{3} = \boxed{288\pi}$

$SA = 4 \cdot \pi \cdot 6 \cdot 6 = \boxed{144\pi}$

9. Find the reciprocal in the given mod.

reciprocal of 4 mod 9 $\underline{\quad 7 \quad}$ 4x7=28, 28–27=1

reciprocal of 3 mod 4 $\underline{\quad 3 \quad}$ 3x3=9, 9–8=1

reciprocal of 2 mod 8 $\underline{\text{none}}$

reciprocal of 0 mod 2 $\underline{\text{none}}$

5. How many positive perfect squares are less than 360,025?

360,000 = 600^2

$1^2 \longrightarrow 600^2$

$\boxed{600}$

10. Find the sum of the angles of each polygon.

heptagon $\underline{\quad 900 \quad}$ 180 x 5

13-gon $\underline{\quad 1980 \quad}$ 180 x 11

32-gon $\underline{\quad 5400 \quad}$ 180 x 30

Level 7	Number 20

1. The angles of a triangle are in the ratio 2:3:5. Find the angles.

sum angles = 180

2 + 3 + 5 = 10

180 / 10 = 18

angles = (36, 54, 90)

---

6. Operate and simplify.

$$12\frac{3}{10} + 3\frac{4}{15} - 6\frac{5}{12}$$

$$12\frac{18}{60} + 3\frac{16}{60} - 6\frac{25}{60}$$

$$9\frac{9}{60} \quad \left(9\frac{3}{20}\right)$$

---

2. Operate.

−28 − −27	−1	42 − −39	81
−69 + −52	−121	−21 + 87	66
36 − 81	−45	23 − −57	80
−16 − 58	−74	24 − 79	−55

---

7. Find the area of the circle inscribed in a square of area 121 square units.

A_{sq} = 121

S = 11

D = 11

R = 11/2

A_{circ} = $\left(\dfrac{121\pi}{4}\right)$

---

3. Find the simple interest on $16,000 at 4.5% annually for 30 months.

I = PRT

$$I = \frac{\overset{40}{\cancel{16,000}}}{1} \cdot \frac{4.5}{\cancel{100}} \cdot \frac{\overset{\underset{4}{10}}{\cancel{30}}}{\cancel{12}}$$

I = 40 x 45 = ($1800)

---

8. Convert to base ten.

444_{five}	4 x 25 + 4 x 5 + 4	124
133_{thirty}	1 x 900 + 2 x 30 + 3	963
374_{nine}	3 x 81 + 7 x 9 + 4	310
$21T_{twelve}$	2 x 144 + 1 x 12 + 10	310

---

4. Find the diagonal of a rectangle with area 660 square units and length 11 units.

660/11 = 60

11, 60, 61

D = (61)

---

9. Find the average using the arithmetic sequence method.

8.71, 8.83, 8.95, 9.07, 9.19, 9.31	9.01
7.75, 8.77, 9.79, 10.81, 11.83	9.79
41.41, 45.49, 49.57, 53.65	47.53

---

5. Solve.

−5x − 7 > 11	4x + 3 ≥ 77	−3x − 4 < 5
−5x > 18	4x ≥ 74	−3x < 9
$\left(x < \dfrac{-18}{5}\right)$	2x ≥ 37	(x > −3)
	$\left(x \ge \dfrac{37}{2}\right)$	

---

10. Find the volume in cubic units of a square pyramid with edge 15 and altitude 11 units.

$$\frac{15 \times 15 \times 11}{3} = 5 \times 15 \times 11 = 75 \times 11$$

$$= (825)$$

Level 7	Number 21

1. How many palindromes occur on a 12-hour digital clock? Ignore the colon, so that 1:11 as well as 11:11 are both palindromes.

x0x, x1x, x2x, x3x, x4x, x5x
  where x can be 1 through 9
Plus 1001, 1111, 1221

$$54 + 3 = \boxed{57}$$

6. Multiply by 25 mentally.

Divide by 4 (cut in half twice). Tack on two 0s.

6004 x 25	150,100	608 x 25 15,200
1212 x 25	30,300	124 x 25 3,100
2048 x 25	51,200	160 x 25 4,000
4008 x 25	100,200	872 x 25 21,800

2. Find the missing sides of the similar triangles. (NTS)

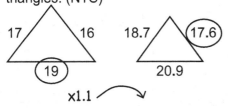

x1.1

7. Operate.

60 ÷ −12	−5	4 x (−13)	−52
−3 x (−15)	45	8 x (−11)	−88
−39 ÷ 3	−13	42 ÷ (−3)	−14
−11 x 25	−275	−13 x 6	−78

3. Subtract the rectangular matrices.

$$\begin{bmatrix} 22 & 6 \end{bmatrix} - \begin{bmatrix} 8 & 11 \end{bmatrix} = \begin{bmatrix} \mathbf{14} & \mathbf{-5} \end{bmatrix}$$

$$\begin{bmatrix} -3 & 12 \\ 0 & 9 \\ -8 & 14 \end{bmatrix} - \begin{bmatrix} 4 & 15 \\ -1 & -8 \\ -6 & -5 \end{bmatrix} = \begin{bmatrix} \mathbf{-7} & \mathbf{-3} \\ \mathbf{1} & \mathbf{17} \\ \mathbf{-2} & \mathbf{19} \end{bmatrix}$$

8. Check if the row number is divisible by the column factor.

	3	7	8	9	21	72
37,856	–	✓	✓	–	–	–
52,488	✓	–	✓	✓	–	✓
66,528	✓	✓	✓	✓	✓	✓

4. Name the lattice points in the picture that are equidistant from A and B.
One box = one unit.

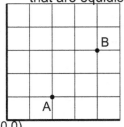

(3, 2)   (2, 3)
(4, 1)   (1, 4)
(5, 0)   (0, 5)

(0,0)

9. Find the supplement of the angle.

Supp (108°)	72°	Supp (163°)	17°
Supp (59°)	121°	Supp (99°)	81°
Supp (141°)	39°	Supp (128°)	52°
Supp (32°)	148°	Supp (77°)	103°

5. Operate on the sets.

A = {10, 20, 30, 40}
B = {5, 10, 25, 35, 40, 45}
U = {5, 10, 15, 20, 25, 30, 35, 40, 45}

A′ = {5, 15, 25, 35, 45}
B′ = {15, 20, 30}
(A ∩ B)′ = {5, 15, 20, 25, 30, 35, 45}

10. Find the area of the parallelogram.

3, 4, 5 right Δ

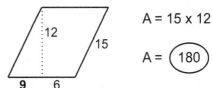

A = 15 x 12

A = $\boxed{180}$

| Level 7 | Number 22 |

1. Find the day of the week given each separate condition.

569 days ago if today is Thursday

−567 = Thurs   Tuesday

354 days ago if today is Friday

−357 = Fri   Monday

6. Mentally calculate the midline of each trapezoid.

base	BASE	midline
8.25	19.75	**14**
19	42	**30.5**
39.2	51.6	**45.4**

2. Find the equal number of nickels, dimes, and quarters that together make $11.20.

$5x + 10x + 25x = 1120$

$40x = 1120$

$4x = 112$

$x = \boxed{28}$

7. Find the average using the rightmost digit methods.

312, 313, 315, 317, 319 $26 \div 5$ 315.2
keep 310

602, 614, 621, 632 $69 \div 4$ 617.25
keep 600

702, 709, 711, 717 $39 \div 4$ 709.75
keep 700

3. Find the volume of the cone in cubic units.

radius = 5
height = 24

$V = \dfrac{5 \cdot 5 \cdot \pi \cdot 24}{3}$

$= 25 \cdot 8 \cdot \pi$

$= \boxed{200\pi}$

8. Find the complement of the angle.

Comp (17°)   73°   Comp (23°)   67°

Comp (39°)   51°   Comp (41°)   49°

Comp (62°)   28°   Comp (54°)   36°

Comp (88°)   2°   Comp (76°)   14°

4. 10! contains how many factors of 4?

$10 \cdot \cancel{9} \cdot 8 \cdot \cancel{7} \cdot 6 \cdot \cancel{5} \cdot 4 \cdot \cancel{3} \cdot 2$

$\boxed{4}$

9. Convert 239 base ten to base four.

239
− 192
____
47
− 32
____
15
− 12
____
3

3	2	3	3
64	16	4	1

$\boxed{3233_{four}}$

5. Find the least natural number n such that kn is a perfect cube.

k = 12   n = ___18___   12=4x3, 2x9=18

k = 18   n = ___12___   18=2x9, 4x3=12

k = 20   n = ___50___   20=4x5, 2x25=50

Three of each prime create a perfect cube.

10. Solve.

$|\,x + 7\,| = 20$ $x = $ ___13, −27___

$|\,x - 8\,| = 22$ $x = $ ___30, −14___

$|\,x - 1\,| = 12$ $x = $ ___13, −11___

$|\,x + 9\,| = 18$ $x = $ ___9, −27___

Level 7	Number 23

1. Find the number of arrangements of the letters PRIME if the first letter must be a vowel.

$\underline{2}$ · $\underline{4}$ · $\underline{3}$ · $\underline{2}$ · $\underline{1}$ ⟨48⟩

↑
1st

6. Find the area of parallelogram ABCD with vertices A(–5, 9), B(6, 9), C(8, –9), and D(–3, –9).

B = 11, H = 18

A = 11 x 18 = ⟨198⟩ sq un

---

2. Answer ALWAYS, SOMETIMES, or NEVER.

Vertical angles are adjacent.	N
Two right angles are complementary.	N
Alternate interior angles are acute.	S
A straight angle is acute.	N
Alternate exterior angles are obtuse.	S

7. Answer YES or NO as to whether the numbers are valid sides of a triangle.

8.1, 8.2, 8.3	Y	8.1, 9.5, 17.2	Y
2.2, 3.3, 5.5	N	8.1, 8.1, 16.3	N
3.5, 5.3, 6.4	Y	7.9, 9.7, 17.9	N

---

3. List all subsets of {a, b, c}.

{ }   {a, b}

{a}   {a, c}

{b}   {b, c}

{c}   {a, b, c}

8. Find the volume in cubic units of a rectangular prism with the areas of its noncongruent faces 15, 42, and 70 square units.

15 = 3 x 5
42 = 2 x 3 x 7 = 3 x 14
70 = 2 x 5 x 7 = 5 x 14

V = 3 x 5 x 14 = 3 x 70 = ⟨210⟩

---

4. Graph on the coordinate plane. One box equals one unit.

y = 2x – 2   y = –2x – 1

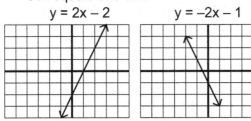

9. Operate in the given mod.

8 + (5 x 7)	≡	7	(mod 9)
7 x 5 x 6	≡	2	(mod 8)
(6 x 6) + (5 x 4)	≡	0	(mod 7)
4 + 3 + 4 + 2	≡	3	(mod 5)

7 x 6 = 42 or 2 (mod 8).
2 x 5 = 10 or 2 (mod 8).

---

5. Operate using mental math.

7 – 4.49	2.51	15 ÷ .375	40
16 – .53	15.47	72 ÷ .6	120
4 – .395	3.605	99 ÷ .11	900
31 – .52	30.48	60 ÷ .75	80

10. Solve.

6 – 4(2x – 3) = 78
–8x + 12 = 72
–8x = 60

x = ⟨$\frac{-15}{2}$⟩

6(4x – 3) + 2 = 14
24x – 18 = 12
24x = 30

x = ⟨$\frac{5}{4}$⟩

Level 7	Number 24

1. Arrange the digits 1, 3, 4, 6, 7, and 8 into two 3-digit numbers with the:

greatest positive difference	least positive difference	
876 Max L to R.	**713**	**416**
− **134** Min L to R.	− **684**	− **387**
742	**29**	**29**

813−764=49

2. If the product of the digits of a 2-digit number is odd, is the sum of the digits even or odd?

Only odd x odd = odd.

Then odd + odd = (even)

3. Write each fraction as a repeating decimal.

$\frac{2}{45}$ .04̅

$\frac{2}{300}$ .006̅

$\frac{7}{900}$ .007̅

$\frac{6}{90}$ .06̅

$\frac{8}{330}$ .024̅

$\frac{23}{990}$ .023̅

4. Find the sum of all 2-digit factors of 200.

1, 2, 4, 5, 8, 10

20, 25, 40, 50, 100, 200

10 + 20 + 25 + 40 + 50 = (145)

5. Find the probability of selecting a point from the shaded area within the square.

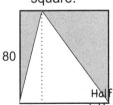

$\frac{A_{sh}}{A_{squ}} = \frac{(80)(80)/2}{6400}$

$= \frac{3200}{6400}$ ($\frac{1}{2}$)

Half may be seen by drawing the dotted line without doing any work.

6. Bob's age now is half of Joe's age 6 years ago and one third what Hank's age will be in 7 years. If Hank is 17 now, how old is Joe now?

	−6	now	+7
H		17	24
J	16	(22)	
B		8	

7. Find the volume of a wedge cut at a 45° central angle from a cylinder with radius 12 and height 5.

$\frac{45}{360} = \frac{1}{8}$

$V = \frac{12 \cdot 12 \cdot 5 \cdot \pi}{8}$

$V = 3 \cdot 6 \cdot 5 \cdot \pi$

(90π)

8. Complete the chart for a circle.

radius	diameter	area	circumference
$\frac{3}{2}$	3	$\frac{9\pi}{4}$	3π
$\frac{7}{4}$	$\frac{7}{2}$	$\frac{49\pi}{16}$	3.5π
2.5	5	6.25π	5π

9. Of 7 2-digit multiples of 10, the mode is 90, the median is 70, the mean is 60, and the range is 70. Find the numbers.

20   30   40   70   80   90   90

The low and high cannot be 10 and 80 because the mode is 90. Place 20, 70, 90, and 90. Sum = 60 x 7 = 420. Then 420 − 270 = 150. 30, 30, 90 won't work.

10. Evaluate.

$\sqrt{.0081}$ .09

$\sqrt{.0064}$ .08

$\sqrt{.0001}$ .01

$\sqrt{.0036}$ .06

$\sqrt{.49}$ .7

$\sqrt{1.44}$ 1.2

$\sqrt{2.93^2}$ 2.93

$\sqrt{.0121}$ .11

MAVA Math: Middle Reviews Solutions Copyright © 2013 Marla Weiss

Level 7	**Number 25**

1. Hal boxed 1/6 of his 120 books and gave 1/4 away. Of those left, 1/7 went to charity. Of those then left, he sold half. The remaining number of books is what fractional part of the number of starting books?

120
20 30 70
10 60
30 30
$\frac{30}{120}$ $\left(\frac{1}{4}\right)$

6. Find the area and perimeter of the rectangle with vertices:

(9, 6), (9, –2), (–4, –2), and (–4, 6)
b = 13   A = 104
h = 8   P = 42

(2, –2), (12, –2), (12, –9), and (2, –9)
b = 10   A = 70
h = 7   P = 34

2. Find the shaded area in square units.

4 4π $(64 - 16\pi)$
16
16
4 (32)

7. Name the property exemplified.

abc = bac   CPM

abc = abc   RPE

a = b implies b = a   SPE

(ab)c = a(bc)   APM

3. Find the perimeter of the figure on the unit grid.

17 (56)
3
5
8
17
6

8. Convert to scientific notation.

53,895 5.3895×10^4

0.00385 3.85×10^{-3}

0.026 2.6×10^{-2}

634.1 6.341×10^2

4. Find the area of the rectangle bounded by the lines y = –6, x = –6, x = 9, and y = 3.

NTS 9
15
(135) sq un

9. Operate.

$8 - 2 \div 3 \times 9 - 7 \div 10$
$8 - 6 - .7$ (1.3)

$7 \div 3 \times 6 \div 7 - 4 \div 8$
$2 - .5$ (1.5)

$7 \div 5 + 9 - 1 \div 4 - 4$
$1.4 + 9 - .25 - 4$ (6.15)

$-(5 + 7) \div 5 \times 2 + .2$
$- 2.4 \times 2 + .2$ (-4.6)

5. Find the face diagonal of a cube with

volume = 125 $5\sqrt{2}$ e = 5

area of face = 1 $\sqrt{2}$ e = 1

volume = 64 $4\sqrt{2}$ e = 4

edge = 8 $8\sqrt{2}$ e = 8

10. How many perfect squares are between 101 and 10,001?

$10^2 = 100$
$100^2 = 10,000$
11 to 100
$100 - 11 + 1$ (90)

Do not list the perfect squares but rather the numbers that generate them. Subtract and add 1 to get the number of numbers.

Level 7	Number 26

1. The average of 12 numbers is 15. Add 5 to three of these numbers. Find the new average.

sum = 12 x 15 = 180
new sum = 180 + 15 = 195
new average = 195/12 = $\boxed{16.25}$

6. Solve.

$7\sqrt{x} - 5 = 555$	$8\sqrt{x} + 11 = 131$
$7\sqrt{x} = 560$	$8\sqrt{x} = 120$
$\sqrt{x} = 80$	$\sqrt{x} = 15$
x = $\boxed{6400}$	x = $\boxed{225}$

2. 14 pounds of chips serve 64 people. How many ounces of chips serve 18 people proportionally?

$$\frac{x}{18} = \frac{14 \cdot 16}{64} \qquad x = 7 \cdot 9$$
$$x = \boxed{63}$$
$$x = \frac{14 \cdot 16 \cdot 18}{64}$$

7. For a trapezoid:

A = 320, b = 11, h = 16. Find B.

M = 320 / 16 = 20

11, 20, $\boxed{29}$

A = 375, B = 36, h = 15. Find b.

M = 375 / 15 = 25

$\boxed{14}$, 25, 36

Calculate by arithmetic sequence.

3. Verify the first 3 perfect numbers as the sum of their proper divisors.

6 = 1 + 2 + 3

28 = 1 + 2 + 4 + 7 + 14

496 = 1 + 2 + 4 + 8 + 16 +
31 + 62 + 124 + 248

8. Complete the table of values for 4x + 7y = 28.

x	−7	−2	−1	0	1	2	3	5	7	14
y	8	$\frac{36}{7}$	$\frac{32}{7}$	4	$\frac{24}{7}$	$\frac{20}{7}$	$\frac{16}{7}$	$\frac{8}{7}$	0	−4

4. Find x in degrees. (NTS)

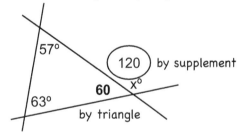

120 by supplement

60

$x°$

by triangle

57°

63°

9. Find the remainder without dividing.

369,054 ÷ 18   369,054 div by 2, 9   R 0

679,938 ÷ 10   679,930 div by 10   R 8

426,304 ÷ 14   426,300 div by 2, 7   R 4

271,367 ÷ 15   271,365 div by 3, 5   R 2

5. Operate.

$$3165_{seven} + 2643_{seven}$$
$$6141_{seven}$$

$$7615_{eight} - 1726_{eight}$$
$$5667_{eight}$$

$$3532_{six} + 1154_{six}$$
$$5130_{six}$$

10. A rectangular prism has sides 2, 3, and 7. Find the 3 distinct face diagonals.

2, 3, $\boxed{\sqrt{13}}$ 4 + 9 = 13

2, 7, $\boxed{\sqrt{53}}$ 4 + 49 = 53

3, 7, $\boxed{\sqrt{58}}$ 9 + 49 = 58

1. Answer as indicated. 27 is what % of 90? $\frac{27}{90} = \frac{3}{10}$ (30%) 12 is 150% of what number? $12 = \frac{3}{2}(x)$ (8) 36 is what % of 60? $\frac{36}{60} = \frac{6}{10}$ (60%)	**6.** Translate the cipher with each letter as its own digit. BAAL B557 **2557** + SRAA + S155 **+ 8155** RELRB 1E71B **10712** R=1, the most that can be carried from 2 addends. In 10s, A=0 or A=5. But A≠0 by 1s. In 100s, L=7 (with carry). In 1s, B=2. S=8. E=0.
2. Answer as indicated. What % of 20 is 12? $\frac{12}{20} = \frac{60}{100}$ (60%) 20% of what number is 12? $12 = \frac{1}{5}(x)$ (60) What is 12% of 20? 10% is 2 1% is .2 2% is .4 (2.4)	**7.** Give the simplified fractional part that the first decimal is of the second. 8.6/60.2 $\frac{86}{602}$ (1/7) 1.12/7.2 $\frac{112}{720}$ (7/45) 9.1/10.5 $\frac{91}{105}$ (13/15) 2.25/8.1 $\frac{225}{810}$ (5/18)
3. Estimate the fraction to the nearest whole number. $\frac{0.5 \; 100 \; 20 \; 0.4 \; 7}{0.51 \times 98.74 \times 0.39 \times 7.06}$ (8) $\frac{6.98 \times 5.09 \times 0.491}{7 \; 5 \; 0.5}$ 4/10 of 20 is 8.	**8.** Divide by 5 mentally. Double; move decimal point 1 place to left. 4209 ÷ 5 841.8 947 ÷ 5 189.4 3538 ÷ 5 707.6 818 ÷ 5 163.6 2746 ÷ 5 549.2 729 ÷ 5 145.8 3930 ÷ 5 786 536 ÷ 5 107.2
4. Without graphing on the number line, identify each as 2 points, 2 rays, line, open ray, or open/half-open segment. −3 ≤ x < 10 half-open segment x = 4 AND x = 8 IMPOSSIBLE x ≥ 7 OR x ≤ 3 2 rays	**9.** Find the angle in degrees formed by clock hands at 10:30. 360°/12 = 30° between neighboring numbers 30 × 4.5 = (135) Hour hand is halfway between 10 and 11.
5. Find the perimeter of a rectangle with area 190 square units and width 10 units. A = 190 W = 10 L = 19 SP = 29 (P = 58)	**10.** Copy the prime factorizations from page 39. Then find the GCF and LCM. 220 / 187 / 165 $2^2 \times 5 \times 11$ / 11 × 17 / 3 × 5 × 11 GCF (220, 187, 165) = (11) LCM (220, 187, 165) = $2^2 \times 3 \times 5 \times 11 \times 17$ = 220 × 51 = (11,220)

Level 7	Number 28

1. A line segment connecting any two points on a circle is a _____ chord.

The number of prime numbers is _____ infinite.

Points on the coordinate plane not in any quadrant lie on the _____ axes.

6. Find each term for the repeating sequence ABCDEABCDE... .

cycle length 5

631st   A   5 divides 630; R1

1950th   E   5 divides 1950; R0

478th   C   5 divides 475; R3

2. Of 150 women:
18 like pink, blue, & red ;
40 like pink & blue;
44 like blue & red;
38 like pink & red;
100 like pink;
70 like blue;   150 – 132
66 like red.
How many like none? 18

P=100   B=70
40 22 4
18
20 26
2
R=66

7. Find AB if BC = 48, CD = 30, and AD = 40. (NTS)

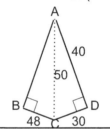

3, 4, 5 x10
30, 40, 50

7, 24, 25 x2
14, 48, 50

14

3. Find the area of an equilateral triangle with edge 16.

$$\frac{s^2 \cdot \sqrt{3}}{4} = \frac{16 \cdot 16 \cdot \sqrt{3}}{4} = 64\sqrt{3} \text{ sq un}$$

8. Answer as a simplified fraction.

The two 100s in each % simplify.

14% is what fractional part of 77%? $\frac{14}{77}$ $\frac{2}{11}$

28% is what fractional part of 72%? $\frac{28}{72}$ $\frac{7}{18}$

24% is what fractional part of 90%? $\frac{24}{90}$ $\frac{4}{15}$

4. Calculate the point on the number line that is:

$\frac{1}{3}$ of the way from 7 to 8.5   1.5/3 = 0.5   7.5

$\frac{3}{5}$ of the way from 11 to –1   12/5 = 2.4   3.8

$\frac{3}{4}$ of the way from –5 to 5   10/4 = 2.5   2.5

9. Find the slope and intercepts of the lines.

line	slope	y-intercept	x-intercept
x = y	1	(0, 0)	(0, 0)
x = 0	undef	y-axis	(0, 0)
y = 0	0	(0, 0)	x-axis

Picture 0 inserted to find intercepts mentally.

5. Calculate the area of each shape.

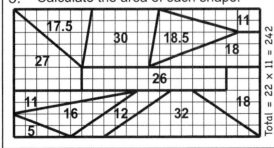

Total = 22 × 11 = 242

10. A book cost $35.02 including the 3% tax. What was the price before the tax?

1.03P = 35.02

$$\frac{3502}{103} = \$34.00$$

Level 7	Number 29

1. Write the inverse of the conditional statement. Are they logically equivalent? (YES)

If a number is prime, then it is odd.

2 is prime.   Statement FALSE

If a number is not prime, then it is not odd.

33 is not prime, but it is odd.   Inverse FALSE

6. Operate and simplify the continued fraction.

$$4 + \cfrac{6}{3 + \cfrac{3}{2 + \cfrac{3}{4}}}$$

$\dfrac{11}{7}$ $\dfrac{11}{3}$ $\dfrac{9}{11}$ $\dfrac{42}{11}$ $\left(5\dfrac{4}{7}\right)$

2. Count the total number of triangles in the picture.

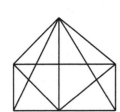

1: 10
2: 13
3: 6
4: 4
5: 0
6: 1
7–12: 0   (34)

7. Show Goldbach's Conjecture: Even integers > 2 are the sum of 2 primes. Answers may vary.

18	5 + 13 7 + 11	76	3 + 73 5 + 71	98	19 + 79 31 + 67
70	3 + 67	82	3 + 79	100	47 + 53
72	5 + 67	88	5 + 83	102	43 + 59
74	3 + 71 7 + 67	92	3 + 89 13 + 79	106	3 + 103 5 + 101

3. Find the probability when drawing once from a standard deck of cards.

P(face OR red) $\dfrac{8}{13}$ 6 black JQK + 26 red = 32

P(8 OR king OR red) $\dfrac{15}{26}$ 26 red + 2 bl 8 + 2 bl K = 30

P(8 OR king OR diamond) $\dfrac{19}{52}$ 13 dia + 3 8s + 3 Ks = 19

8. By what fractional part of:

$\dfrac{7}{9}$ does $\dfrac{5}{9}$ exceed $\dfrac{2}{9}$? $\left(\dfrac{3}{7}\right)$

$\dfrac{6}{7}$ does $\dfrac{5}{7}$ exceed $\dfrac{2}{7}$? $\left(\dfrac{1}{2}\right)$

$\dfrac{3}{5}$ does $\dfrac{2}{5}$ exceed $\dfrac{1}{5}$? $\left(\dfrac{1}{3}\right)$

4. Write each repeating decimal as a simplified fraction.

$.\overline{54}$ $\dfrac{54}{99}$ $\left(\dfrac{6}{11}\right)$ $.\overline{42}$ $\dfrac{42}{99}$ $\left(\dfrac{14}{33}\right)$

$.0\overline{54}$ $\dfrac{6}{110}$ $\left(\dfrac{3}{55}\right)$ $.0\overline{42}$ $\dfrac{14}{330}$ $\left(\dfrac{7}{165}\right)$

9. Find the point equidistant from points (−3, 4), (3, 10), (9, 4), and (3, −2).

Points are vertices of a rhombus or opposite ends of perpendicular diameters of a circle.
vertical = 12; 10 − 6 = 4
horizontal = 12; 9 − 6 = 3   (3, 4)

5. Find the percent increase. $\dfrac{\text{change}}{\text{original}}$

From 16 to 28 $\dfrac{12}{16}$ = (75%) $\dfrac{3}{4}$

From 15 to 18 $\dfrac{3}{15}$ = (20%) $\dfrac{1}{5}$

From 30 to 33 $\dfrac{3}{30}$ = (10%) $\dfrac{1}{10}$

From 24 to 60 $\dfrac{36}{24}$ = (150%) $\dfrac{3}{2}$

10. Multiply and simplify by regrouping.

9 • (5 DAYS 8 HR 9 MIN 7 SEC)

45 DAYS 72 HR 81 MIN 63 SEC

45 DAYS 72 HR 82 MIN 3 SEC

45 DAYS 73 HR 22 MIN 3 SEC

(48 DAYS 1 HR 22 MIN 3 SEC)

Level 7	Number 30

1. Tami bought 10.5 yards of fabric for $47.25. What was the cost per yard?

$$\frac{\text{cost}}{\text{yard}} = \frac{47.25}{10.5} = \boxed{\$4.50}$$

The value to the left of "per" is the numerator.
The value to the right of "per" is the denominator.

6. Find the area of an isosceles triangle with sides 25, 25, 48.

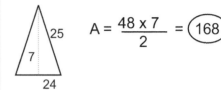

$$A = \frac{48 \times 7}{2} = \boxed{168}$$

2. Reflect each point over the line y = x. Then translate as specified.

(2, –9)	(–9, 2)	down 7	(–9, –5)
(–8, 4)	(4, –8)	left 8	(–4, –8)
(3, 7)	(7, 3)	up 5	(7, 8)

7. Multiply by 11 mentally.

627 x 11	6897
1714 x 11	18,854
263 x 11	2893
4425 x 11	48,675

Write from R to L. For 4425: Write the 5. Then 5+2=7; 2+4=6; 4+4=8. Write the 4.

3. Multiply by 11 mentally.
Leave left and right digits. Insert pairwise digit sums in order.

271 x 11	2981	635 x 11	6985
154 x 11	1694	333 x 11	3663
542 x 11	5962	711 x 11	7821
443 x 11	4873	907 x 11	9977

8. Operate and simplify.

$$\frac{35}{66} \div \frac{49}{22} \quad \frac{35}{66} \times \frac{22}{49} \quad \frac{5}{21}$$

$$\frac{25}{26} \div \frac{10}{39} \quad \frac{25}{26} \times \frac{39}{10} \quad \frac{15}{4} \qquad \frac{5}{21} \times \frac{4}{15}$$

$$\boxed{\frac{4}{63}}$$

4. Define operation ✦ as:

$W ✦ Z = 5WZ + W^2 - Z$. For example, $2 ✦ 5 = 50 + 4 - 5 = 49$.
Find the values.

1.5 ✦ 2	15.25	15 + 2.25 – 2
6 ✦ 1.1	67.9	33 + 36 – 1.1

9. Distribute.

–z(2x – 3y + 7)	–2xz + 3yz – 7z
6(2b – 4c + 6e)	12b – 24c + 36e
4x(–x – 5y + 3z)	$-4x^2 - 20xy + 12xz$
–2e(–6a + 3c + 2d)	12ae – 6ce – 4de

5. Find the least value of x ÷ y.

8 ≤ x ≤ 40	4 ≤ y ≤ 16	$\frac{1}{2}$
–1 ≤ x ≤ 15	–3 ≤ y ≤ 30	–5
–9 ≤ x ≤ –3	3 ≤ y ≤ 9	–3
7 ≤ x ≤ 14	–7 ≤ y ≤ 28	–2

10. Find the principal for a time of 1 year.

principal	rate	interest
$1700	3%	$51
$1550	2%	$31
$4500	4%	$180
$3300	5%	$165

Get 1%. Then x 100.

3% is 51
1% is 17
100% is 1700

| Level 7 | Number 31 |

1. Given the equation below and that a, b, c, and d are different positive integers. Find the value:
$5d - 3c + 4a - 9b$

$1 \times 10^a + 5 \times 10^b + 3 \times 10^c + 8 \times 10^d$
$= 53,810$

$a = 1$
$b = 4$
$c = 3$
$d = 2$

$10 - 9 + 4 - 36$ ⟨-31⟩

6. Find the probability of tossing a sum greater than 16 with 3 standard dice.

6, 6, 6
6, 6, 5
6, 5, 6 $\dfrac{4}{6 \times 6 \times 6} = $ ⟨$\dfrac{1}{54}$⟩
5, 6, 6

$5 + 5 + 5 = 15$
$5 + 5 + 6 = 16$

2. Graph on the number line.

$x = 1$ OR $x > 3$

$x > -3$ AND $x < 4$

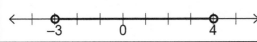

7. Combine like terms.

$4(2x + 5y) - 2(x - 4) - 5y - 8$	$6x + 15y$
$-2(-3x + 6) + 7(3x - 2) - 4x$	$23x - 26$
$6(5x - 2) + 5(3 - 6x) + x - 3$	x
$-9(2x + 4) + 3(1 + 7x) - 5x$	$-2x - 33$

3. Small roses cost $0.95. Small pansies cost $1.15. How many of each does one buy to get 100 plants for $100?

$0.95 $4 \times 25 = 100$
$0.95
$0.95   ⟨75 roses
$1.15   25 pansies⟩
─────
$4.00

8. Write as an equation.

Half a number minus three is five times the number. $\dfrac{x}{2} - 3 = 5x$

Triple the quantity two minus a number is eight. $3(2 - x) = 8$

A number squared plus the number is nine. $x^2 + x = 9$

A number cubed increased by five is the number. $x^3 + 5 = x$

4. Find the percent mentally.

16% of 100	16	12% of 325	39
16% of 200	32	8% of 125	10
16% of 150	24	2% of 650	13
16% of 25	4	4% of 350	14

9. Find the missing sides of the triangles with $m\angle A = 45°$.

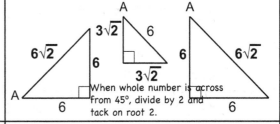

When whole number is across from 45°, divide by 2 and tack on root 2.

5. Find the height of an isosceles right triangle with each given area.

$A = 12.5$ $A = 40.5$
$s^2 = 25$ $s^2 = 81$

$s = h = $ ⟨5⟩ $s = h = $ ⟨9⟩

An isosceles right triangle is half of a square.

10. A 4 by 4 by 4 cube is painted on the outside and then cut into 64 1 by 1 by 1 cubes. How many of the little cubes have each number of faces painted?

| 3 faces? | 8 | 1 face? | 24 |
| 2 faces? | 24 | 0 faces? | 8 |

Level 7	Number 32

1. Two and one half hours are what fraction of time between noon on Sunday and noon of Friday of the same week?

$$\frac{\frac{5}{2}}{5 \times 24} = \frac{5}{2} \times \frac{1}{5 \times 24} = \boxed{\frac{1}{48}}$$

6. Operate and simplify.

$$7\frac{7}{15} + 4\frac{17}{30} + 5\frac{16}{45}$$

$$7\frac{42}{90} + 4\frac{51}{90} + 5\frac{32}{90}$$

$$16\frac{125}{90} \qquad 17\frac{35}{90} \qquad \boxed{17\frac{7}{18}}$$

2. Answer YES or NO as to whether the numbers may be expressed as the product of 2 consecutive even integers.

24	Y	42	N	72	N
4 x 6		6 x 7		8 x 9	
36	N	60	N	80	Y
6 x 6		6 x 10		8 x 10	

7. Find the next letter in each pattern.

Skip 3, 4, 5, 6

A   e   J   p   **W**
___ ___ ___ ___ ___

Skip 0, 1, 5, 0, 1, 5, 0

A   B   d   J   K   m   S   **T**
___ ___ ___ ___ ___ ___ ___ ___

Skip 0, 1, 0, 3, 0, 5, 0

z   Y   w   V   r   Q   k   **J**
___ ___ ___ ___ ___ ___ ___ ___

3. Multiply the rectangular matrix by the constant (scalar multiplication).

$$7\begin{bmatrix} 11 & 8 & -1 \\ 9 & 0 & 6 \end{bmatrix} = \begin{bmatrix} \mathbf{77} & \mathbf{56} & \mathbf{-7} \\ \mathbf{63} & \mathbf{0} & \mathbf{42} \end{bmatrix}$$

$$3\begin{bmatrix} -3 & 8 & 9 & 11 \\ 16 & -2 & 0 & 14 \end{bmatrix} = \begin{bmatrix} \mathbf{-9} & \mathbf{24} & \mathbf{27} & \mathbf{33} \\ \mathbf{48} & \mathbf{-6} & \mathbf{0} & \mathbf{42} \end{bmatrix}$$

8. Find the statistics for the first 11 whole perfect squares.

0, 1, 4, 9, 16, 25, 36, 49, 64, 81, 100

n	11	median	25
sum	385	mode	none
mean	35	range	100

4. Solve by cross multiplication.

$$\frac{3}{7} = \frac{10}{x} \qquad \frac{7}{9} = \frac{x}{7} \qquad \frac{8}{x} = \frac{11}{8}$$

$$3x = 70 \qquad 9x = 49 \qquad 11x = 64$$

$$x = \boxed{\frac{70}{3}} \qquad x = \boxed{\frac{49}{9}} \qquad x = \boxed{\frac{64}{11}}$$

9. Find the sum of the edges of a cube with SA 294 cubic units.

SA = 294
A face = 49
e = 7
sum edges = 12 x 7
= $\boxed{84}$ un

Find the volume of a cube with surface area 726 square units.

SA = 726
A face = 121
e = 11
V = 11^3 = $\boxed{1331}$ cu un

5. Find the midpoint of the line segment with the given endpoints.

(1.5, 2) and (8.5, −18) → (5, −8)

(−1, 3.4) and (−9, 6.6) → (−5, 5)

(1.8, −7) and (1.6, 21) → (1.7, 7)

(3.5, −3) and (5.3, −9) → (4.4, −6)

Average the xs, average the ys.

10. Simplify.

$$\frac{16!}{13!\,4!\,2!} \qquad \frac{14 \cdot 15 \cdot 16}{4 \cdot 3 \cdot 2 \cdot 2} \qquad 14 \cdot 5 \quad \boxed{70}$$

$$\frac{17!}{14!\,3!} \qquad \frac{15 \cdot 16 \cdot 17}{3 \cdot 2} \qquad 5 \cdot 8 \cdot 17 \quad \boxed{680}$$

Level 7	Number 33

1. The time on a 12-hour circular clock is 7:00 AM. What will be the time after the hour hand goes around once?

(7:00 PM)

A full circle of the hour hand is 12 hours.

6. Count the number of paths from A to B moving only right and/or up.

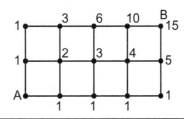

(15)

2. Find the prime factorization. Work down. In the answer use exponents with primes in ascending order.

2412	1870	1001
4 x 603	187 x 10	**7 x 11 x 13**
4 x 9 x 67	11 x 17 x 2 x 5	*Memorize.*
2^2 x 3^2 x 67	**2 x 5 x 11 x 17**	

7. Evaluate for x = 4.

$(x - 6)^4 - (x - 5)^3$	16 – –1	17
$x^{-1} - x + (x - 1)^2$	.25 – 4 + 9	5.25
$(x - 2)^2 - (5x)^{-1}$	4 – .05	3.95
$(5x - 5)^0 - 3x^2 + x$	1 – 48 + 4	–43

3. Find the volume and surface area of a sphere with radius 1 units.

$$V = \frac{4 \cdot \pi \cdot 1 \cdot 1 \cdot 1}{3} = \left(\frac{4\pi}{3} \right) \text{cu un}$$

$$SA = 4 \cdot \pi \cdot 1 \cdot 1 = \left(4\pi \right) \text{sq un}$$

8. Answer YES or NO as to whether the 3 numbers form sides of a right triangle.

11, 22, 25	N	14, 48, 50	Y
		7, 24, 25 x 2	
12, 60, 71	N	48, 48, 48	N
11, 60, 61	Y	30, 40, 50	Y
		3, 4, 5 x 10	

4. Find the area in square units of a rectangle with diagonal 34 and length 16 units.

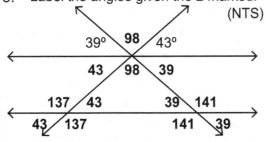

8, 15, 17
x2 blow-up
A = 16 x 30
A = (480)

9. Find the opposite in the given mod.

opposite of 5 mod 8 ___ 3 5+3=8, 8–8=0

opposite of 2 mod 4 ___ 2 2+2=4, 4–4=0

opposite of 3 mod 9 ___ 6 3+6=9, 9–9=0

opposite of 1 mod 7 ___ 6 1+6=7, 7–7=0

5. Label the angles given the 2 marked. (NTS)

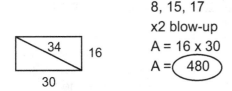

10. Find the area of the parallelogram.

5, 12, 13 right △ (NTS)

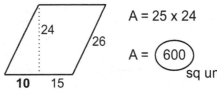

A = 25 x 24

A = (600) sq un

Level 7	Number 34

1. Find the rate in miles per hour for traveling 2.5 miles in 24 minutes.

D = RT

$\dfrac{5}{2}$ = R x $\dfrac{24}{60}$

R = $\dfrac{5}{2}$ x $\dfrac{60}{24}$ = $\dfrac{25}{4}$ R = (6.25)

6. The sum of two whole numbers is 50. If their ratio is 3:7, find their positive difference.

3:7 sum = 10
15:35 sum = 50

35 – 15 = (20)

2. Operate.

–19 – –77	58	49 – –56	105
–44 + –31	–75	–66 + 24	–42
–38 – 27	–65	11 – 92	–81
18 – 36	–18	71 – –83	154

7. Express the reciprocal of each decimal as a simplified fraction.

9.6   9 $\dfrac{6}{10}$ 9 $\dfrac{3}{5}$ $\dfrac{48}{5}$ ($\dfrac{5}{48}$)

7.8   7 $\dfrac{8}{10}$ 7 $\dfrac{4}{5}$ $\dfrac{39}{5}$ ($\dfrac{5}{39}$)

3. Find the missing terms of the arithmetic sequences.

1.5, __4__, 6.5, 9, __11.5__, 14, __16.5__

94, 91, __88__, 85, 82, __79__, 76, 73

75, 68, 61, __54__, 47, __40__, 33

8. Evaluate.

–6 x |4 x 5| ÷ 3 – 9 x |–3 – 7| ÷ 3 + 1

–40 – 30 + 1   __–69__

– 3 x |6 – 8| ÷ 2 x |–6 – 4| – |0 – 9| x 3

–3 x 10 – 27   __–57__

4. Draw the reflection of each rectangle over the line y = –1.

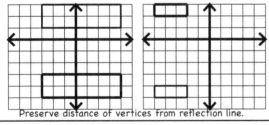

Preserve distance of vertices from reflection line.

9. Find the mean, median, mode, and range of the data in the frequency table.   n = 6

number	frequency
100	3
98	1
94	1
90	1

sum = 582

(mean = 97
median = 99
mode = 100
range = 10)

5. Solve.

5 – 4x ≤ 41	4x – 3 ≥ 19	9 – 6x < 48
–4x ≤ 36	4x ≥ 22	–6x < 39
(x ≥ –9)	(x ≥ $\dfrac{11}{2}$)	(x > $\dfrac{–13}{2}$)

10. Find the next term in the sequence by successive differences.

5   6   14   32   64   115   (**191**)

  1   **8**   **18**   **32**   **51**   **76**

    7   **10**   **14**   **19**   **25**

      3   **4**   **5**   **6**

Compute the differences moving down in rows until a pattern emerges. Then build back up.

| Level 7 | Number 35 |

---

1. The number of unique digits in base nine is _____ nine.

   A side of a right triangle across from an acute angle is called a _____ leg.

   The side of a right triangle across from the right angle is called the _____ hypotenuse.

6.

	R	T	D
To	60	4	240
Fr	48	5	240
RT	(53.$\overline{3}$)	9	480

A bus went 240 miles from City A to City B in 4 hours. The bus made the return trip in 5 hours. What was its average rate in mph for the round trip?

"Average" means "overall" in this type of problem. Rates do not average.

---

2. Find the missing sides of the similar triangles. (NTS)

   2.8 / 6.4 / (8.6)   14 / (32) / 43

   ×5

7. How many dots are in the 20th figure if the pattern continues?

   1 + 2   2 + 3   3 + 4

   20 + 21 = (41)

---

3. Find the probability when drawing once from a standard deck of cards.

   P(red OR 1-digit even) $\dfrac{17}{26}$ 2,4,6,8 2 suits + 26 red = 34

   P(heart OR 1-digit even) $\dfrac{25}{52}$ 2,4,6,8 3 suits + 13 hearts = 25

   P(red OR 1-digit) $\dfrac{11}{13}$ 1-9 2 suits + 26 red = 44

8. Divide by 5 mentally.

   Double; move decimal point 1 place to left.

$1928 \div 5$	385.6	$737 \div 5$	147.4
$2945 \div 5$	589	$848 \div 5$	169.6
$3827 \div 5$	765.4	$939 \div 5$	187.8
$4136 \div 5$	827.2	$626 \div 5$	125.2

---

4. Find the area of the triangle bounded by both axes and the line $y = -2x - 6$.

   3   6   (9) sq un

9. Find the average using the arithmetic sequence method.

   4.9, 5.9, 6.9, 7.9, 8.9, 9.9   7.4

   4.9, 6.0, 7.1, 8.2, 9.3, 10.4, 11.5   8.2

   4.1, 4.2, 4.3, 4.4, 4.5, 4.6   4.35

---

5. The product of 3 consecutive integers is 54834. What is their sum?

   30 x 31 x 32 = 27,000+
   40 x 41 x 42 = 64,000+   1st, estimate

   7 x 8 x 9 ends in 4
   37 x 38 x 39   2nd, use ones digits

   37 + 38 + 39 = (114)

10. A punch recipe is 3 parts juice, 5 parts seltzer, and 2 parts sherbet. The sherbet is what fractional part of the beverage?

   $\dfrac{S}{P} = \dfrac{2}{10} = \left(\dfrac{1}{5}\right)$

   On a half hour TV show, the 6-minute commercials were what fractional part of the show?

   $\dfrac{C}{S} = \dfrac{6}{30} = \left(\dfrac{1}{5}\right)$

Level 7	Number 36

1. The angles of a pentagon are in the ratio 2:3:4:5:6. Find the angles.

sum angles = 540

2 + 3 + 4 + 5 + 6 = 20

540 / 20 = 27

angles = ⟨ 54, 81, 108, 135, 162 ⟩

6. The area of △ABC is 32 sq un. The total area of the shaded regions is 16. Find the side of square DEFG.

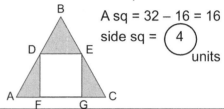

A sq = 32 − 16 = 16

side sq = ⟨ 4 ⟩ units

2. What time is 307 minutes before 1:25 AM?

307 = 60 x 5 + 7   1 − 5 → 8   25 − 7 → 18

⟨ 8:18 PM ⟩

What time is 235 minutes before 2:50 PM?

235 = 60 x 4 − 5   2 − 4 → 10   50 + 5 → 55

⟨ 10:55 AM ⟩

7. Convert to base ten.

$ABE_{sixteen}$ 10 x 256 + 11 x 16 + 14   2750

$A5F_{fifteen}$ 10 x 225 + 5 x 15 + 15   2340

$22C_{thirteen}$ 2 x 169 + 2 x 13 + 12   376

$2T8_{eleven}$ 2 x 121 + 10 x 11 + 8   360

3. Find the volume of the cone in cubic units.

radius = 8
height = 6

$V = \dfrac{8 \cdot 8 \cdot \pi \cdot 6}{3}$

$= 64 \cdot 2 \cdot \pi$

$= ⟨ 128\pi ⟩$

8. Evaluate for n = 100. Answer as a decimal.

$\dfrac{(n + 1)!}{(n - 1)!} \quad \dfrac{101!}{99!} \quad \dfrac{101 \cdot \cancel{100}}{100 \cdot \cancel{100}}$

$\dfrac{}{n^2} \qquad \dfrac{}{100 \cdot 100}$

Dividing by 100 moves the decimal point 2 places to the left.

⟨ 1.01 ⟩

4. The arc is a semi-circle. The bottom portion is a trapezoid with H = 3, b = 8, and B = 18. Find the area of the figure.

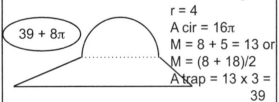

⟨ 39 + 8π ⟩

r = 4
A cir = 16π
M = 8 + 5 = 13 or
M = (8 + 18)/2
A trap = 13 x 3 = 39

9. Evaluate.

$\sqrt{18}$ 9 · 2 $3\sqrt{2}$ $\sqrt{60}$ 4 · 15 $2\sqrt{15}$

$\sqrt{27}$ 9 · 3 $3\sqrt{3}$ $\sqrt{72}$ 36 · 2 $6\sqrt{2}$

$\sqrt{48}$ 16 · 3 $4\sqrt{3}$ $\sqrt{75}$ 25 · 3 $5\sqrt{3}$

$\sqrt{50}$ 25 · 2 $5\sqrt{2}$ $\sqrt{90}$ 9 · 10 $3\sqrt{10}$

5. Simplify.

$\dfrac{28!}{26! + 26!} \quad \dfrac{28!}{2 \cdot 26!} \quad \dfrac{27 \cdot \overset{14}{\cancel{28}}}{2} \quad ⟨ 378 ⟩$

$\dfrac{17! + 17!}{19!} \quad \dfrac{2 \cdot 17!}{19!} \quad \dfrac{\cancel{2}}{\underset{9}{\cancel{18} \cdot 19}} \quad ⟨ \dfrac{1}{171} ⟩$

10. Find the volume in cubic units of a square pyramid with edge 6 and altitude 4 units.

$\dfrac{6 \times 6 \times 4}{3} = 2 \times 6 \times 4 = ⟨ 48 ⟩$

Level 7	Number 37

1. Write the inverse of the conditional statement. Are they logically equivalent? (NO)

If a number is divisible by 10, then the number is not a prime.   statement TRUE
                                              inverse FALSE
If a number is not divisible by 10, then the number is a prime.   25 is one of countless counterexamples.

6. Multiply by 25 mentally.

Divide by 4 (cut in half twice). Tack on two 0s.

6480 x 25	162,000	308 x 25	7,700
4452 x 25	111,300	728 x 25	18,200
4036 x 25	100,900	912 x 25	22,800
7672 x 25	191,800	516 x 25	12,900

2. What is the ratio of the final price of an item to the original price if the price first rises by 25%, then rises by 50%, and then drops by 40%?

$$\frac{5}{4} \cdot \frac{3}{2} \cdot \frac{3}{5} = \left(\frac{9}{8}\right)$$

9/8 to 1 is 9/8. Using $100 as the original price is another method but involves more calculations.

7. Operate.

30^{-2}	$\frac{1}{900}$	29^0	$\frac{1}{1}$
43^{-1}	$\frac{1}{43}$	16^2	256
		199^1	199
14^{-2}	$\frac{1}{196}$	2^9	512

15^{-2}	$\frac{1}{225}$
8^{-3}	$\frac{1}{512}$
10^{-2}	$\frac{1}{100}$

3. Find the formula for the sequence −2, 3, 8, 13, 18, . . . in functional notation.

0 1 2 3 4
−2 3 8 13 18 +5 . . .

$$f(n) = 5n - 2$$

8. Answer as a simplified fraction.

The two 100s in each % simplify.

27% is what fractional part of 45%? $\frac{27}{45}$ $\frac{3}{5}$

13% is what fractional part of 52%? $\frac{13}{52}$ $\frac{1}{4}$

42% is what fractional part of 60%? $\frac{42}{60}$ $\frac{7}{10}$

4. Find the reciprocal of the square root of the cube of .25.

$\frac{1}{4}$ $\frac{1}{64}$ $\frac{1}{8}$ (8)

Find the reciprocal of the cube of the square root of 1/9.

$\frac{1}{9}$ $\frac{1}{3}$ $\frac{1}{27}$ (27)

9. Find the maximum number of rectangular boxes measuring 4 by 6 by 8 inches that can fit into a box measuring 1 by 1.5 by 2 feet.

$$\frac{12}{4} \cdot \frac{18}{6} \cdot \frac{24}{8} = 3 \cdot 3 \cdot 3 = (27)$$

5. Find the area in square units of a rectangle with perimeter 52 and length 9 units.

P = 52
SP = 26
L = 9
W = 17   9 x 17 = 90 + 63
              by DPMA
(A = 153)

10. How many individual squares are in the 13th figure if the pattern continues?

13 x 13 + 4 x 13 = 169 + 52 = (221)

Level 7	Number 38

1. Let a "cut-prime" be a prime number that remains prime when its unit digit is removed. For example, both 173 and 17 are prime, so 173 is a cut-prime. Find all 2-digit cut-primes.

$$\boxed{23, 29, 31, 37, 53, 59, 71, 73, 79}$$

6. Find the statistics for the factors of 72.

1, 2, 3, 4, 6, 8, 9, 12, 18, 24, 36, 72

n	12	median	8.5
sum	195	mode	none
mean	16.25	range	71

2. Solve by cross multiplication.

$\dfrac{4}{x} = \dfrac{x}{9}$ | $\dfrac{x}{36} = \dfrac{4}{x}$ | $\dfrac{4}{x} = \dfrac{x}{25}$

$x^2 = 36$ | $x^2 = 144$ | $x^2 = 100$

$x = \boxed{\pm 6}$ | $x = \boxed{\pm 12}$ | $x = \boxed{\pm 10}$

7. Find the probability of tossing a sum of 15 with 3 standard dice.

6, 6, 3   6, 5, 4
6, 3, 6   6, 4, 5
3, 6, 6   5, 6, 4
          5, 4, 6
5, 5, 5   4, 6, 5
          4, 5, 6

$$\dfrac{10}{6 \times 6 \times 6} = \boxed{\dfrac{5}{108}}$$

The entire list need not be written if one can see the 3 + 1 + 6 options.

3. Find the simple interest on $24,000 at 2.5% annually for 22 months.

I = PRT

$$I = \dfrac{\overset{20}{\cancel{24,000}}}{1} \cdot \dfrac{2.5}{\cancel{100}} \cdot \dfrac{22}{\cancel{12}}$$

$I = 2 \times 25 \times 2 \times 11 = 100 \times 11 = \boxed{\$1100}$

8. Divide by 5 mentally.

Double; move decimal point 1 place to left.

$1{,}802{,}411 \div 5$	360482.2
$2{,}734{,}532 \div 5$	546906.4
$4{,}817{,}328 \div 5$	963465.6
$3{,}925{,}483 \div 5$	785096.6

4. Find x + y in degrees if y is one third of a straight angle and x is one third of y. (NTS)

y = 60
x = 20
x + y = $\boxed{80}$

9. Convert 981 base ten to base six.

```
 981
-864
 117
-108
   9
-  6
   3
```

4	3	1	3
216	36	6	1

$$\boxed{4313_{six}}$$

5. Simplify.

$$\sqrt{1\dfrac{3}{4} + 1\dfrac{11}{18}} \qquad \sqrt{\dfrac{121}{36}}$$

$$\sqrt{\dfrac{7}{4} + \dfrac{29}{18}}$$

$$\sqrt{\dfrac{63}{36} + \dfrac{58}{36}} \qquad \boxed{\dfrac{11}{6}}$$

10. Solve.

$|x + 11| = 0$	$|x - 4| = 30$
x = -11	x = $34, -26$
$|x - 6| = 17$	$|x + 6| = -7$
x = $23, -11$	x = impossible

Level 7	Number 39

1. Find the number of ways to arrange 3 boys and 2 girls in a row, alternating boys and girls.

$$\underline{\frac{3}{\uparrow} \cdot \frac{2}{\uparrow} \cdot \frac{2}{\uparrow} \cdot \frac{1}{\uparrow} \cdot \frac{1}{\uparrow}} \; (12)$$

1st B 1st G 2nd B 2nd G 3rd B

6. Translate the cipher with each letter as its own digit.

TAME	8AME	**8594**
+ SAME	+ 1AME	**+ 1594**
SYSTT	1Y188	**10188**

S=1, the most that can be carried from 2 addends. In 1000s, T must be great enough to carry. T≠9 by 1s. T=8. E=4. M=9. A=5. Y=0.

2. Answer as indicated.

35 is what % of 56?	35% of what number is 56?	What is 35% of 56?
$\dfrac{35}{56} = \dfrac{5}{8}$	$56 = \dfrac{7}{20}(x)$	10% is 5.6 30% is 16.8 5% is 2.8
(62.5%)	(160)	(19.6)

7. Answer YES or NO as to whether the numbers are valid sides of a triangle.

4.4, 4.6, 9.9	N	5.9, 6.9, 12.6	Y
3.4, 7.5, 10.8	Y	5.7, 7.8, 13.5	N
4.8, 6.9, 11.7	N	7.7, 7.7, 14.4	Y

3. Find the area of the rhombus in square units given both diagonals.

D = 24 d = 15

$$A = \frac{24 \times 15}{2}$$
$$= 12 \times 15$$
$$= 150 + 30 = (180)$$

D = 60 d = 17

$$A = \frac{60 \times 17}{2}$$
$$= 30 \times 17 = (510)$$

8. Rolling 3 octahedra, each numbered 1 to 8, what is the probability that the sum of the up faces is 5?

1, 1, 3
1, 2, 2
1, 3, 1
2, 1, 2
2, 2, 1
3, 1, 1

$$\frac{\overset{3}{\cancel{6}}}{\underset{4}{\cancel{8 \times 8 \times 8}}} \quad \left(\frac{3}{256}\right)$$

4. Find the 100th digit of the decimal for

$$\frac{1}{27}.$$

$$27\overline{)1.000000} \quad (0)$$
$$\underline{81}$$
$$190$$
$$\underline{189}$$
$$1$$
.037037

99th is 7. 100th is 0.

9. Find the point equidistant from points (−9, 1), (−9, 9), (−1, 9), and (−1, 1).

Points are vertices of a square.
b = 8; −9 + 4 = −5
h = 8; 9 − 4 = 5

(−5, 5)

5. The area of the top face of the pentagonal prism is 63. Find the prism's volume.

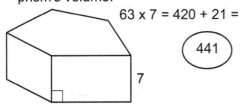

63 x 7 = 420 + 21 =

441

7

10. Solve.

9 + 5(3x − 7) = 54	7(3x − 7) + 9 = 8
15x − 35 = 45	21x − 49 = −1
15x = 80	21x = 48
x = $\dfrac{16}{3}$	x = $\dfrac{16}{7}$

Level 7	Number 40

1. Find the equal number of dimes, quarters, and half dollars that together make $29.75.

$10x + 25x + 50x = 2975$

$85x = 2975$

$17x = 595$

$x = \boxed{35}$

6. Mentally calculate the midline of each trapezoid.

base	BASE	midline
22.8	36.2	**29.5**
15.31	45.31	**30.31**
55	90	**72.5**

2. Evaluate.

$\sqrt{12}$	$4 \cdot 3$	$2\sqrt{3}$	$\sqrt{45}$	$9 \cdot 5$	$3\sqrt{5}$
$\sqrt{20}$	$4 \cdot 5$	$2\sqrt{5}$	$\sqrt{63}$	$9 \cdot 7$	$3\sqrt{7}$
$\sqrt{24}$	$4 \cdot 6$	$2\sqrt{6}$	$\sqrt{80}$	$16 \cdot 5$	$4\sqrt{5}$
$\sqrt{32}$	$16 \cdot 2$	$4\sqrt{2}$	$\sqrt{98}$	$49 \cdot 2$	$7\sqrt{2}$

7. Find the area of the circle inscribed in a square of area 49 square units.

$A_{sq} = 49$

$S = 7$

$D = 7$

$R = 7/2$

$A_{circ} = \boxed{\dfrac{49\pi}{4}}$

3. Find the area of an equilateral triangle with edge 12.

$$\frac{s^2 \cdot \sqrt{3}}{4} = \frac{12 \cdot 12 \cdot \sqrt{3}}{4} = \boxed{36\sqrt{3}}$$

8. Find the complement of the angle.

Comp (74°)	16°	Comp (84°)	6°
Comp (11°)	79°	Comp (51°)	39°
Comp (68°)	22°	Comp (22°)	68°
Comp (33°)	57°	Comp (46°)	44°

4. 10! contains how many factors of 2?

$10 \cdot 9 \cdot 8 \cdot 7 \cdot 6 \cdot 5 \cdot 4 \cdot 3 \cdot 2$

$\;1\quad\;\;3\quad\;1\quad\;\;2\quad\;1$

$\boxed{8}$

9. Find the principal for a time of 1 year.

principal	rate	interest	Get 1%. Then x 100.
$2500	8%	$200	8% is 200
$2600	5%	$130	1% is 25
$2800	2%	$56	100% is 2500
$3500	3%	$105	

5. Find the least natural number n such that kn is a perfect cube.

$k = 44 \quad n = \underline{\;242\;}$ 44=4x11, 2x121=242

$k = 45 \quad n = \underline{\;75\;}$ 45=9x5, 3x25=75

$k = 50 \quad n = \underline{\;20\;}$ 50=2x25, 4x5=20

Three of each prime create a perfect cube.

10. A square with area 49 square units and an equilateral triangle have equal perimeters. Find the side in units of the triangle.

$A\square = 49$

$S\square = 7$

$P\square = 28$

$P\triangle = 28$

$S\triangle = \boxed{9.\overline{3}}$

Level 7	Number 41

1. Of 100 girls, 10% have red hair and 20% have blond hair. Of those left, 10% have curls, 30% have waves, and the rest have straight hair. The number with straight hair is what fractional part of the total number of girls?

100
10   20   70
7   21   42
$\frac{42}{100}$ $\boxed{\frac{21}{50}}$

6. Find the area of parallelogram ABCD with vertices A(10, 10), B(0, 10), C(–6, –9), and D(4, –9).

B = 10, H = 19

A = 10 x 19 = $\boxed{190}$ sq un

2. Find the shaded area in square units.

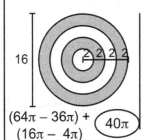
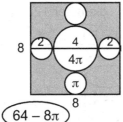

$(64\pi - 36\pi) +$ $\boxed{40\pi}$ $\boxed{64 - 8\pi}$
$(16\pi - 4\pi)$

7. Operate.

–242 ÷ 11	–22	–12 x (–4)	48
–3 x (–17)	51	19 ÷ (–19)	–1
–77 ÷ 7	–11	80 ÷ (–5)	–16
–21 x 4	–84	–19 x 3	–57

3. Find the missing terms of the geometric sequences.

–10, 20, __–40__, 80, __–160__, 320, __–640__

8100, __2700__, 900, __300__, 100

800, __–400__, 200, –100, __50__, –25

8. Convert to scientific notation.

32.3	3.23×10^1
0.0967	9.67×10^{-2}
2,058	2.058×10^3
0.0005	5.0×10^{-4}

4. Answer as indicated.

What % of 150 is 48?
$\frac{48}{150} = \frac{16}{50}$
$\boxed{32\%}$

36 is 30% of what number?
$36 = \frac{3}{10}$ x
$\boxed{120}$

What % of 225 is 90?
$\frac{90}{225} = \frac{10}{25}$
$\boxed{40\%}$

9. Operate in the given mod.

3 + 3 + 3 + 3 + 3 ≡ __3__ (mod 4)

5 x 5 x 8 ≡ __2__ (mod 9)

(4 x 4) + (3 x 4) ≡ __4__ (mod 6)

(7 x 7) – 6 ≡ __3__ (mod 8)

Can simplify value partway rather than at end.

5. How many dots are in the 15th figure if the pattern continues?

225 + 1 = $\boxed{226}$

10. Find the sum of the angles of each polygon.

hexagon	720	180 x 4 = 360 + 360
11-gon	1620	180 x 9 = 900 + 720
17-gon	2700	180 x 15 = 1800 + 900

| Level 7 | Number 42 |

1. Given the equation below and that a, b, c, and d are different positive integers. Find the value:
$2b - 4c + 5a - 7d$

$1 \times 10^a + 2 \times 10^b + 3 \times 10^c + 4 \times 10^d$
$= 24{,}130$

$a = 2$
$b = 4$
$c = 1$
$d = 3$

$8 - 4 + 10 - 21$ $\boxed{-7}$

6. Rob's age now is half of Jan's age 7 years ago and one third of Tim's age 8 years ahead. If Jan is 31 now, how old was Tim 7 years ago?

	−7	now	+8
J	24	31	
R		12	
T	㉑	28	36

2. If the product of the digits of a 3-digit number is odd, is the sum of the digits even or odd?

Only odd x odd x odd = odd.

Then odd + odd + odd = $\boxed{\text{odd}}$

7. Find the volume of a wedge cut at a 90° central angle from a cylinder with diameter 12 and height 11.

$\dfrac{90}{360} = \dfrac{1}{4}$ $\quad V = \dfrac{6 \cdot 6 \cdot 11 \cdot \pi}{4}$

$V = 3 \cdot 3 \cdot 11 \cdot \pi$

$\boxed{99\pi}$

3. Given two sides of a triangle, find the range of values for the perimeter (s for unknown side, p for perimeter).

9, 14	$5 < s < 23$	$28 < p < 46$
14, 16	$2 < s < 30$	$32 < p < 60$
20, 24	$4 < s < 44$	$48 < p < 88$

Add 2 sides to each of LO & HI of 3rd side.

8. The price of making a certain item goes down proportionally the more are made. If making 100 items costs 5200 dimes, find the cost in dollars of making 325 items.

$I \times D = I \times D$
$100 \times 520 = 325 \times D$
$4 \times 520 = 13D$
$D = 4 \times 4 \times 10$ $\boxed{160}$

4. Find the sum of all 2-digit factors of 150.

1, 2, 3, 5, 6, 10
15, 25, 30, 50, 75, 150

$10 + 15 + 25 + 30 + 50 + 75 = \boxed{205}$

9. Find the supplement of the angle.

Supp (87°)	93°	Supp (122°)	58°
Supp (115°)	65°	Supp (133°)	47°
Supp (143°)	37°	Supp (106°)	74°
Supp (159°)	21°	Supp (71°)	109°

5. Calculate the area of each shape.

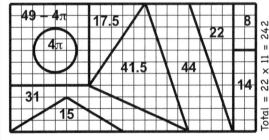

49 − 4π   17.5   8
4π
41.5   44
31   14
15

Total = 22 × 11 = 242

10. A rectangular prism has sides 5, 12, and 13. Find the 3 distinct face diagonals.

5, 12, $\boxed{13}$ Pythagorean primitive

5, 13, $\boxed{\sqrt{194}}$ $169 + 25 = 194$

12, 13, $\boxed{\sqrt{313}}$ $169 + 144 = 313$

| Level 7 | Number 43 |

---

1. Find the 6-digit mystery number.
Clue 1: The number is divisible by 15.
Clue 2: The number is a palindrome.
Clue 3: The number has 3 distinct digits.
Clue 4: The digits are odd and sum to 30.
Clue 5: The middle two digits sum to 18.

Must start and end with 5.
Middle digits = 99

$\boxed{519{,}915}$

---

6. Operate and simplify the continued fraction.

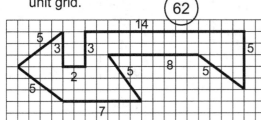

$$5 - \cfrac{3}{5 + \cfrac{7}{2 + \cfrac{4}{5}}}$$

$\dfrac{14}{5}$ $\dfrac{5}{2}$ $\dfrac{15}{2}$ $\boxed{4\,\dfrac{3}{5}}$

---

2. One child can eat a pizza with diameter 6 inches. Four similar children can eat a pizza with a diameter of how many inches? $\boxed{12}$

$d = 6$ $r = 3$ $A = 9\pi$ for 1 child
$A = 36\pi$ for 4 children
$A = 36\pi$ $r = 6$ $d = 12$

One eats area, not diameter.

---

7. Show Goldbach's Conjecture: Even integers > 2 are the sum of 2 primes.
Answers may vary.

	3 + 23		**3 + 43**		**3 + 61**
26	**13 + 13**	46	**23 + 23**	64	**17 + 47**
34	**17 + 17**	58	**29 + 29**	68	**31 + 37**
38	**19 + 19**	60	**29 + 31**	86	**3 + 83**
42	**5 + 37**	62	**3 + 59**	104	**3 + 101**
	11 + 31		**31 + 31**		**43 + 61**

---

3. Find the perimeter of the figure on the unit grid. $\boxed{62}$

---

8. Combine like terms.

$2(x + 8) - 5(7 - x) + 11x - 6$ $\quad \underline{18x - 25}$

$6(x + y) - 9x - 7(x + 1) - 6y$ $\quad \underline{-10x - 7}$

$-3(x + 6) - 9(5 - x) + 5x - 12$ $\quad \underline{11x - 75}$

$16 + 5(x - 5) - 11x - 3(x + 4)$ $\quad \underline{-9x - 21}$

---

4. Find the percent mentally.

6% of 300	18	4% of 175	7
9% of 200	18	2% of 850	17
18% of 100	18	12% of 650	78
8% of 150	12	19% of 200	38

---

9. Complete the unit conversions.

3 cm^2	300 mm^2	2.5 cm^2	250 mm^2
6 cm^3	6000 mm^3	59 mm^3	.059 cm^3
.71 cm^3	710 mm^3	(3 cm)^3	27,000 mm^3

---

5. Find the face diagonal of a cube with

edge = 9	$9\sqrt{2}$	e = 9
area of face = 16	$4\sqrt{2}$	e = 4
volume = 27	$3\sqrt{2}$	e = 3
volume = 512	$8\sqrt{2}$	e = 8

---

10. How many perfect squares are between 624 and 4901?

$25^2 = 625$
$70^2 = 4900$
25 to 70; $\boxed{46}$ numbers

Do not list the perfect squares but rather the numbers that generate them. Subtract and add 1 to get the number of numbers.

| Level 7 | Number 44 |

1. The average of 8 numbers is 12. Subtract 3 from two of these numbers. Find the new average.

sum = 12 x 8 = 96
new sum = 96 – 6 = 90
new average = 90/8 = (11.25)

6. Find the area of an isosceles triangle with sides 16, 17, 17.

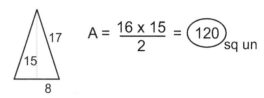

$A = \dfrac{16 \times 15}{2} =$ (120) sq un

2. Of 175 men:
61 like games, cars, & TV;
70 like games & cars;
75 like cars & TV;
80 like games & TV;
95 like games;
99 like cars; 175 – 155
125 like TV.
How many like none? _20_

G=95 C=99
6 9 15
61
19 14
31
T=125

7. For a trapezoid:

A = 225, b = 17, h = 9. Find B.

M = 225 / 9 = 25

17, 25, (33)

A = 360, B = 42, h = 12. Find b.

M = 360 / 12 = 30

(18), 30, 42

Calculate by arithmetic sequence.

3. Simplify.

$18.\overline{18}$ + $81.\overline{81}$ $99.\overline{99}$ $\underline{100}$

$13.\overline{32}$ + $66.\overline{67}$ $79.\overline{99}$ $\underline{80}$

$42.\overline{45}$ + $36.\overline{54}$ $78.\overline{99}$ $\underline{79}$

8. Complete the chart for a circle.

radius	diameter	area	circumference
$\dfrac{13}{2}$	**13**	$\dfrac{169\pi}{4}$	**13π**
0.2	**0.4**	**0.04π**	**0.4π**
1.4	2.8	**1.96π**	**2.8π**

4. Define operation ▲ as:

A ▲ B = 2B + 5A – 4AB. For example,
5 ▲ 2 = 4 + 25 – 40 = –11.
Find the values.

1.2 ▲ 4.5 $\underline{-6.6}$ 9 + 6 – 18(1.2)

2.2 ▲ 1.5 $\underline{0.8}$ 3 + 11 – 6(2.2)

9. Find the missing sides of the triangles with m∠A = 45°.

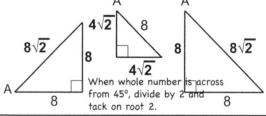

When whole number is across from 45°, divide by 2 and tack on root 2.

5. Find the height of an isosceles right triangle with each given area.

A = 60.5
$s^2 = 121$
s = h = (11)

A = 200
$s^2 = 400$
s = h = (20)

An isosceles right triangle is half of a square.

10. A savings account now has $334.50. If the bank paid 11.5% interest annually, and no money was deposited or withdrawn, how much money was in the account 1 year ago?

1.115P = 334.50

$\dfrac{334500}{1115}$ = ($300.00)

Level 7	Number 45

1. Find the number of arrangements of the letters HOUSE if the consonants must be first and last.

$$\underline{2} \cdot \underline{3} \cdot \underline{2} \cdot \underline{1} \cdot \underline{1} \quad \boxed{12}$$

↑ Do 1st.　　　　　↑ Do 2nd.

6. Find the area and perimeter of the triangle with vertices:

(–4, 3), (8, 3), and (–4, –2)	(–3, 7), (9, 7), and (9, –2)
legs = 5 & 12	legs = 9 & 12
hypotenuse = 13	hypotenuse = 15
A = 30 P = 30	A = 54 P = 36

2. Graph on the number line.

x > –4 AND x < 5　　entire number line

←——+——+——+——+——+——+——+——+——→
　　　　　　　0

x ≤ –2 OR x ≥ 0

←——+——●——+——●——+——+——+——→
　　　–2　　0

7. Give the simplified fractional part that the first decimal is of the second.

2.64 3.84	$\dfrac{264}{384}$ $\boxed{\dfrac{11}{16}}$	3.96 7.48	$\dfrac{396}{748}$ $\boxed{\dfrac{9}{17}}$
15.4 23.1	$\dfrac{154}{231}$ $\boxed{\dfrac{2}{3}}$	1.95 4.35	$\dfrac{195}{435}$ $\boxed{\dfrac{13}{29}}$

3. Write an expression in 3 ways with value 100 using only four 9s as digits and any operational symbols.

$99 + (9 \div 9)$　　$99\dfrac{9}{9}$　　$\dfrac{9 \times 9 + 9}{.9}$

8. Operate and simplify.

$$\dfrac{9}{22} \div \dfrac{48}{55} \quad \dfrac{9}{22} \times \dfrac{55}{48} \quad \dfrac{15}{32}$$

$$\dfrac{25}{24} \div \dfrac{50}{27} \quad \dfrac{25}{24} \times \dfrac{27}{50} \quad \dfrac{9}{16}$$

$$\dfrac{15}{32} \times \dfrac{16}{9}$$

$$\boxed{\dfrac{5}{6}}$$

4. Find the area of the triangle bounded by both axes and the line y = 2x + 4.

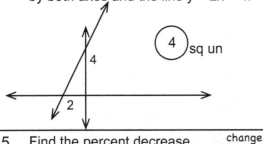

$\boxed{4}$ sq un

9. Operate.

$1 - 1 \div 2 + 2 \div 8 + 8$ $1 - .5 + .25 + 8$ $\boxed{8.75}$	$9 \div 2 + 2 \div 9 \times 27$ $4.5 + 6$ $\boxed{10.5}$
$4 \div 6 \times 9 \times 2 \div 10 + 1$ $12 \div 10 + 1$ $\boxed{2.2}$	$4 + 3 \div (3 + 7) - .7$ $4.3 - .7$ $\boxed{3.6}$

5. Find the percent decrease. $\dfrac{change}{original}$

From 60 to 45	From 20 to 9
$\dfrac{15}{60} = \boxed{25\%}$ $\dfrac{1}{4}$	$\dfrac{11}{20} = \boxed{55\%}$
From 50 to 37	From 40 to 16
$\dfrac{13}{50} = \boxed{26\%}$	$\dfrac{24}{40} = \boxed{60\%}$ $\dfrac{6}{10}$

10. Copy the prime factorizations from page 61. Then find the GCF and LCM.

120	125	200
$2^3 \times 3 \times 5$	5^3	$2^3 \times 5^2$

GCF (120, 125, 200) = **1**

LCM (120, 125, 200) = $2^3 \times 3 \times 5^3$

= 200 x 15 = **3000**

Level 7	Number 46

1. Two hours and twenty-four minutes are what fraction of time between noon on Wednesday and noon of Saturday of the same week?

$$\dfrac{\frac{12}{5}}{3 \times 24} = \dfrac{12}{5} \times \dfrac{1}{3 \times 24} = \boxed{\dfrac{1}{30}}$$

6. Find each term for the repeating sequence LMNLMNLMN... .

cycle length 3

452nd <u>M</u> 3 divides 450; R2

334th <u>L</u> 3 divides 333; R1

513rd <u>N</u> 3 divides 513; R0

2. Reflect each point over the line y = x. Then translate as specified.

(5, 0)	<u>(0, 5)</u>	left 6	<u>(−6, 5)</u>
(8, −1)	<u>(−1, 8)</u>	down 5	<u>(−1, 3)</u>
(−9, 2)	<u>(2, −9)</u>	up 3	<u>(2, −6)</u>

7. Find the next letter in each pattern.

a <u>e</u> <u>i</u> b <u>f</u> <u>j</u> **c**

+4, +4, −7 or 3 consecutive intertwining

a <u>e</u> <u>d</u> h <u>g</u> **k**

+4, −1 or 2 intertwining +3, +3

z <u>d</u> <u>w</u> <u>g</u> <u>t</u> <u>**j**</u>   −3 and +3 intertwining

3. Simplify.

$$\sqrt{2\frac{1}{2} + 2\frac{9}{16}} \quad \sqrt{\frac{81}{16}}$$

$$\sqrt{\frac{5}{2} + \frac{41}{16}}$$

$$\sqrt{\frac{40}{16} + \frac{41}{16}} \qquad \boxed{\dfrac{9}{4}}$$

8. Complete the table of values for 2x + y = 4.

x	−9	−5	−3	−2	0	1	2	3	4	6
y	22	14	10	8	4	2	0	−2	−4	−8

4. Solve by cross multiplication.

$$\dfrac{12}{7} = \dfrac{x}{6} \quad\Big|\quad \dfrac{8}{9} = \dfrac{9}{x} \quad\Big|\quad \dfrac{x}{11} = \dfrac{7}{10}$$

$$7x = 72 \quad\Big|\quad 8x = 81 \quad\Big|\quad 10x = 77$$

$$x = \boxed{\dfrac{72}{7}} \quad\Big|\quad x = \boxed{\dfrac{81}{8}} \quad\Big|\quad x = \boxed{\dfrac{77}{10}}$$

9. Find the remainder without dividing.

714,212 ÷ 21   714,210 div by 3, 7   R <u>2</u>

735,955 ÷ 11   by 11 rule 17 − 17   R <u>0</u>

182,791 ÷ 18   182,790 div by 2, 9   R <u>1</u>

222,154 ÷ 15   222,150 div by 3, 5   R <u>4</u>

5. Find the mean, median, mode, and range of the data in the frequency table.

n = 12
sum = 960

number	frequency
100	4
95	2
90	1
80	1
50	4

mean = 80
median = 92.5
modes = 100, 50
range = 50

10. Evaluate.

$\sqrt{275}$ 25 • 11 <u>$5\sqrt{11}$</u> $\sqrt{117}$ 9 • 13 <u>$3\sqrt{13}$</u>

$\sqrt{252}$ 36 • 7 <u>$6\sqrt{7}$</u> $\sqrt{120}$ 4 • 30 <u>$2\sqrt{30}$</u>

$\sqrt{294}$ 49 • 6 <u>$7\sqrt{6}$</u> $\sqrt{132}$ 4 • 33 <u>$2\sqrt{33}$</u>

$\sqrt{243}$ 81 • 3 <u>$9\sqrt{3}$</u> $\sqrt{175}$ 25 • 7 <u>$5\sqrt{7}$</u>

| | Level 7 | Number 47 |

1. Answer as indicated.

75 is what % of 120?	75% of what number is 120?	120% of 75 is what number?
$\dfrac{75}{120} = \dfrac{5}{8}$	$120 = \dfrac{3}{4}(x)$	100% is 75 10% is 7.5 20% is 15
(62.5%)	(160)	(90)

6. Count the number of paths from A to B moving only right and/or up.

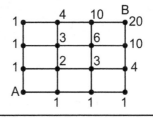

(20)

2. Solve by cross multiplication.

$\dfrac{3}{x} = \dfrac{x}{27}$	$\dfrac{x}{16} = \dfrac{4}{x}$	$\dfrac{9}{x} = \dfrac{x}{25}$
$x^2 = 81$	$x^2 = 64$	$x^2 = 225$
(x = ± 9)	(x = ± 8)	(x = ± 15)

7. Operate.

74 ÷ (−2)	37	−16 x 3	−48
−2 x 49	−98	18 x (−4)	−72
−13 ÷ 0	undef	52 ÷ (−4)	−13
−9 x (−12)	108	−3 x (−32)	96

3. Find the volume and surface area of a sphere with diameter 4 units.

$$V = \frac{4 \cdot \pi \cdot 2 \cdot 2 \cdot 2}{3} = \boxed{\frac{32\pi}{3}} \text{ cu un}$$

$$SA = 4 \cdot \pi \cdot 2 \cdot 2 = \boxed{16\pi} \text{ sq un}$$

8. Check if the row number is divisible by the column factor.

	2	9	11	18	22	99
19,314	✓	✓	−	✓	−	−
49,500	✓	✓	✓	✓	✓	✓
75,746	✓	−	✓	−	✓	−

4. Find the area in square units of a rectangle with diagonal 26 and length 10 units.

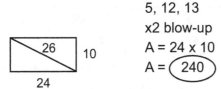

26   10

24

5, 12, 13
x2 blow-up
A = 24 x 10
A = (240)

9. Find the greatest possible product for two whole numbers that sum to:

45	22 · 23 =	(506)
69	34 · 35 =	(1190)
97	48 · 49 =	(2352)

Imagine the product as the area of a rectangle. For a fixed semi-perimeter, a square has the greatest area. So select 2 numbers closest together.

5. When a woman had driven 32 km, she had completed 40% of her car trip. How many km remained?

40% is 32

20% is 16

60% is (48)

10. Evaluate.

$\sqrt{28}$	4 · 7	$2\sqrt{7}$	$\sqrt{56}$	4 · 14	$2\sqrt{14}$
$\sqrt{40}$	4 · 10	$2\sqrt{10}$	$\sqrt{68}$	4 · 17	$2\sqrt{17}$
$\sqrt{52}$	4 · 13	$2\sqrt{13}$	$\sqrt{92}$	4 · 23	$2\sqrt{23}$
$\sqrt{54}$	9 · 6	$3\sqrt{6}$	$\sqrt{96}$	16 · 6	$4\sqrt{6}$

Level 7	Number 48

1. Dana paid $132 for 35.2 gallons of gas. What was the cost per gallon?

$$\frac{\text{cost}}{\text{gallon}} = \frac{132.00}{35.2} = \boxed{\$3.75}$$

The value to the left of "per" is the numerator.
The value to the right of "per" is the denominator.

6. Find the probability of tossing 3 different numbers with 3 standard dice.

$$\frac{6}{6} \cdot \frac{5}{6} \cdot \frac{4}{6} = \boxed{\frac{5}{9}}$$

2. Typing 250 words in 20 minutes, how many hours are needed to type 7500 words proportionally?

$$\frac{1/3}{250} = \frac{x}{7500} \qquad x = \boxed{10}$$

$$250x = 2500$$

7. Find the average using the rightmost digit methods.

401, 405, 407, 409, 410 $32 \div 5$ 406.4
keep 400

180, 180, 180, 181 $1 \div 4$ 180.25
keep 180

212, 231, 234, 240, 251 $168 \div 5$ 233.6
keep 200

3. Find the probability when drawing once from a standard deck of cards.

P(perfect square OR club) $\frac{11}{26}$ 13 club + 1,4,9 x 3 suits = 22

P(perfect square OR black) $\frac{8}{13}$ 26 black + 1,4,9 x 2 suits = 32

P(ace OR face OR black) $\frac{17}{26}$ 26 black + 2 ace + 6 JQK = 34

8. Find the sum in degrees of the complement of $(2w)°$ and the supplement of $(4w)°$.

$$90 - 2w$$
$$180 - 4w$$
$$\boxed{270 - 6w}$$

4. Find x in degrees. (NTS)

by congruent vertical angles

$$4y = 180$$
$$y = 45$$

9. Find the slope and intercepts of the lines.

line	slope	y-intercept	x-intercept
$y = 6x - 3$	**6**	**(0, −3)**	**(0.5, 0)**
$x + y = 6$	**−1**	**(0, 6)**	**(6, 0)**
$2x - y = 11$	**2**	**(0, −11)**	**(5.5, 0)**

Picture 0 inserted to find intercepts mentally.

5. Find the perimeter of a rectangle with area 280 square units and length 35 units.

$$A = 280$$
$$L = 35$$
$$W = 8$$
$$SP = 43$$
$$\boxed{P = 86}$$

10. Find the next term in the sequence by successive differences.

6   11   20   35   60   101   166

5   9   15   25   41   65

4   6   10   16   24

2   4   6   8

Compute the differences moving down in rows until a pattern emerges. Then build back up.

| Level 7 | Number 49 |

1. If a chicken and a half lays an egg and a half in a day and a half, 12 chickens will lay how many eggs in 12 days?

chickens	eggs	days
$\frac{3}{2}$	$\frac{3}{2}$	$\frac{3}{2}$
1	1	$\frac{3}{2}$
12	12	$\frac{3}{2}$
12	(96)	12

6. Translate the cipher with each letter as its own digit.

$$\begin{array}{r} \text{SEND} \\ + \text{SUUD} \\ \hline \text{DESUE} \end{array} \qquad \begin{array}{r} \text{S2N1} \\ + \text{SUU1} \\ \hline \text{12SU2} \end{array} \qquad \begin{array}{r} \textbf{6201} \\ + \textbf{6441} \\ \hline \textbf{12642} \end{array}$$

D=1, the most that can be carried from 2 addends. In 1s, E=2. S+S=12 (even) means no carry. S=6. U=4. N=0.

2. Find the missing sides of the similar triangles. (NTS)

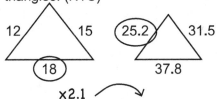

12, 15, (25.2), 31.5, (18), 37.8

×2.1

7. If numbering a book's pages starting with 1 consecutively takes 852 digits, how many pages does the book have?

PAGES	DIGITS USED
1 – 9	9
10 – 99	180 (90 numbers x 2 digits)
100 – 320	663 (221 numbers x 3 digits)
	852

(320)

3. List all subsets of {1, 2, 3, 4}.

{ }	{4}	{2, 3}	{1, 2, 4}
{1}	{1, 2}	{2, 4}	{1, 3, 4}
{2}	{1, 3}	{3, 4}	{2, 3, 4}
{3}	{1, 4}	{1, 2, 3}	{1, 2, 3, 4}

8. In problem #7, how many times does the digit 7 appear?

PAGES	7s USED
7 – 67, 87, 97	9 without 77
70 – 79	11 1 extra for 77
101 – 200	20
201 – 300	20
307	1
317	1

(62)

4. Write each repeating decimal as a simplified fraction.

$.\overline{36} \quad \frac{36}{99} \quad \left(\frac{4}{11}\right) \qquad .\overline{12} \quad \frac{12}{99} \quad \left(\frac{4}{33}\right)$

$.0\overline{36} \quad \frac{4}{110} \quad \left(\frac{2}{55}\right) \qquad .0\overline{12} \quad \frac{4}{330} \quad \left(\frac{2}{165}\right)$

9. Find the volume in cubic units of a rectangular prism with the areas of its noncongruent faces 18, 75, and 150 square units.

18 = 2 x 3 x 3 = 3 x 6
75 = 3 x 5 x 5 = 3 x 25
150 = 2 x 3 x 5 x 5 = 6 x 25

V = 3 x 6 x 25 = 6 x 75 = (450)

5. Operate using mental math.

5 – .637	4.363	25 ÷ .625	40
9 – 1.112	7.888	8 ÷ .4	20
8 – .345	7.655	99 ÷ .33	300
12 – .34	11.66	42 ÷ .7	60

10. Answer YES or NO as to whether the 3 numbers form sides of a right triangle.

21, 72, 75 7, 24, 25 x 3	Y	36, 48, 60 3, 4, 5 x 12	Y
8, 11, 15	N	7, 8, 9	N
9, 9, 9	N	25, 60, 65 5, 12, 13 x 5	Y

Level 7	Number 50

1. Arrange the digits 1, 3, 4, 5, 7, and 9 into two 3-digit numbers with the:

greatest positive difference	least positive difference
975 Max L to R.	**513** Min L to R.
– 134 Min L to R.	**– 497** Max L to R.
841	**16** First, 100s 1 apart.

415 – 397 = 18

6. If a drum strikes 6 times evenly in 5 seconds, starting and ending the time, how often will it similarly strike in 10 seconds?

(11)

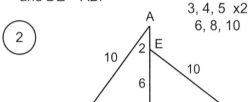

X   X   X   X   X   X
1   2   3   4   5

"fencepost" problem

2. Operate.

–13 – –24	11	59 – –69	128
–32 + –39	–71	–72 + 86	14
–43 – 47	–90	61 – 82	–21
54 – 91	–37	34 – –97	131

7. Find AE if AB = 10, BC = 6, CD = AC, and DE = AB.

(2)

3, 4, 5 x2
6, 8, 10

3. Find the volume of the cone in cubic units.

radius = 7
height = 12

$$V = \frac{7 \cdot 7 \cdot \pi \cdot 12}{3}$$

$$= 49 \cdot 4 \cdot \pi$$

$$= (196\pi)$$

Two mental math options: double twice or 50x4 – 4.

8. Operate and simplify.

$$\frac{\frac{12}{35} \times \frac{14}{15}}{\frac{18}{15} \div \frac{27}{20}} \quad \frac{\frac{12}{35} \times \frac{14}{15}}{\frac{18}{15} \times \frac{20}{27}} \quad \frac{\frac{8}{25}}{\frac{8}{9}} \quad \frac{8}{25} \times \frac{9}{8}$$

($\frac{9}{25}$)

4. Name the specified lattice point in the picture. One box = one unit.

three times as far from A as from B   (5, 4)

three times as far from B as from A   (1, 0)

(0,0)

9. Of 7 2-digit multiples of 10, the modes are 10 and 90, the median is 60, the mean is 50, and the range is 80. Find the numbers.

10   10   20   60   70   90   90

Place 10, 10, 60, 90, and 90. Sum = 50 x 7 = 350. Then 350 – 260 = 90. Two numbers add to 90. Only 20 and 70 will work.

5. Solve.

8 – 7x ≤ 57	9x – 5 ≥ 43	5 – 8x < 33
–7x ≤ 49	9x ≥ 48	–8x < 28
(x ≥ –7)	(x ≥ $\frac{16}{3}$)	(x > $\frac{-7}{2}$)

10. Find the sum of the angles of each polygon.

nonagon	1260	180 x 7 = 700 + 560
42-gon	7200	180 x 40
27-gon	4500	180 x 25

Level 7	Number 51

1. In a bag of 200 marbles, 15% are blue and 12% are red. Of the rest, 1/2 plus 1 are dotted; the rest are striped. Of the striped, 1/3 are chipped. The non-chipped striped marbles are what fractional part of all the marbles?

200
30   24   146
48/200   74   72
24   48
6/25

6. Place one of 1, 2, 3, 4, 5, and 6 in each circle of Figure A to make the sum along each line 11. Do the same for Figure B with each line sum 12.

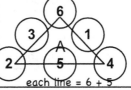
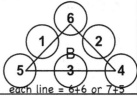

each line = 6 + 5   each line = 6+6 or 7+5

2. Find the prime factorization. Work down. In the answer use exponents with primes in ascending order.

2808	3640	4125
8 x 351	364 x 10	25 x 165
8 x 9 x 39	4 x 91 x 10	25 x 15 x 11
$2^3 \times 3^3 \times 13$	$2^3 \times 5 \times 7 \times 13$	$3 \times 5^3 \times 11$

7. Operate.

13^{-2}	$\dfrac{1}{169}$	53^1	53	4^{-4}	$\dfrac{1}{256}$
95^{-1}	$\dfrac{1}{95}$	19^2	361	7^{-3}	$\dfrac{1}{343}$
		38^0	1		
40^{-2}	$\dfrac{1}{1600}$	5^4	625	50^{-2}	$\dfrac{1}{2500}$

3. Estimate the fraction to the nearest whole number.

0.5   ~~8~~   ~~600~~   20
$\dfrac{0.498 \times 8.11 \times 601.067}{31.017 \times 7.92}$
   ~~30~~   ~~8~~

⟨10⟩

8. Write as an equation.

The quotient of a number divided by five plus 1 is eight. $\dfrac{x}{5} + 1 = 8$

Nine times the quantity of a number decreased by 6 is 2. $9(x - 6) = 2$

Nine times a number decreased by six is two. $9x - 6 = 2$

The positive difference of x and 1 is triple x. $|x - 1| = 3x$

4. Graph on the coordinate plane. One box equals one unit.

y = 2x – 3   y = –3x

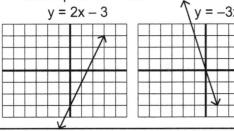

9. Find the angle in degrees formed by clock hands at 6:45.

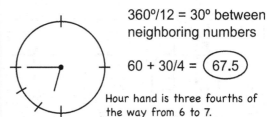

360°/12 = 30° between neighboring numbers

60 + 30/4 = ⟨67.5⟩

Hour hand is three fourths of the way from 6 to 7.

5. Operate on the sets.

A = {2, 5, 6, 12, 14, 16}
B = {12, 13, 14, 15, 16}
U = {2, 3, 4, 5, 6, 12, 13, 14, 15, 16}

A′ = {3, 4, 13, 15}
B′ = {2, 3, 4, 5, 6}
(A ∩ B)′ = {2, 3, 4, 5, 6, 13, 15}

10. A soccer team lost 10 games and won 20. What fraction of its games did the team win?

$\dfrac{W}{G} = \dfrac{20}{30} = \left\langle \dfrac{2}{3} \right\rangle$

In a class of 60 students, 42 are boys. What fraction of the students are girls?

$\dfrac{G}{S} = \dfrac{18}{60} = \left\langle \dfrac{3}{10} \right\rangle$

Level 7	Number 52

1. Find the day of the week given each separate condition.

421 days ago if today is Wednesday

–420 = Wed   Tuesday

285 days ago if today is Saturday

–287 = Sat   Monday

6. Operate and simplify.

$$9\frac{5}{6} - 2\frac{7}{9} + 8\frac{9}{10}$$

$$9\frac{75}{90} - 2\frac{70}{90} + 8\frac{81}{90}$$

$$15\frac{86}{90} \quad \boxed{15\frac{43}{45}}$$

2. If 210 is written as the product of 3 positive integers, each greater than 1, how many of them are even?

21 x 10   ①
2 x 3 x 5 x 7
The 2 may be multiplied by the 3, 5, or 7.
But only 1 of the 3 factors would be even.

7. Convert to base ten.

555_{seven} 5 x 49 + 5 x 7 + 5   285

$15D_{sixteen}$ 1 x 256 + 5 x 16 + 13   349

$5T9_{twelve}$ 5 x 144 + 10 x 12 + 9   849

$A3B_{fourteen}$ 10 x 196 + 3 x 14 + 11   2013

3. Find the sum of each arithmetic sequence using the "First plus Last" method.

–35 + –31 + –27 + –23 + –19 + –15 + –11 + – 7

–50 + –45 + –40 + –35 + –30 + –25 + –20

$$\frac{(-42)(8)}{2} = \boxed{-168} \quad \frac{(-70)(7)}{2} = \boxed{-245}$$

8. Multiply by 11 mentally.

354 x 11   3894   Write from R
                         to L. For 7263:
7263 x 11   79,893   Write the 3.
                         Then 3+6=9;
517 x 11   5687   6+2=8; 2+7=9.
                         Write the 7.
8072 x 11   88,792

4. Draw the reflection of each trapezoid over the line x = 2.

Preserve distance of vertices from reflection line.

9. Distribute.

c(9x – 7y + 5z)   9cx – 7cy + 5cz

–8(–x – 2y – 8z)   8x + 16y + 64z

3x(–7x – 5y + 3) $-21x^2 - 15xy + 9x$

–7b(–3b + 4c – d) $21b^2 - 28bc + 7bd$

5. Find the probability of selecting a point from the shaded area within the square given an inscribed circle.

s = 10 d = 10 r = 5

$$\frac{A_{sh}}{A_{squ}} = \frac{100 - 25 \cdot \pi}{100}$$

Factor 25 in numerator.
Simplify. $\boxed{\frac{4 - \pi}{4}}$

10. How many perfect squares are between 903 and 250,001?

$30^2 = 900$
$500^2 = 250,000$ ④⑦⓪
31 to 500; 470 numbers

Do not list the perfect squares, but rather the numbers that generate them. Subtract and add 1 to get the number of numbers.

Level 7	Number 53

1. The number of unique numbers in mod six is ___six___.

For a line, the change in y divided by the change in x is its ___slope___.

The horizontal bar symbol is used to mark the pattern in a ___repeating decimal___.

6. Operate and simplify the continued fraction.

$$6 + \cfrac{2}{4 + \cfrac{5}{4 + \cfrac{1}{6}}}$$

$\dfrac{5}{13}$ $\boxed{6\frac{5}{13}}$

$4 + \dfrac{1}{6} \quad \dfrac{25}{6} \quad \dfrac{6}{5} \quad \dfrac{26}{5}$

2. Count the total number of triangles in the picture.

1: 8
2: 8
3: 0
4: 2
5–8: 0

$\boxed{18}$

7. Name the property exemplified.

If a = 6 and 6 = b, then a = b. ___TPE___

167 x 1 = 167 ___IdPM___

6 + 7 = 7 + 6 ___CPA___

6 + 7 = 6 + 7 ___RPE___

3. Answer ALWAYS, SOMETIMES, or NEVER.

The sum of 2 acute angles is < 90°. ___S___
A central angle is right. ___S___
Vertical angles are obtuse. ___S___
Two adjacent angles are congruent. ___S___
Two right angles are congruent. ___A___

8. Divide by 5 mentally.

Double; move decimal point 1 place to left.

2346 ÷ 5	469.2	194 ÷ 5	38.8
1835 ÷ 5	367	733 ÷ 5	146.6
4418 ÷ 5	883.6	624 ÷ 5	124.8
3712 ÷ 5	742.4	835 ÷ 5	167

4. Without graphing on the number line, identify each as 2 points, 2 rays, line, open ray, or open/half-open segment.

−11 < x < 5 ___open segment___

x = −3 OR x = 0 ___2 points___

x > 4 OR x < 9 ___line___

9. Find the reciprocal in the given mod.

reciprocal of 3 mod 4 ___3___ 3x3=9, 9–8=1

reciprocal of 7 mod 11 ___8___ 7x8=56, 56–55=1

reciprocal of 5 mod 7 ___3___ 5x3=15, 15–14=1

reciprocal of 4 mod 5 ___4___ 4x4=16, 16–15=1

5. Copy the prime factorizations from page 51. Then find the GCF and LCM.

135	190	315
$3^3 \times 5$	$2 \times 5 \times 19$	$3^2 \times 5 \times 7$

GCF (135, 190, 315) = **5**
LCM (135, 190, 315) = $2 \times 3^3 \times 5 \times 7 \times 19$
= 135 x 266 = **35,910**

10. A 5 by 5 by 5 cube is painted on the outside and then cut into 125 1 by 1 by 1 cubes. How many of the littlest cubes have each number of faces painted?

3 faces? ___8___   1 face? ___54___

2 faces? ___36___   0 faces? ___27___

Level 7	Number 54

1. The angles of a quadrilateral are in the ratio 2:3:4:6. Find the angles.

sum angles = 360

2 + 3 + 4 + 6 = 15

360 / 15 = 24

angles = $\boxed{48, 72, 96, 144}$

6. Solve.

$3\sqrt{x} + 9 = 219$	$9\sqrt{x} - 8 = 100$
$3\sqrt{x} = 210$	$9\sqrt{x} = 108$
$\sqrt{x} = 70$	$\sqrt{x} = 12$
$\boxed{x = 4900}$	$\boxed{x = 144}$

2. What time is 189 minutes before 5:12 AM?

$189 = 60 \times 3 + 9 \quad 5 - 3 \rightarrow 2 \quad 12 - 9 \rightarrow 3$

$\boxed{\text{2:03 AM}}$

What time is 118 minutes before 9:34 PM?

$118 = 60 \times 2 - 2 \quad 9 - 2 \rightarrow 7 \quad 34 + 2 \rightarrow 36$

$\boxed{\text{7:36 PM}}$

7. Find the area of the circle inscribed in a square of area 25 square units.

$A_{sq} = 25$

$S = 5$

$D = 5$

$R = 5/2$

$A_{circ} = \boxed{\dfrac{25\pi}{4}}$

3. Find the simple interest on $3200 at 3.5% annually for 33 months.

$I = PRT$

$I = \dfrac{\overset{8}{\cancel{3200}}}{1} \cdot \dfrac{3.5}{\cancel{100}} \cdot \dfrac{\overset{11}{\cancel{33}}}{\underset{4}{\cancel{12}}}$

$I = 4 \times 7 \times 11 = 28 \times 11 = \boxed{\$308}$

8. Answer as a simplified fraction.

The two 100s in each % simplify.

26% is what fractional part of 65%? $\dfrac{26}{65}$ $\dfrac{2}{5}$

16% is what fractional part of 72%? $\dfrac{16}{72}$ $\dfrac{2}{9}$

30% is what fractional part of 75%? $\dfrac{30}{75}$ $\dfrac{2}{5}$

4. Find the sum of all 2-digit factors of 160.

1, 2, 4, 5, 8, 10

16, 20, 32, 40, 80, 160

10 + 16 + 20 + 32 + 40 + 80 = $\boxed{198}$

9. Convert 993 base ten to base nine.

$\begin{array}{r} 993 \\ -729 \\ \hline 264 \\ -243 \\ \hline 21 \\ -18 \\ \hline 3 \end{array}$

1	3	2	3
729	81	9	1

$\boxed{1323_{nine}}$

5. Find the midpoint of the line segment with the given endpoints.

(5.5, 17) and (6.7, −1)	(6.1, 8)	
(−1, 5.2) and (−7, 5.8)	(−4, 5.5)	Average the xs, average the ys.
(7.3, −8) and (3.7, 20)	(5.5, 6)	
(−4, −3) and (−5, −7)	(−4.5, −5)	

10. A rectangular prism has sides 4, 7, and 9. Find the 3 distinct face diagonals.

4, 7, $\boxed{\sqrt{65}}$ 16 + 49 = 65

4, 9, $\boxed{\sqrt{97}}$ 16 + 81 = 97

7, 9, $\boxed{\sqrt{130}}$ 49 + 81 = 130

Level 7	Number 55

1. Find the number of 5-digit even numbers selecting from the digits 1, 2, 3, 4, 5 without repeating digits.

$$\underline{4} \cdot \underline{3} \cdot \underline{2} \cdot \underline{1} \cdot \underline{2} \quad \textcircled{48}$$

↑ Do first.

6.

	R	T	D	Jane ran 18 miles twice to prepare for a marathon. The first time she ran at 3 mph. The second time she ran at 4 mph. What was her total time in hours for the two practices?
#1	3	6	18	
#2	4	4.5	18	
tot		⑩.5	36	

2. Find the probability when drawing once from a standard deck of cards.

P(ace OR red) $\dfrac{7}{13}$ 26 red + 2 black aces = 28

P(jack OR heart) $\dfrac{4}{13}$ 13 hearts + 3 jacks = 16

P(face OR spade) $\dfrac{11}{26}$ JQK of 3 suits + 13 spades = 22

7. Combine like terms.

$8(x - 2y) - 2x + 3y - (x - y)$ $5x - 12y$

$-6(6 - 4x) + 4x + 10(3 - x)$ $18x - 6$

$5(7y - 2x) - 7(x + y) + 18x$ $x + 28y$

$9x - 9 + 9(9 - x) + 9(x + 1)$ $9x + 81$

3. Find the area of the rectangle bounded by the lines y = 1, x = –2, x = 10, and y = –10.

NTS   11   ⑬2   sq un
12

8. By what fractional part of:

$\dfrac{8}{11}$ does $\dfrac{7}{11}$ exceed $\dfrac{3}{11}$? $\textcircled{\dfrac{1}{2}}$

$\dfrac{7}{12}$ does $\dfrac{5}{12}$ exceed $\dfrac{2}{12}$? $\textcircled{\dfrac{3}{7}}$

$\dfrac{8}{13}$ does $\dfrac{7}{13}$ exceed $\dfrac{5}{13}$? $\textcircled{\dfrac{1}{4}}$

4. Find the percent mentally.

12% of 50	6	4% of 125	5
19% of 100	19	11% of 600	66
4% of 150	6	13% of 200	26
8% of 25	2	18% of 250	45

9. Find the missing sides of the triangles with m∠A = 30°. (NTS)

24 ... 12√3 ... 12

6√3 ... 6 ... 12

12 ... 8√3 ... 4√3

When whole number is across from 60°, divide by 3 and tack on root 3.

5. Find the face diagonal of a cube with

area of face = 400 $20\sqrt{2}$ e = 20

area of face = 121 $11\sqrt{2}$ e = 11

volume = 1000 $10\sqrt{2}$ e = 10

edge = 12 $12\sqrt{2}$ e = 12

10. Find the area of the parallelogram.

7, 24, 25 right △   (NTS)

24 ... 25

7 ... 13

A = 20 x 24

A = ⑭80

| Level 7 | Number 56 |

1. Find the number of miles traveled in 4 minutes 48 seconds at an avergae rate of 10 miles per hour.

D = RT   48 sec = 4/5 min

$D = 10 \times \dfrac{24}{5} \times \dfrac{1}{60}$

$D = \boxed{\dfrac{4}{5}}$

6. Find the statistics for the factors of 200.

1, 2, 4, 5, 8, 10, 20, 25, 40, 50, 100, 200

n	12	median	15
sum	465	mode	none
mean	38.75	range	199

2. Given the equation below and that a, b, c, and d are different positive integers. Find the value:
6b − 9a + 8d − 5c

$7 \times 10^a + 9 \times 10^b + 3 \times 10^c + 5 \times 10^d$
= 53,790

a = 2
b = 1
c = 3
d = 4

6 − 18 + 32 − 15   (5)

7. Find the next letter in each pattern.

+3 and +1 intertwining

A F D G G H **J**

+3 and +4 alternating

A D H K O R **V**

+4 and +1 intertwining

c c g d k e **o**

3. Write each fraction as a repeating decimal.

| $\dfrac{1}{45}$ $.0\overline{2}$ | $\dfrac{9}{330}$ $.0\overline{27}$ | $\dfrac{6}{900}$ $.00\overline{6}$ |
| $\dfrac{1}{90}$ $.0\overline{1}$ | $\dfrac{4}{330}$ $.0\overline{12}$ | $\dfrac{19}{990}$ $.0\overline{19}$ |

8. Evaluate.

$-5 \times |4 - 9| - 20 \div 2 \times |2 - 7| + 7 \times 5$
−25 − 50 + 35
−40

$-8 \div |2 - 4| - |-3 - 4| \div |1 - 8| - 1 \div 2$
− 4 − 1 − 0.5
−5.5

4. Calculate the point on the number line that is:

$\dfrac{1}{2}$ of the way from −5 to 6   11/2 = 5.5   .5

$\dfrac{2}{5}$ of the way from 9 to −9   18/5 = 3.6   1.8

$\dfrac{3}{4}$ of the way from 4 to −2   6/4 = 1.5   −.5

9. If pizza with a 20 cm diameter serves 2, then 2 pizzas each with a 30 cm diameter should serve how many?

d = 20   r = 10   A = 100π for 2 people
50π for 1 person
d = 30   r = 15   A = 225π
225π + 225π = 450π   450π ÷ 50π = (9)

One eats area, not diameter.

5. Operate.

$\begin{array}{r} 1012_{three} \\ + 1122_{three} \\ \hline \mathbf{2211_{three}} \end{array}$
$\begin{array}{r} 8043_{nine} \\ - 1256_{nine} \\ \hline \mathbf{6676_{nine}} \end{array}$
$\begin{array}{r} 1233_{four} \\ + 1132_{four} \\ \hline \mathbf{3031_{four}} \end{array}$

10. Evaluate.

$\sqrt{.0016}$ .04	$\sqrt{.000004}$ .002
$\sqrt{.0009}$ .03	$\sqrt{.000121}$ .011
$\sqrt{.0169}$ .13	$\sqrt{.000144}$ .012
$\sqrt{6.25^2}$ 6.25	$\sqrt{.000625}$ .025

MAVA Math: Middle Reviews Solutions Copyright © 2013 Marla Weiss

| Level 7 | Number 57 |

---

1. The time on a 12-hour circular clock is 3:00 PM. What will be the time after the hour hand goes around once?

    (3:00 AM)

    A full circle of the hour hand is 12 hours.

6. Find the area of parallelogram ABCD with vertices A(1, 10), B(13, 10), C(10, –2), and D(–2, –2).

    B = 12, H = 12

    A = 12 x 12 = (144)

---

2. Find the prime factorization. Work down. In the answer use exponents with primes in ascending order.
    24,750
    2475 x 10
    5 x 495 x 5 x 2   ($2 \times 3^2 \times 5^3 \times 11$)
    5 x 5 x 99 x 5 x 2
    5 x 5 x 9 x 11 x 5 x 2

7. Evaluate for x = 5.

$x^3 - 2x^2 + x - 3$	125 – 50 + 2	77
$x + 7 + (2x + 5)^0$	5 + 7 + 1	13
$x^{-1} + 2(x - 2)^2 - 1$	.2 + 18 – 1	17.2
$(x - 1)^3 - 4x^2 + x^0$	64 – 100 + 1	–35

---

3. Find the area of an equilateral triangle with edge 8.

    $$\frac{s^2 \cdot \sqrt{3}}{4} = \frac{8 \cdot 8 \cdot \sqrt{3}}{4} = (16\sqrt{3})$$

8. Write as an equation.
    One third a number minus two is four times the number. $\frac{x}{3} - 2 = 4x$
    Five times the quantity twice a number minus one is six. $5(2x - 1) = 6$
    A number squared plus triple the number is ten. $x^2 + 3x = 10$
    A number cubed decreased by nine is double the number. $x^3 - 9 = 2x$

---

4. Find the diagonal of a rectangle with area 1080 square units and length 45 units.

    1080/45 =
    216/9 = 24
    8, 15, 17
    x3 blow-up
    D = (51) units

    51 │ 24
    45

9. Operate in the given mod.

    (6 x 5) + 4   ≡ _6_ (mod 7)
    4 x 3 x 2 x 3   ≡ _2_ (mod 5)
    (3 x 2) + (3 x 3)   ≡ _3_ (mod 4)
    5 + 4 + 3 + 2 + 1 ≡ _3_ (mod 6)

    Can simplify value partway rather than at end.

---

5. Label the angles given the 2 marked.   (NTS)

    59° 95 26°
    26 95 59
    154 26   59 121
    26 154   121 59

10. Divide and simplify by regrouping.

    (15 YD 2 FT 8 IN) ÷ 4
    (12 YD 11 FT 8 IN) ÷ 4
    (12 YD 8 FT 44 IN) ÷ 4

    (3 YD 2 FT 11 IN)

Level 7	Number 58

1. Find the value of AC – AD given the system of 4 equations.

AB = 24   A = 4, B = 6   By T & E
CD = 56   C = 7, D = 8
BD = 48   B = 6, D = 8
BC = 42   B = 6, C = 7

AC – AD = A(C – D) = (4)(–1) = $\boxed{-4}$

6. Four years ago Mary was the same age as Jill will be in 6 years. Now Mary's age is 4 times Mike's age who is 6. How old is Jill now?

	–4	now	+6
Ma	20	24	
Mi		6	
J		$\boxed{14}$	20

2. Of 200 students:

MA=150   PE=110
47, 13, 18, 75, 15, 4, 6
MU=100

75 like math, PE, & music;
88 like math & PE;
90 like math & music;
79 like PE & music;
150 like math;
110 like PE;   200 – 178
100 like music.
How many like none? 22

7. Find the average using the rightmost digit methods.

61, 61, 62, 64, 69, 70   27 ÷ 6   64.5
keep 60

135, 135, 136, 136, 141   8 ÷ 5   136.6
keep 135

971, 973, 977, 980   21 ÷ 4   975.25
keep 970

3. Find the height of an isosceles right triangle with each given area.

A = 40   A = 45
$s^2 = 80$ $s^2 = 90$
$s = h = \sqrt{80}$ $s = h = \sqrt{90}$

An isosceles right triangle is half of a square.

$\boxed{4\sqrt{5}}$ $\boxed{3\sqrt{10}}$

8. Find the complement of the angle.

Comp (61°) 29°   Comp (14°) 76°
Comp (53°) 37°   Comp (38°) 52°
Comp (79°) 11°   Comp (86°) 4°
Comp (42°) 48°   Comp (27°) 63°

4. Find the least natural number n such that kn is a perfect cube.

k = 14   n = 196   14=2x7, 4x49=196
k = 16   n = 4   16=8x2, 2x2=4
k = 22   n = 484   22=2x11, 4x121=484

Three of each prime create a perfect cube.

9. Operate and simplify.

$$\frac{16}{27} \div \frac{44}{15} \quad \frac{16}{27} \times \frac{15}{44} \quad \frac{20}{99}$$
$$\frac{11}{6} \div \frac{99}{48} \quad \frac{11}{6} \times \frac{48}{99} \quad \frac{8}{9} \quad \frac{20}{99} \times \frac{9}{8}$$

$\boxed{\dfrac{5}{22}}$

5. Find the least value of x ÷ y.

–15 ≤ x ≤ 0   –20 ≤ y ≤ –5   0
–9 ≤ x ≤ 36   –3 ≤ y ≤ 12   –12
–12 ≤ x ≤ –2   –2 ≤ y ≤ 12   –1
5 ≤ x ≤ 45   –5 ≤ y ≤ 9   –9

10. A square with area 25 square units and an equilateral triangle have equal perimeters. Find the side in units of the triangle.

A □ = 25
S □ = 5
P □ = 20
P △ = 20
S △ = $\boxed{6.\overline{6}}$

Level 7	Number 59

1. The time on a 12-hour circular clock is 10:00 AM. What will be the time after the minute hand goes around 4 times?

(2:00 PM)

A full circle of the minute hand is one hour.

6. Count the number of paths from A to B moving only right and/or up.

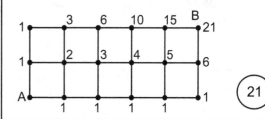

(21)

2. Find the missing sides of the similar triangles. (NTS)

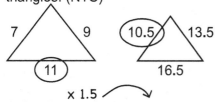

(11)   16.5

x 1.5

7. Give the simplified fractional part that the first decimal is of the second.

4.9
‾‾‾‾ $4\frac{9}{10} \div 12\frac{1}{4}$ $\frac{49}{10} \times \frac{4}{49}$ $\left(\frac{2}{5}\right)$
12.25

4.25
‾‾‾‾ $4\frac{1}{4} \div 12\frac{3}{4}$ $\frac{17}{4} \times \frac{4}{51}$ $\left(\frac{1}{3}\right)$
12.75

3. Find the area of the rhombus in square units given both diagonals.

D = 24 d = 12

$A = \dfrac{24 \times 12}{2}$

= 12 x 12 = (144)

D = 30 d = 15

$A = \dfrac{30 \times 15}{2}$

= 15 x 15 = (225)

8. Answer YES or NO as to whether the 3 numbers form sides of a right triangle.

13, 14, 15	N	12, 41, 45	N
33, 44, 55	Y	11, 19, 29	N
3, 4, 5 x 11			
20, 48, 52	Y	20, 30, 40	N
5, 12, 13 x 4			

4. Identify as a cube or not a cube.

NO   YES   NO

9. Find the point equidistant from points (15, 13), (15, 1), (1, 1), and (1, 13).

Points are vertices of a rectangle.
b = 14; 1 + 7 = 8
h = 12; 1 + 6 = 7

(8, 7)

5. Write each repeating decimal as a simplified fraction.

$.\overline{72}$ $\frac{72}{99}$ $\left(\frac{8}{11}\right)$ $.\overline{15}$ $\frac{15}{99}$ $\left(\frac{5}{33}\right)$

$.0\overline{72}$ $\frac{8}{110}$ $\left(\frac{4}{55}\right)$ $.0\overline{15}$ $\frac{5}{330}$ $\left(\frac{1}{66}\right)$

10. The meat in a recipe has 3 parts beef, 2 parts lamb, 1 part pork, and 2 parts veal. The lamb is what fractional part of all the meat?

$\dfrac{L}{M} = \dfrac{2}{8} = \left(\dfrac{1}{4}\right)$

The vowels including Y are what fractional part of the letters in the alphabet?

$\dfrac{V}{L} = \dfrac{6}{26} = \left(\dfrac{3}{13}\right)$

Level 7	Number 60

1. Two and one third hours are what fraction of time between noon on Monday and noon of Monday of the next week?

$$\dfrac{\frac{7}{3}}{7 \times 24} = \dfrac{7}{3} \times \dfrac{1}{7 \times 24} = \boxed{\dfrac{1}{72}}$$

2. Find the equal number of nickels, dimes, and half dollars that together make $27.30.

$5x + 10x + 50x = 2730$

$65x = 2730$

$13x = 546$

$x = \boxed{42}$

3. Given two sides of a triangle, find the range of values for the perimeter (s for unknown side, p for perimeter).

6, 17	$11 < s < 23$	$34 < p < 46$
16, 21	$5 < s < 37$	$42 < p < 74$
18, 28	$10 < s < 46$	$56 < p < 92$

Add 2 sides to each of LO & HI of 3rd side.

4. 10! contains how many factors of 3?

$10 \cdot 9 \cdot 8 \cdot 7 \cdot 6 \cdot 5 \cdot 4 \cdot 3 \cdot 2$

    2   1   1

$\boxed{4}$

5. Calculate the area of each shape.

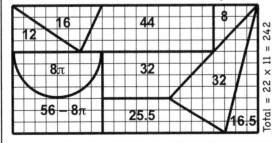

16	44	8
12		
8π	32	32
$56 - 8\pi$	25.5	16.5

Total = 22 x 11 = 242

6. Operate and simplify the continued fraction.

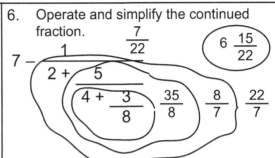

$$7 - \cfrac{1}{2 + \cfrac{5}{4 + \cfrac{3}{8}}}$$

$\dfrac{7}{22}$ $\dfrac{35}{8}$ $\dfrac{8}{7}$ $\dfrac{22}{7}$ $\boxed{6\dfrac{15}{22}}$

7. Find the probability of tossing a sum of 6 with 3 standard dice.

4, 1, 1	3, 2, 1
1, 4, 1	3, 1, 2
1, 1, 4	2, 1, 3
	2, 3, 1
2, 2, 2	1, 2, 3
	1, 3, 2

$\dfrac{10}{6 \times 6 \times 6} = \boxed{\dfrac{5}{108}}$

Use Multiplication Principle to see the 3 + 1 + 6 options without writing the entire list.

8. Find the statistics for the first 5 positive perfect cubes.

1, 8, 27, 64, 125

n	5	median	27
sum	225	mode	none
mean	45	range	124

9. Find the principal for a time of 1 year.

principal	rate	interest
$1400	6%	$84
$1900	7%	$133
$3100	8%	$248
$4200	5%	$210

Get 1%. Then x 100.

6% is 84
1% is 14
100% is 1400

10. Find the volume in cubic units of a square pyramid with edge 11 and altitude 9 units.

$$\dfrac{11 \times 11 \times 9}{3} = 11 \times 11 \times 3$$

$$= \boxed{363}$$

Level 7	Number 61

1. 1500 bees make 36 pounds of honey in 1 year. How much honey do 1100 bees make in two and a half years?

bees	honey lb	years
1500	36	1
100	$\frac{12}{5}$	1
1100	$\frac{11 \cdot 12}{5}$	1
1100	⟨66⟩	$\frac{5}{2}$

6. Multiply by 25 mentally.
Divide by 4 (cut in half twice). Tack on two Os.

7204 x 25	180,100	836 x 25	20,900
8828 x 25	220,700	472 x 25	11,800
4864 x 25	121,600	960 x 25	24,000
6808 x 25	170,200	660 x 25	16,500

2. Find the prime factorization. Work down. In the answer use exponents with primes in ascending order.

9009	5120	1313
9 x 1001	512 x 10	13 x 101
3^2 x 7 x 11 x 13	2^9 x 10	**13 x 101**
	2^{10} x 5	

7. Answer YES or NO as to whether the numbers are valid sides of a triangle.

5.6, 5.6, 6.5	Y	6.5, 7.5, 14.2	N
6.5, 6.5, 6.5	Y	7.9, 8.1, 15.6	Y
3.3, 4.4, 7.6	Y	9.6, 9.8, 19.5	N

3. Find the perimeter of the figure on the unit grid.
⟨60⟩

8. Convert to scientific notation.

0.0054	5.4×10^{-3}
76,000	7.6×10^4
0.00006	6.0×10^{-5}
0.719	7.19×10^{-1}

4. Without graphing on the number line, identify each as 2 points, 2 rays, line, open ray, or open/half-open segment.

$x \le 8$ OR $x \ge 9$	2 rays
$-6 \le x < 12$	half-open segment
$0 < x < 35$	open segment

9. Find the average using the arithmetic sequence method.
Add half of 1.2 to 4.7.

2.3, 3.5, 4.7, 5.9, 7.1, 8.3	5.3
11.4, 13.7, 16, 18.3, 20.6	16
13.5, 15.5, 17.5, 19.5	16.5

5. Find the percent change. $\frac{change}{original}$

From 80 to 20	From 18 to 27
$\frac{60}{80}$ = ⟨75% D⟩	$\frac{9}{18}$ = ⟨50% I⟩
From 40 to 12	From 20 to 31
$\frac{28}{40}$ = ⟨70% D⟩	$\frac{11}{20}$ = ⟨55% I⟩

10. Find the sum of the angles of each polygon.

octagon	1080	180 x 6 = 600 + 480
20-gon	3240	180 x 18
dodecagon	1800	180 x 10

| Level 7 | Number 62 |

1. The average of 15 numbers is 10. Add 6 to five of these numbers. Find the new average.

sum = 15 x 10 = 150
new sum = 150 + 30 = 180
new average = 180/15 = (12)

6. For a trapezoid:

A = 165, b = 12, h = 11. Find B.

M = 165/11 = 15

12, 15, (18)

A = 286, B = 30, h = 13. Find b.

M = 286/13 = 22

(14), 22, 30

Calculate by arithmetic sequence.

2. If 330 is written as the product of 3 positive integers, each greater than 1, how many of them are even?

33 x 10
2 x 3 x 5 x 11
The 2 may be multiplied by the 3, 5, or 11.
But only 1 of the 3 factors would be even.

(1)

7. Find the volume of a wedge cut at a 40° central angle from a cylinder with radius 15 and height 8.

$$\frac{40}{360} = \frac{1}{9}$$

$$V = \frac{15 \cdot 15 \cdot 8 \cdot \pi}{9}$$

$$V = 5 \cdot 5 \cdot 8 \cdot \pi$$

(200π)

3. Reflect each point over the line y = x. Then translate as specified.

(7, –8) ___(–8, 7)___ right 5 ___(–3, 7)___

(0, 9) ___(9, 0)___ up 4 ___(9, 4)___

(4, 6) ___(6, 4)___ down 6 ___(6, –2)___

8. Complete the table of values for x + 5y = 15.

x	–5	–1	0	1	5	10	15	20	30	45
y	4	$\frac{16}{5}$	3	$\frac{14}{5}$	2	1	0	–1	–3	–6

4. Define operation ❋ as:

X ❋ Y = XY + 4Y – 6X. For example, 3 ❋ 2 = 6 + 8 – 18 = –4.
Find the values.

3 ❋ 2.25 ___–2.25___ 6.75 + 9 – 18

6 ❋ 7.5 ___39___ 45 + 30 – 36

9. Find the supplement of the angle.

Supp (132°) ___48°___ | Supp (94°) ___86°___

Supp (158°) ___22°___ | Supp (146°) ___34°___

Supp (167°) ___13°___ | Supp (113°) ___67°___

Supp (109°) ___71°___ | Supp (121°) ___59°___

5. Solve.

11x – 7 ≤ 48 | 7 – 9x ≥ 73 | 5 – 6x < 31

11x ≤ 55 | –9x ≥ 66 | –6x < 26

(x ≤ 5) | (x ≤ $\frac{-22}{3}$) | (x > $\frac{-13}{3}$)

10. If an item at 23% off costs $508.20, what is its original price?

.77P = 508.20

$$\frac{50820}{77} = \frac{7260}{11} = (\$660.00)$$

Level 7	Number 63

1. Write the inverse of the conditional statement. Are they logically equivalent? (NO)

If a shape is a polygon, then it has sides.

statement TRUE
inverse FALSE

If a shape is not a polygon, then it does not have sides.

The shape could have open or crossing sides.

6. Find each term for the repeating sequence KLMNOPKLMNOP... .

cycle length 6

129th   M   6 divides 126; R3

656th   L   6 divides 654; R2

144th   P   6 divides 144; R0

---

2. Graph on the number line.

$x > 2$ OR $x \le -3$

$x \le 3$ AND $x \le 1$

7. Show Goldbach's Conjecture: Even integers > 2 are the sum of 2 primes. Answers may vary.

16	3 + 13	28	11 + 17	66	19 + 47
20	3 + 17	30	13 + 17	80	7 + 73
52	5 + 47	48	19 + 29	84	5 + 79
54	7 + 47	50	3 + 47	90	11 + 79
	23 + 31		23 + 27		43 + 47

Top of columns: 5 + 11 | 5 + 23 | 5 + 61

---

3. Estimate the fraction to the nearest whole number.

0.6   ~~50~~ 10   ~~3~~

$$\frac{0.618 \times 51.1 \times 2.987}{3.009 \times 4.931}$$ (6)

~~3~~   ~~5~~

8. Find the volume in cubic units of a rectangular prism with the areas of its noncongruent faces 40, 45, and 72 square units.

$40 = 5 \times 8$
$45 = 5 \times 9$
$72 = 8 \times 9$

$V = 5 \times 8 \times 9 = 40 \times 9 = $ (360)

---

4. Find the volume and surface area of a sphere with radius 5 units.

$$V = \frac{4 \cdot \pi \cdot 5 \cdot 5 \cdot 5}{3} = \left(\frac{500\pi}{3}\right)$$

$$SA = 4 \cdot \pi \cdot 5 \cdot 5 = (100\pi)$$

9. Operate.

$3 \div 2 \times 3 + 2 \div 3 \times 3$ $4.5 + 2$	$4 \div 14 \times 7 - 4 \div 5$ $2 - .8$
(6.5)	(1.2)
$6 \times 5 \div 3 \times 9 \div 10 \div 5$ $18 \div 10$	$(4 \times 3) \div (6 \times 4) - .8$ $.5 - .8$
(1.8)	(−0.3)

---

5. Copy the prime factorizations from page 17. Then find the GCF and LCM.

345	253	180
$3 \times 5 \times 23$	11×23	$2^2 \times 3^2 \times 5$

GCF (345, 253, 180) = **1**

LCM (345, 253, 180) = $2^2 \times 3^2 \times 5 \times 11 \times 23$

$= 180 \times 253 = $ **45,540**

10. Solve.

$2(7x + 6) - 8 = 39$ $14x + 12 = 47$ $14x = 35$	$3(3x - 8) - 7 = 11$ $9x - 24 = 18$ $9x = 42$
$x = \left(\dfrac{5}{2}\right)$	$x = \left(\dfrac{14}{3}\right)$

Level 7	Number 64

1. Dawn bought 13 pounds of beef for $120.90. What was the cost per pound?

$$\frac{\text{cost}}{\text{pound}} = \frac{120.90}{13} = \boxed{\$9.30}$$

The value to the left of "per" is the numerator.
The value to the right of "per" is the denominator.

6. Mentally calculate the midline of each trapezoid.

base	BASE	midline
12.36	36.54	**24.45**
17.73	18.27	**18**
10.18	18.32	**14.25**

2. Operate.

$-53 - -11$	-42	$36 - -19$	55
$-48 + -56$	-104	$-26 + 84$	58
$-84 - 25$	-109	$17 - 71$	-54
$33 - 74$	-41	$56 - -34$	90

7. Find AC if BD = 42, AB = 26, and AD = 40. The 26 is the key to finding the 10 in 42.

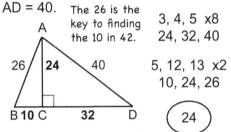

3, 4, 5 x8
24, 32, 40

5, 12, 13 x2
10, 24, 26

$\boxed{24}$

3. Draw the reflection of each polygon over the line y = 1.

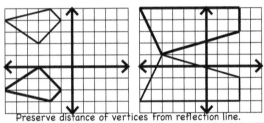

Preserve distance of vertices from reflection line.

8. Complete the chart for a circle.

radius	diameter	area	circumference
$\frac{4}{9}$	$\frac{8}{9}$	$\frac{16\pi}{81}$	$\frac{8\pi}{9}$
0.8	**1.6**	**0.64π**	**1.6π**
1.2	2.4	**1.44π**	**2.4π**

4. Solve by cross multiplication.

$\frac{15}{8} = \frac{4}{x}$	$\frac{5}{6} = \frac{x}{5}$	$\frac{10}{x} = \frac{13}{9}$
$15x = 32$	$6x = 25$	$13x = 90$
$x = \boxed{\frac{32}{15}}$	$x = \boxed{\frac{25}{6}}$	$x = \boxed{\frac{90}{13}}$

9. Distribute.

$-0.5(2x - 8y + 6z)$	$-x + 4y - 3z$
$0.8x(5y + 10z + 15)$	$4xy + 8xz + 12x$
$9a(-2 - 3y + z)$	$-18a - 27ay + 9az$
$-a(-6a - 6c - 6e)$	$6a^2 + 6ac + 6ae$

5. How many perfect squares are between 220 and 1598?

$15^2 = 225$
$40^2 = 1600$
15 to 39; 25 numbers

$\boxed{25}$

Do not list the perfect squares, but rather the numbers that generate them. Subtract and add 1 to get the number of numbers.

10. Solve.

$\|x - 5\| = 34$ $\quad x = \underline{39, -29}$	$\|x + 2\| = 29$ $\quad x = \underline{27, -31}$
$\|x + 3\| = 23$ $\quad x = \underline{20, -26}$	$\|x - 9\| = 33$ $\quad x = \underline{42, -24}$

Level 7	Number 65

1. Answer as indicated.

24 is what % of 40? $\dfrac{24}{40} = \dfrac{6}{10}$ (60%)	24 is 40% of what number? $24 = \dfrac{2}{5}x$ (60)	What is 24% of 40? 10% is 4 20% is 8 1% is .4 4% is 1.6 (9.6)

6.

	R	T	D
D1	4	2.5	10
D2		7.5	20
tot	(3)	10	30

Maya jogged a total of 30 miles over 2 days, doing 1/3 on the 1st day. Her time the 2nd day was triple her 1st time. Her rate the 1st day was 4 mph. Find her overall rate in mph.

2. If 24 people can do a job in 5 full weeks, then how many people are needed to do the same job in 5 days?

W x T = W x T
24 • 5 • 7 = x • 5
x = 24 • 7
(168)

Workers and time are inversely proportional-- as one goes up, the other goes down. Their product is constant.

7. Operate.

86 ÷ (−2)	−43	−15 x 6	−90
−2 x (−37)	74	16 x (−4)	−64
0 ÷ (−43)	0	88 ÷ (−22)	−4
−11 x 17	−187	−2 x (−39)	78

3. Count the total number of triangles in the picture.

1: 10
2: 10
3: 5 + 3 = 8
4: 0
5: 5
6–11: 0
(33)

8. Check if the row number is divisible by the column factor.

	3	4	11	12	33	44
15,840	✓	✓	✓	✓	✓	✓
38,016	✓	✓	✓	✓	✓	✓
96,944	−	✓	−	−	−	−

4. Find the area in square units of a rectangle with diagonal 25 and length 15 units.

3, 4, 5
x5 blow-up
A = 20 x 15
A = (300)

9. Evaluate.

$\sqrt{44}$	$4 \cdot 11$ $2\sqrt{11}$	$\sqrt{99}$	$9 \cdot 11$ $3\sqrt{11}$
$\sqrt{76}$	$4 \cdot 19$ $2\sqrt{19}$	$\sqrt{180}$	$36 \cdot 5$ $6\sqrt{5}$
$\sqrt{84}$	$4 \cdot 21$ $2\sqrt{21}$	$\sqrt{125}$	$25 \cdot 5$ $5\sqrt{5}$
$\sqrt{88}$	$4 \cdot 22$ $2\sqrt{22}$	$\sqrt{147}$	$49 \cdot 3$ $7\sqrt{3}$

5. Name the lattice points in the picture that are equidistant from A and B.
One box = one unit.

(0, 4)
(1, 3)   (2, 2)
(3, 1)   (4, 0)

10. Find the angle in degrees formed by clock hands at 8:15.

360°/12 = 30° between neighboring numbers

150 + 30/4 = (157.5)

Hour hand is one fourth of the way between 8 and 9.

Level 7	Number 66

1. Arrange the digits 2, 3, 4, 6, 8, and 9 into two 3-digit numbers with the:

greatest positive difference	least positive difference
986 Max L to R.	**426** Min L to R.
− 234 Min L to R.	**− 398** Max L to R.
752	First, 100s 1
	28 apart.

346−298=48; 923−864=59

6. Find the area of an isosceles triangle with sides 26, 26, 48.

$A = \dfrac{48 \times 10}{2} = \boxed{240}$ sq un

26, 10, 24

---

2. Subtract the sum of the odd numbers 1 through 39 from the sum of the even numbers 2 through 40.

$$2 + 4 + \ldots + 38 + 40$$
$$-\ 1 + 3 + \ldots + 37 + 39$$
$$\overline{1 + 1 + \ldots + 1 + 1}$$

20 even numbers
20 odd numbers

$\boxed{20}$

7. Convert to base ten.

598_{twenty}	$5 \times 400 + 9 \times 20 + 8$	2188
345_{eleven}	$3 \times 121 + 4 \times 11 + 5$	412
$5BE_{fifteen}$	$5 \times 225 + 11 \times 15 + 14$	1304
$15B_{thirteen}$	$1 \times 169 + 5 \times 13 + 11$	245

---

3. Find the volume of the cone in cubic units.

radius = 4
height = 18

$V = \dfrac{4 \cdot 4 \cdot \pi \cdot 18}{3}$

$= 16 \cdot 6 \cdot \pi$

$= \boxed{96\pi}$

8. Multiply by 11 mentally.

725 x 11	7975	Write from R to L. For 6351:
6351 x 11	69,861	Write the 1. Then 1+5=6;
816 x 11	8976	5+3=8; 3+6=9.
9027 x 11	99,297	Write the 6.

---

4. Find the sum of each arithmetic sequence using the "First plus Last" method.

$-52 + -45 + -38 + -31 + -24 + -17 + -10 + -3$

$-50 + -44 + -38 + -32 + -26 + -20 + -14 + -8$

$\dfrac{(-55)(8)}{2} = \boxed{-220}$ $\dfrac{(-58)(8)}{2} = \boxed{-232}$

9. Find the slope and intercepts of the lines.

line	slope	y-intercept	x-intercept
y = x + 10	1	**(0, 10)**	**(−10, 0)**
4x − y = −8	4	**(0, 8)**	**(−2, 0)**
2x + 2y = 7	−1	**(0, 3.5)**	**(3.5, 0)**

Picture 0 inserted to find intercepts mentally.

---

5. Operate.

3726_{eight}
$+\ 3654_{eight}$
$\overline{7602_{eight}}$

4102_{five}
$-\ 2223_{five}$
$\overline{1324_{five}}$

2053_{six}
$+\ 2524_{six}$
$\overline{5021_{six}}$

10. Find the remainder without dividing.

345,879 ÷ 10	345,870 div by 10	R 9
431,546 ÷ 11	431,541 div by 11	R 5
763,045 ÷ 14	763,042 div by 2, 7	R 3
821,731 ÷ 15	821,730 div by 3, 5	R 1

Level 8	Number 1

1. Find the least perfect square divisible by the first three prime numbers.

$(2 \times 3 \times 5)(2 \times 3 \times 5) = 30 \times 30 =$ (**900**)

6. Find twin primes with sum equal to:

24	11, 13
120	59, 61
84	41, 43
384	191, 193

Because twin primes are 2 apart, halve the sum to get the median.

2. Find the measures of 2 supplementary angles, the greater of which measures 5 times the lesser.

x + 5x = 180
6x = 180
x = 30

(**30 and 150**)

7. A swimming pool 25 by 40 feet has a uniform walkway all around. Find the width and area of the walkway if the total area is 1750 square feet.

25 x 40 = 1000
35 x 50 = 1750
A = 750
W = 5

Use T&E and ones digit instead of algebra.

3. Find the area of the rhombus in square units given perimeter P = 60 and diagonal D = 24.

P = 60   A = 54 x 4
s = 15   or 24 x 9
D = 24
D/2 = 12   A = (**216**)
9, 12, 15
d =18

8. Write in scientific notation.

$(7 \times 10^8) \div (8 \times 10^{-4})$ | $(1 \times 10^7) \div (4 \times 10^2)$

0.875×10^{12} | 0.25×10^5

(**8.75×10^{11}**) | (**2.5×10^4**)

4. Find the endpoint of the line segment.

Endpoint	Midpoint	Endpoint
(5, 10)	(5.5, 14)	**(6, 18)**
(−1, 5.1)	(−3, 6.2)	**(−5, 7.3)**
(7, −8)	(11, −4)	(15, 0)
(−6, −3)	(−1, 4)	(4, 11)

9. Complete.

39 ≡ 6 (mod 11) | 29 ≡ 14 (mod 15)
40 ≡ 1 (mod 13) | 40 ≡ 2 (mod 19)
84 ≡ 4 (mod 20) | 51 ≡ 3 (mod 24)
36 ≡ 11 (mod 25) | 77 ≡ 7 (mod 10)

5. Solve.

$(x − 3)(x + 6) = 0$ x = (**3, −6**) | $(2x − 3)(4x + 5) = 0$ x = (**$\frac{3}{2}, \frac{-5}{4}$**)
$(x + 5)(x − 1) = 0$ x = (**−5, 1**) | $(3x − 4)(5x + 7) = 0$ x = (**$\frac{4}{3}, \frac{-7}{5}$**)

10. Find the distance between the points.

(2, 7) and (5, 11) 3, 4, (**5**) | (3, −8) and (−2, 4) 5, 12, (**13**)
(4, 8) and (−4, −7) 8, 15, (**17**) | (−2, −8) and (5, 16) 7, 24, (**25**)

MAVA Math: Middle Reviews Solutions Copyright © 2013 Marla Weiss

Level 8	Number 2

1. A line that crosses two parallel lines is called a ___transversal.___

The multiplier from term to term in a geometric sequence is the ___common ratio.___

A counting number is also called ___natural.___

6. Find all digits d so that the 5-digit number 25,d83 is divisible by 3.

2 + 5 + 8 = 15   ignore the 3   (0, 3, 6, 9)

Find all digits x so that the 4-digit number 1x46 is divisible by 6. already divisible by 2

6 + 4 + 1 = 11   (1, 4, 7)

2. Find the area of the parallelogram bounded by the lines y = –3, y = 2, y = x + 3, and y = x.

b = 3, h = 5

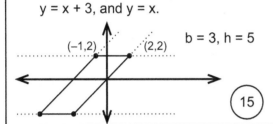

(–1,2)   (2,2)

(15)

7. Point B divides segment AC such that AC:BC = 3:1. AB = 60. Find BC.

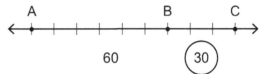

A   B   C

60   (30)

AB is 2 parts or 60. BC is 1 part or 30.
AC is 3 parts or 90.

3. Find x degrees in the crook picture. (NTS)

130° **50** by supp

50 by corr
x
40 by alt int
40°

(90)

8. Five years from now the sum of Jim and Tim's ages is 36. Tim's age now is one less than Jim's age in 5 years. Find Tim's age 2 years ago.

	–2	now	+5
J		J	J+5
T	(13)	J+4	J+9

J+5+J+9 = 36
2J + 14 = 36
2J = 22
J = 11
Tim now 15

4. Find x in degrees. (NTS)

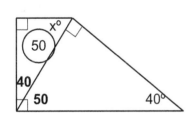

x°
(50)
40
50   40°

9. Operate.

$(-2)^3 - 3^2 - (-3)^3 + 2^2$	$-4^3 - 3^4 - (-4)^2 + 4^4$
$-8 - 9 + 27 + 4$ (14)	$-64 - 81 - 16 + 256$ (95)
$(-2)^4 \times 2^2 - (-2)^5 - 2^6$	$(-5)^3 - 5^2 + (-5)^4 - 5$
$16 \times 4 + 32 - 64$ (32)	$-125 - 25 + 625 - 5$ (470)

5. Find the probability of selecting a point from the shaded area within the square given an inscribed circle.

s = 60 d = 60 r = 30

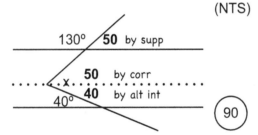

$$\frac{A_{sh}}{A_{squ}} = \frac{60 \cdot 60 - 30 \cdot 30 \cdot \pi}{60 \cdot 60}$$

60

Factor 30 • 30 in numerator.
Simplify. See p. 118.
Probability is the same.

$\left(\dfrac{4 - \pi}{4}\right)$

10. Find the volume in cubic units of a rectangular pyramid with length 12, width 8, and altitude 11 units.

$$\frac{12 \times 8 \times 11}{3} = 4 \times 8 \times 11 = 32 \times 11$$

$$= (352)$$

Level 8	Number 3

1. If b bees need 1 year to make p pounds of honey, e bees will make how many pounds of honey in 1 century?

bees	years	pounds
b	1	p
1	1	$\dfrac{p}{b}$
e	1	$\dfrac{ep}{b}$
e	100	$\boxed{\dfrac{100ep}{b}}$

6. Find the number of 5-digit even numbers with the middle digit 7, the leftmost digit an odd prime, and no repeated digits.

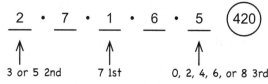

2 · 7 · 1 · 6 · 5   (420)

3 or 5 2nd   7 1st   0, 2, 4, 6, or 8 3rd

2. Solve.

$16^x = 2^{12}$ $3^8 = 9^x$ $4^{3x} = 8^{3x+1}$

$2^{4x} = 2^{12}$ $3^8 = 3^{2x}$ $2^{6x} = 2^{9x+3}$

$4x = 12$ $2x = 8$ $6x = 9x + 3$

$\boxed{x = 3}$ $\boxed{x = 4}$ $\boxed{x = -1}$

7. Estimate the specified root to the nearest whole number.

square root of 41 — 6   36 vs. 49

square root of 91 — 10   100 vs. 81

cube root of 110 — 5   125 vs. 64

cube root of 40 — 3   27 vs. 64

3. The area of rectangle ABCD is 6 times the area of rectangle BCFE. Find the coordinates of E. (NTS) $\boxed{(18, 8)}$

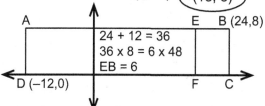

A   E   B (24,8)

24 + 12 = 36
36 x 8 = 6 x 48
EB = 6

D (−12,0)   F   C

8. Find the angle in degrees formed by clock hands at 7:10.

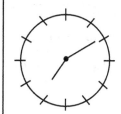

360°/12 = 30° between neighboring numbers

150 + 5 = (155)

Hour hand is 1/6 of way from 7 to 8.

4. Evaluate the functions.

SQRT(64)	8	TRUNC(1.2)	1
SQR(9)	81	INT(3.5)	3
ABS(−12)	12	SIGMA(4)	3
SGN(−3)	−1	INT(−2.1)	−3

9. Add.

$\dfrac{1}{x}^y + \dfrac{1}{y}^x$ $\dfrac{1}{a}^{bc} + \dfrac{1}{b}^{ac} + \dfrac{1}{c}^{ab}$

$\boxed{\dfrac{y + x}{xy}}$ $\boxed{\dfrac{bc + ac + ab}{abc}}$

5. Find the points of trisection of the line segment with the given endpoints.

(7, 15) & (31, 9)	(15, 13)	(23, 11)
(−9, 30) & (0, 9)	(−6, 23)	(−3, 16)
(8, −11) & (44, 1)	(20, −7)	(32, −3)
(14, −5) & (38, −8)	(22, −6)	(30, −7)

10. How many triangles with whole sides have greatest side 5?

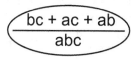   (9)

5, 5, 5   5, 4, 4
5, 5, 4   5, 4, 3
5, 5, 3   5, 4, 2
5, 5, 2   5, 3, 3
5, 5, 1

Other options do not form a triangle. The sum of 2 sides is greater than the 3rd.

| Level 8 | Number 4 |

1. In a windowless, cube-shaped closet, the ceiling, walls, and door, but not the floor, are painted. The painted area equals 180 square feet. What is the volume of the room in cubic units?

$5e^2 = 180$

$e^2 = 36$

$e = 6$

$V = 6 \times 6 \times 6 = \boxed{216}$

6. Two boats leave from the same dock at 11:00 AM. One travels north at 60 mph; the other travels east at 80 mph. When will they be 150 miles apart?

3, 4, 5 x 30

90, 120, 150

D = RT =   D =150

60 x 1.5 = 90

11:00 + 1.5 hr

D = RT = 80 x 1.5 = 120 $\boxed{12:30\ PM}$

2. If n is an odd integer, find:

a.) the next greater odd integer. $n + 2$

b.) the next greater even integer. $n + 1$

c.) the even integer 8 times greater. $8n$

d.) the next lesser odd integer. $n - 2$

7. Simplify using DPMA.

435 x 198 + 435 x 2   435 x (198 + 2)

435 x 200

$\boxed{87{,}000}$

25% of 376 + 25% of 124   25% (376 + 124)

25% of 500

$\boxed{125}$

$11^2 + 11 \times 74$ 11(11 + 74)

11 x 85 $\boxed{935}$

3. Multiply using the difference of 2 squares.

53 x 47   2491   (50 + 3)(50 − 3) 2500−9

32 x 28   896   (30 + 2)(30 − 2) 900−4

91 x 89   8099   (90 + 1)(90 − 1) 8100−1

65 x 75   4875   (70 − 5)(70 + 5) 4900−25

8. Find the circumference of a circle given its area is π^3.

$A = \pi r^2 = \pi^3$

$r^2 = \pi^2$

$r = \pi$

$C = 2r\pi$

$C = \boxed{2\pi^2}$

4. The measures of the interior angles of a triangle are 5x, 4x + 8, and 3x − 2. Find the median of the angles.

$5x + 4x + 8 + 3x - 2 = 180$

$12x + 6 = 180$

$12x = 174$

$x = 14.5$

$4(14.5) + 8 = \boxed{66}$

That 4x + 8 is the middle one in the list is irrelevant. It is the middle of the 3 values once x is found.

9. Find the value using fractions.

$33\frac{1}{3}$ % of $62\frac{1}{2}$ % of 75% of 864

$\frac{1}{3} \times \frac{5}{8} \times \frac{3}{4} \times \overset{\overset{27}{108}}{864}$

27 x 5 = 100 + 35 = $\boxed{135}$

5. If f(x) = 3x + 1 and g(x) = 4x − 3, find g(f(2)) + f(g(1)).

$f(2) = 6 + 1 = 7$

$g(7) = 28 - 3 = 25$

$g(1) = 4 - 3 = 1$

$f(1) = 3 + 1 = 4$

$25 + 4 = \boxed{29}$

10. Multiply as two binomials.

$8\frac{2}{3} \times 6\frac{9}{10}$ $3\frac{4}{5} \times 5\frac{1}{2}$

F =   48   F =   15

O =   7.2   O =   1.5

I =   4   I =   4

L =   0.6 $\boxed{59.8}$ L =   0.4 $\boxed{20.9}$

Level 8	Number 5

1. Find the time in minutes needed for 2 people to do a job together given that: Person A can do the job in 1 hour; Person B can the same job in 2 hours.

Rate A = 1
Rate B = 1/2
Total Rate = 3/2
Total Time = 2/3

In work problems, rates and times are reciprocals.

$\boxed{40}$

6. Find the volume of each prism.

pentagonal base with area 27, altitude 11   27 x 11   297

octagonal base with area 45, altitude 20   45 x 20   900

heptagonal base with area 12, altitude 15   12 x 15   180

2. Is each the graph of a function?

YES   NO

7. Ring the fictitious operations that are commutative.

$\boxed{x \text{❋} y = xy + x + y}$ $x \text{❀} y = x - y + 1$

$\boxed{x \text{❦} y = x + y - 5}$ $x \text{❁} y = 3x + 2y$

3. Find the area of an isosceles trapezoid with legs 13 and bases 13 and 23.

13
13 / 12
5   13   5

M = 18
H = 12
A = 18 x 12
= 180 + 36
= $\boxed{216}$

8. Given the perimeter of a rectangle and the upper bound of the width. Find the range of values for the length.

P = 30; max W < 12   P = 70; max W < 25
SP = 15   SP = 35
0 < W < 12   0 < W < 25
$\boxed{3 < L < 15}$ $\boxed{10 < L < 35}$

4. Find the space diagonal of a cube with

edge = 3 $3\sqrt{3}$ e = 3

area of face = 36 $6\sqrt{3}$ e = 6

volume = 125 $5\sqrt{3}$ e = 5

face diagonal = $9\sqrt{2}$ $9\sqrt{3}$ e = 9

9. Find the angles of parallelogram ABCD given:

m∠A = 7x + 6   m∠B = 6x + 5

7x + 6 + 6x + 5 = 180 $\Big($ m∠A, C = 97°
13x + 11 = 180
13x = 169   m∠B, D = 83° $\Big)$
x = 13   *Consecutive angles are supplementary. Opposite angles are congruent.*

5. Find the sum of all 1-digit numbers with exactly 3 factors.

4 + 9 = $\boxed{13}$

Find the sum of all 2-digit numbers with exactly 3 factors.   *Only the square of a prime has exactly 3 factors.*

25 + 49 = $\boxed{74}$

10. Answer YES or NO as to whether the 3 numbers form sides of a right triangle.

4, 7.5, 8.5 ×2 Y	6.1, 8.1, 10.1 N
3.5, 4.5, 5.5 N	3.5, 12, 12.5 ×2 Y
1.4, 4.8, 5 ×5 Y	5.5, 30, 30.5 ×2 Y

| Level 8 | Number 6 |

1. Find two consecutive integers with the given sum.

35	17, 18	–77	–38, –39
–53	–26, –27	69	34, 35
91	45, 46	–81	–40, –41

2. Write the contrapositive of the conditional statement. Are they logically equivalent? (YES)

If polygon P is a square, then polygon P is a rectangle.   Both are TRUE.

If polygon P is a not a rectangle, then

polygon P is a not a square.

3. Find the area of a rectangle given the perimeter is 90 and the length is twice the width.

P = 90   W = 15
SP = 45   L = 30
L + W = 45   A = 30 x 15
2W + W = 45
3W = 45   A = (450)

4. Define the binary operation ~ as:

~X~ = 5X + 2. For example,
~5~ = 25 + 2 = 27.
Find the values.

~(~4~)~   112   Do ~4~ = 22 first.

~(~0.2~)~   17   Do ~0.2~ = 3 first.

5. Find the space diagonal of the rectangular prism with the given dimensions.

1 by 2 by 2   3 $\sqrt{1 + 4 + 4}$

8 by 9 by 12   17 $\sqrt{64 + 81 + 144}$

6. Find the measure of one angle of each regular polygon.

Interior and exterior angles are supplements.

dodecagon 150   ext < = 360/12 = 30
                  int < = 180 – 30 = 150

18-gon   160   ext < = 360/18 = 20
                  int < = 180 – 20 = 160

pentagon   108   ext < = 360/5 = 72
                  int < = 180 – 72 = 108

Or 16 x 180 / 18 = 160

7. Find the area of the square inscribed in a circle of area 64π square units.

$A_{circ} = 64\pi$
R = 8
D = diag sq = 16
A_{sq} = 16 x 16 / 2 = (128)

A square is a rhombus. Use the area formula: half the product of the diagonals.

8. Calculate the combinations.

$C(6, 3) = \dfrac{6!}{3!\ 3!} = \dfrac{6 \cdot 5 \cdot 4}{3 \cdot 2 \cdot 1} = (20)$

$C(8, 7) = \dfrac{8!}{7!\ 1!} = (8)$

$C(10, 2) = \dfrac{10!}{2!\ 8!} = \dfrac{10 \cdot 9}{2 \cdot 1} = (45)$

9. Answer YES or NO if the matrices may be multiplied. If yes, give the dimensions of the product.

3 by 4 times 3 by 4   NO

4 by 3 times 3 by 4   YES 4 by 4

4 by 4 times 4 by 4   YES 4 by 4

10. A regular hexagon has perimeter 30 units. Find its area in square units.

P = 30
S = 5
$A = \dfrac{6 \cdot 5^2 \sqrt{3}}{4} = \left(\dfrac{75\sqrt{3}}{2} \right)$

Level 8	Number 7

1. Colored streamers are hung on a cord in the repeating pattern: red, yellow, blue, green, purple, red, yellow, blue, green, purple, and so on. What is the color of the 69th streamer?

RYBGP RYBGP
cycle length 5
65th is full sequence ⟶ green

6. Find the number of integer values that satisfy:

$|x + 5| \leq 6$ 13   –11 ⟶ 1

$|x - 3| \leq 8$ 17   –5 ⟶ 11

$|x + 3| \leq 10$ 21   –13 ⟶ 7

2. Find the perimeter of the rhombus with diagonals 24 and 32.

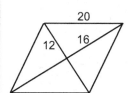

3, 4, 5 x 4
12, 16, 20
P = 4 x 20
P = 80

7. 14 quarters, 15 dimes, and 40 nickels are what percent of $10?

14 x .25 = 3.50   Use mental math for
15 x .10 = 1.50   the multiplications.
40 x .05 = 2.00
sum = 7.00

$\dfrac{7}{10}$ 70%

3. Multiply the matrices.

$$\begin{bmatrix} 10 & 6 & 9 \\ 3 & 0 & 7 \\ 5 & 1 & 11 \end{bmatrix} \times \begin{bmatrix} 2 & 3 & 5 \\ 0 & 4 & 6 \\ 6 & 2 & 1 \end{bmatrix} =$$

20 + 0 + 54   15 + 0 + 7
30 + 24 + 18   10 + 0 + 66
50 + 36 + 9   15 + 4 + 22
6 + 0 + 42   25 + 6 + 11
9 + 0 + 14

$$\begin{bmatrix} 74 & 72 & 95 \\ 48 & 23 & 22 \\ 76 & 41 & 42 \end{bmatrix}$$

8. Determine if the triangle with the 3 sides is acute or obtuse.

2, 4, 5

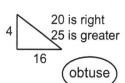 20 is right
25 is greater

obtuse

4, 5, 6

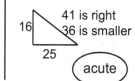 41 is right
36 is smaller

acute

4. Find the perimeter of parallelogram ABCD given AB = 9x – 4, BC = 11, and AD = 10x + 1.

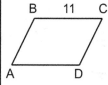

10x + 1 = 11
x = 1
AB = 9 – 4 = 5
P = 2(11 + 5) = 32

9. Complete the unit conversions.

1 foot	12	inches	1 yard	3	feet
1 foot²	144	inches²	1 yard²	9	feet²
1 foot³	1728	inches³	1 yard³	27	feet³

5. Find the mean of 13x + 9, 12x – 6, 5x + 1, 2x – 2, and 3x + 8.

$\dfrac{35x + 10}{5}$ = 7x + 2

10. A, B, C, D, and E are consecutive points on a line. AB:BC = 1:2, BC:CD = 2:3, and CD:DE = 3:4. Find AC:AE.

No adjustments to the ratios are necessary.

AB	1	1
BC	2 2	2
CD	3	3
DE		4

3:10

MAVA Math: Middle Reviews Solutions Copyright © 2013 Marla Weiss

Level 8	Number 8

1. Walking 30 feet east, 10 feet north, 10 feet east, and 20 feet north in that order ends how many feet from the starting point?

$\boxed{50}$

3, 4, 5 x 10

50   30
40

2. Find the formula that generates the pattern 2, 5, 10, 17, 26, ... Then find the 25th term using the formula.

1	2
2	5
3	10
4	17
5	26

$\boxed{n^2 + 1}$

$25^2 + 1$
$625 + 1$

$\boxed{626}$

3. Find the volume of the cone in cubic units.

diameter = 12
slant height = 10

$V = \dfrac{6 \cdot 6 \cdot \pi \cdot 8}{3}$

$= 2 \cdot 48 \cdot \pi$

$= \boxed{96\pi}$

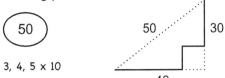

10
8
6

4. Find the mean and range of the 5 different integers x, x − 2, 19, −10, and 6x if their median is 6 and x > 0.

| −10 | 4 | 6 | 19 | 36 |
| −10 | x − 2 | x | 19 | 6x |

range = 36 − (−10) = $\boxed{46}$ Draw 5 slots.

mean = 55/5 = $\boxed{11}$ Try most logical order. x = 6

5. Multiply mentally.

84 x 1001	84,084
77 x 101	7777
95 x 10,001	950,095
258 x 1001	258,258

Picture 101 as (100 + 1), 1001 as (1000 + 1), and 10,001 as (10,000 + 1).

6. Calculate the sum of the arithmetic sequence.

15 + 20 + 25 + . . . + 75 + 80 + 85

5x3 5x4 5x5   5x17
1 to 17 is 17 numbers; omit 1 & 2; n = 15
F + L = 15 + 85 = 100
100 x 15 / 2 = 50 x 15 = $\boxed{750}$

7. Label the lines perpendicular, parallel, or neither.

y = 5x + 7   y = 4x + 3
y = 5x − 9   y = 3x + 4

$\boxed{\text{parallel}}$ $\boxed{\text{neither}}$

8. Convert 1,481 in base ten to base twelve.

```
  1481        T    3    5
− 1440      ───  ───  ───
   41       144   12    1
−  36
    5
```

$\boxed{T35_{twelve}}$

9. Find the surface area in square units of a square pyramid with base edge 16 and lateral edge 10 units.

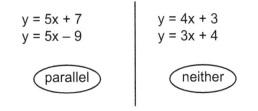

$A_{sq} = 16 \times 16 = 256$

$A_{tri} = 16 \times 6 / 2 = 48$

48 x 4 = 192

256
+ 192
$\boxed{448}$

10
6
8

10. How many right triangles with whole number sides have one side of length 15 units?

3, 4, 5 $\longrightarrow$ 9, 12, 15
          15, 20, 25

5, 12, 13 $\longrightarrow$ 15, 36, 39

8, 15, 17

$\boxed{4}$

Level 8	Number 9

1. Find the area of each rectangle in square inches.

4 feet by 5 feet

48 inches by 60 inches

2400 + 480

(2880)

2 yards by 3 yards

72 inches by 108 inches

7200 + 560 + 16

(7776)

6. Solve.

$X + 3Y = 1$
$-5X + Y = 11$

$5X + 15Y = 5$
$-5X + Y = 11$
_____
$16 Y = 16$ $X + 3 = 1$
$Y = 1$ $X = -2$

$-3Y = 6$
$X - 4Y = 38$

$Y = -2$
$X + 8 = 38$
$X = 30$

2. Find the length of the segment joining the midpoints of segments $\overline{AB}$ and $\overline{BC}$.

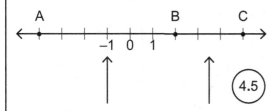

(4.5)

7. Find the area of a regular hexagon with edge 4.

$$\frac{6 \cdot s^2 \cdot \sqrt{3}}{4} = \frac{6 \cdot 4 \cdot 4 \cdot \sqrt{3}}{4} = (24\sqrt{3})$$

Area of regular hexagon is 6 times area of equilateral triangle.

3. Solve for the variables in the matrices.

$$\begin{bmatrix} 19 & 3y - 5 \\ 5x + 4 & 33 \end{bmatrix} = \begin{bmatrix} 6w + 1 & 22 \\ 34 & 9z - 3 \end{bmatrix}$$

$6w + 1 = 19$ $5x + 4 = 34$
$w = 3$ $x = 6$
$3y - 5 = 22$ $9z - 3 = 33$
$y = 9$ $z = 4$

8. Simplify. Answer in exponential form.

$3^8 \times 3^6$	$(9^3)^5$	$(7^4)^4$
(3^{14})	(9^{15})	(7^{16})
$4^8 \div 4^2$	$15^7 \times 15^8$	$8^9 \div 8^3$
(4^6)	(15^{15})	(8^6)

4. The radii of 2 spheres are in the ratio 2:3. Find the ratio of their volumes and surface areas.

$$\frac{v}{V} = \frac{4/3 \cdot \pi \cdot 2 \cdot 2 \cdot 2}{4/3 \cdot \pi \cdot 3 \cdot 3 \cdot 3} = \left(\frac{8}{27}\right)$$

$$\frac{sa}{SA} = \frac{4 \cdot \pi \cdot 2 \cdot 2}{4 \cdot \pi \cdot 3 \cdot 3} = \left(\frac{4}{9}\right)$$

9. Find the number of subsets of each set.

{He, Me} 2^p where p = number of elements   4

{m, a, t, h}   16

{G, O, O, D}   The second O in GOOD is irrelevant. An element only appears once in a set.   8

{P, O, I, N, T}   32

5. Graph on the coordinate plane. One box equals one unit.

$3x + 5y = 15$ $5x - 4y = 20$

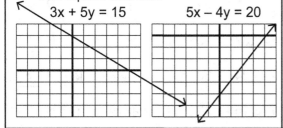

10. How many numbers from 1 to 100 inclusive are multiples of 2 or 3?

$100 \div 2 = 50$
$100 \div 3 = 33$
$100 \div 6 = 16$
$50 + 33 - 16 = (67)$

Subtract the overlap: the multiples of 2 AND 3.

Level 8	Number 10

1. The sum of the dates of all Fridays in a given month is 70. What is the date of the first Friday?

70 = 7 x 10
7 + 14 + 21 + 28 = 70

(7th)

6. A concrete walkway 10 m wide by 15 m long by 4 cm thick contains how many cubic meters of concrete?

4 cm = .04 m = 4/100 m = 1/25 m

$10 \cdot 15 \cdot \dfrac{1}{25} = 2 \cdot 3 =$ (6)

---

2. Find 3 consecutive even integers such that the sum of the least and greatest is 72.

x, x + 2, x + 4
x + (x + 4) = 72
2x = 68
x = 34
(34, 36, 38)

7. Solve.

$5\sqrt{x} = x$
$25x = x^2$
$0 = x^2 - 25x$
$0 = x(x - 25)$
(x = 0, 25)

$4x^2 = 800$
$x^2 = 200$
$x = \pm \sqrt{200}$
($x = \pm 10\sqrt{2}$)

---

3. Rewrite the statements using DeMorgan's Laws.

NOT (A OR B)
$\underline{\hspace{2cm}}$ (NOT A) AND (NOT B) $\underline{\hspace{2cm}}$

NOT (A AND B)
$\underline{\hspace{2cm}}$ (NOT A) OR (NOT B) $\underline{\hspace{2cm}}$

8. Operate and simplify.

$$\dfrac{3\frac{4}{7} \times 8\frac{2}{5}}{3\frac{7}{9} \div 5\frac{2}{3}} \qquad \dfrac{\frac{25}{7} \times \frac{42}{5}}{\frac{34}{9} \times \frac{3}{17}} \qquad \dfrac{30}{\frac{2}{3}}$$

$30 \times \dfrac{3}{2}$ (45)

---

4. Find the surface area of the cone in square units.   LA = "πrl"

radius = 8
slant height = 10

$SA = 8 \cdot 8 \cdot \pi$
$+ 8 \cdot 10 \cdot \pi$
$= 64\pi + 80\pi$
$= $ (144π)

9. Find the total value after compounding annually.

principal	rate	years	value	
$2000	5%	2	**$2205**	2100
$1000	10%	3	**$1331**	1100, 1210
$2500	2%	2	**$2601**	2550
$5000	6%	2	**$5618**	5300

---

5. Solve for n.

$\dfrac{11!}{n!} = 990$

990 = 99 x 10 = 11 x 10 x 9

n = (8)

10. Find the remainder without dividing.

560,472 ÷ 22 div by 2, 11   R 0

129,376 ÷ 25 129,375 div by 25   R 1

723,243 ÷ 30 723,240 div by 3, 10   R 3

773,159 ÷ 33 773,157 div by 3, 11   R 2

143

Level 8	Number 11

1. Find the 6-digit mystery number.
Clue 1: The number is divisible by 15.
Clue 2: The digits are in descending order.
Clue 3: The digits are odd.
Clue 4: The number is divisible by 25.
Clue 5: The middle 2 digits sum to 14.

Ends in 75.
59 and 95 cannot be middle.   ⬭977,775⬭
997775 and 777775 wrong sum.

6. A clock that is correctly set at 4:00 PM loses 4 minutes every hour. What is the correct time when the clock reads 8:00 AM the next day?

4:00 PM to 8:00 AM is 16 hours
real time is extra 4 minutes per hour
4 x 16 = 64 = 60 + 4 minutes
⬭9:04 AM⬭

2. In a set of 5 numbers, the unique mode is less than the median. Four of the numbers are 11, 21, 31, and 41. Find the missing number.

⬭11⬭   11   21   31   41

7. Translate the cipher, with each letter as its own digit.

SAVE   3A71   **3271**
x   S   x   3   x   **3**
PRES   PR13   **9813**

In 1s, E=1. In 1000s, SxS does not carry; also in 10s, SxV ends in 1. S=3. V=7. P=9. In 100s, A needs to be small so as not to carry. A=2. R=8.

3. Graph on the number line, and write the solution to the quadratic inequality.

$(x - 4)(x + 3) \geq 0$

yes   no   yes
←●——————●——→
−3   0   4

⬭x ≥ 4 OR x ≤ −3⬭

8. Simplify.

$8^{\frac{1}{3}}$ ②	$27^{\frac{4}{3}}$ ㉛	$121^{\frac{1}{2}}$ ⑪
$9^{\frac{3}{2}}$ ㉗	$32^{\frac{3}{5}}$ ⑧	$625^{\frac{3}{4}}$ ⑫⑤

4. Find the probability of drawing 2 reds successively without replacement from a bowl containing 10 green, 14 yellow, 16 red, and 25 blue marbles.

$\dfrac{\cancel{16}^{1}}{\cancel{65}_{13}} \cdot \dfrac{\cancel{15}^{3}}{\cancel{64}_{4}}$ ⬭$\dfrac{3}{52}$⬭

9. Find the square root in the given mod.

square root of 3 mod 6 ___3___   3 x 3 = 9 / 9 − 6 = 3

square root of 6 mod 8 ___none___

square root of 0 mod 4 __0, 2__   4 x 4 = 16 / 16 − 9 = 7 and

square root of 7 mod 9 __4, 5__   5 x 5 = 25 / 25 − 18 = 7

5. Write each repeating decimal as a simplified fraction.

$.3\overline{1}$ $3.\overline{1}$ $3\frac{1}{9}$ $\frac{28}{9}$ $\frac{28}{90}$ ⬭$\frac{14}{45}$⬭
x 10   ÷ 10

$.0\overline{43}$ $4.\overline{3}$ $4\frac{1}{3}$ $\frac{13}{3}$ $\frac{13}{3}$ ⬭$\frac{13}{300}$⬭
x 100   ÷ 100

10. Find the area and perimeter of the trapezoid with vertices (6, 8), (11, 8), (14, 4), and (3, 4).   A = ⬭32⬭ 8 x 4
P = ⬭26⬭ 15 + 11

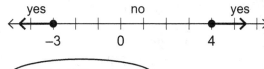

(6,8) — 5 — (11,8)
(3,4)   (14,4)

B = 11   M = 8
b = 5   h = 4

MAVA Math: Middle Reviews Solutions Copyright © 2013 Marla Weiss

Level 8	Number 12

1. The average of 5 tests is 85. If the worst grade is dropped, the new average is 91. What was the worst grade?

$$\frac{425}{5} = 85 \qquad \frac{364}{4} = 91 \qquad \begin{array}{r} 425 \\ -364 \\ \hline \boxed{61} \end{array}$$

6. Of 61 children, 19 drink milk, 33 drink juice, and 12 drink neither. How many drink both?

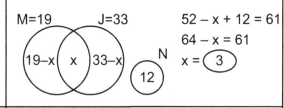

M=19 J=33   52 – x + 12 = 61
64 – x = 61
x = ③

2. X varies directly as Y. X = 60 when Y = 5. Find X when Y = 11.

X = kY
60 = 5k
k = 12
X = 12 • 11
X = ⑬⑫ 132

7. Reflect each point over y = –x. Then translate as specified.

(5, 6)	(–6, –5)	up 9	(–6, 4)
(–8, –2)	(2, 8)	left 5	(–3, 8)
(3, –4)	(4, –3)	down 3	(4, –6)

3. Find the 51st term of each sequence.

11, 17, 23, 29, . . .   d = 6

11 + (50) (6) = 11 + 300 = ⟨311⟩

17, 22, 27, 32, . . .   d = 5

17 + (50) (5) = 17 + 250 = ⟨267⟩

8. Find the volume between 2 cylinders with concentric bases of diameters 14 and 10, both 11 units high.

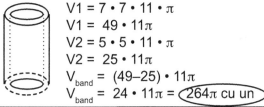

V1 = 7 • 7 • 11 • π
V1 = 49 • 11π
V2 = 5 • 5 • 11 • π
V2 = 25 • 11π
V_{band} = (49–25) • 11π
V_{band} = 24 • 11π = ⟨264π cu un⟩

4. Find the area of a rectangle given the perimeter is 50 and the ratio of the sides is 3:2.

P = 50   A = 15 x 10
SP = 25
ratio sum = 5   A = ⟨150⟩
blow-up = 25/5 = 5   sq un
sides = 15 and 10

9. Solve. $\left| 4x - 3 \right| = 17$

4x – 3 = 17   OR   4x – 3 = –17
4x = 20   4x = –14
⟨x = 5⟩ $x = \dfrac{-7}{2}$

5. Find the principal that yields $22 simple interest at 5.5% for 8 months.

I = PRT

$22 = P \ \dfrac{5.5}{100} \ \dfrac{\overset{2}{\cancel{8}}}{\underset{3}{\cancel{12}}}$

$22 = P \ \dfrac{11}{300}$

$P = 22 \ \dfrac{300}{11}$

P = ⟨$600⟩

10. Name two lattice points C in the picture that would yield ΔABC with area 6.

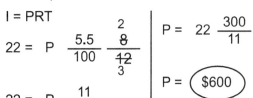

B   1 box = 1 unit

⟨(1, 4) or (5, 1)⟩

A
(0,0)   3-4-5 right triangles

Level 8	Number 13

1. Draw the discards on the left.

A baker gave 1/2 of his cookies away. After he boxed 1/5 of those left, 40 remained. Find the starting number of cookies.

$$C$$

$$\frac{C}{2} \qquad \frac{C}{2}$$

$$\frac{C}{10} \qquad \frac{4C}{10} = 40$$

$$4C = 400 \qquad \boxed{100}$$

6. Find the positive geometric mean between each pair. Show the proportion.

1 and 16	3 and 27	4 and 16
$\dfrac{1}{4} = \dfrac{4}{16}$	$\dfrac{3}{9} = \dfrac{9}{27}$	$\dfrac{4}{8} = \dfrac{8}{16}$
$\boxed{4}$	$\boxed{9}$	$\boxed{8}$

2. Find BD − AC.

A B   C   D

$$-6 \quad 0 \quad 6$$

8 of length 6 − 6 of length 6

$\boxed{12}$

7. Two similar triangles have sides 2, 8, 9 and 6, 24, 27. Find the ratio of their:

sides   1:3

altitudes   1:3

perimeters   1:3

areas   1:9

3. Find the measures of two complementary angles, the greater of which measures 8 times the lesser.

$$x + 8x = 90$$
$$9x = 90$$
$$x = 10$$

$\boxed{10 \text{ and } 80}$

8. Write in scientific notation.

$(6 \times 10^7) + (3 \times 10^6)$ $(5 \times 10^5) \times (7 \times 10^2)$

$(60 \times 10^6) + (3 \times 10^6)$ 35×10^7

63×10^6

Can only add like terms. $\boxed{3.5 \times 10^8}$

$\boxed{6.3 \times 10^7}$

4. Multiply using DPMA.

35 x 46   1610
$(30 + 5)(40 + 6) = 1200 + 180 + 200 + 30$

78 x 45   3510
$(70 + 8)(40 + 5) = 2800 + 350 + 320 + 40$

9. Operate.

$$\frac{2}{wx}^{z} - \frac{3}{xz}^{w} \qquad \frac{3}{2a}^{3bc} + \frac{2}{3b}^{2ac} + \frac{1}{c}^{6ab}$$

$\boxed{\dfrac{2z - 3w}{wxz}} \qquad \boxed{\dfrac{9bc + 4ac + 6ab}{6abc}}$

5. Find the probability that three of the numbers 2, 3, 6, or 8 chosen randomly form the sides of a triangle.

2, 3, 6   NO
2, 3, 8   NO
2, 6, 8   NO
3, 6, 8   YES $\boxed{\dfrac{1}{4}}$

10. How many triangles with whole sides have greatest side 6?

6, 6, 6   6, 5, 5 $\boxed{12}$
6, 6, 5

                6, 5, 2   Other options do not
                6, 4, 4   form a triangle. The
6, 6, 1   6, 4, 3   sum of 2 sides is
                         greater than the 3rd.

| Level 8 | Number 14 |

1. Draw a Venn Diagram to show the relationship between each pair of sets.

A = {3, 5, 7}　　　　A = {2, 6, 8}
B = {2, 3, 6, 7}　　　B = {7, 9}

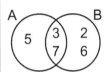　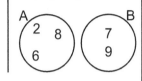

6. Find the next letter(s) in each pattern.

A, b, D, e, H, i, M, n,　__S__

The A, D, H, M pattern is all capitals. A to D advances 3, D to H 4, H to M 5, and M to S 6.

A, w, B, c, w, D, e, f,　__w__　　__G__　　__h__

Small w interleaves with a 2nd pattern, always starting with a capital, with 1, 2, 3, . . . letters.

2. Find the area of the trapezoid bounded by the x-axis and the lines y = x, x = 5, and y = 2.

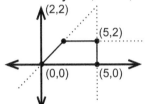

b = 3, B = 5,
M = 4, h = 2

(8)

7. Simplify.

$$\frac{\overline{.216}}{\overline{.72}} \quad \frac{216}{999} \times \frac{99}{72} \quad \frac{3}{111} \times \frac{11}{1} \left(\frac{11}{37}\right)$$

$$\frac{\overline{.435}}{\overline{.585}} \quad \frac{435}{999} \times \frac{999}{585} \quad \left(\frac{29}{39}\right)$$

3. Multiply using the difference of 2 squares.

31 x 29　　__899__　(30 + 1)(30 − 1)　900−1

65 x 55　　__3575__　(60 + 5)(60 − 5)　3600−25

73 x 67　　__4891__　(70 + 3)(70 − 3)　4900−9

36 x 44　　__1584__　(40 − 4)(40 + 4)　1600−16

8. Show that primes of the form 4k+1 equal the sum of 2 perfect squares.

5 (k = 1)	1 + 4	37 (k = 9)	1 + 36
13 (k = 3)	4 + 9	41 (k = 10)	16 + 25
17 (k = 4)	1 + 16	53 (k = 13)	4 + 49
29 (k = 7)	4 + 25	61 (k = 15)	25 + 36

4. Draw the reflection of each triangle over the line y = x.　Reflect vertices first. (x,y) becomes (y,x).

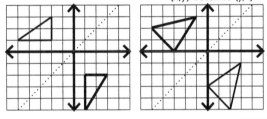

9. Jack paid $2.50 each for small books. He sold all but 30 of them for $5.00 each. His profit was $50. How many books did he buy?

$IN − $OUT = PROFIT
5(B − 30) − 2.5B = 50
5B − 150 − 2.5B = 50
2.5B = 200　　　　　　B = (80)
25B = 2000

5. The ratio of an angle to its supplement is 2:7. Find the ratio of the angle to its complement.

2 + 7 = 9
180 ÷ 9 = 20
angle = 40 (supplement = 140)
complement = 50　(4:5)

10. Multiply.

$$2745_{eight} \qquad 4322_{five} \qquad 752T_{eleven}$$
$$\times \quad 3_{eight} \qquad \times \quad 4_{five} \qquad \times \quad 6_{eleven}$$
$$\mathbf{10657}_{eight} \qquad \mathbf{33343}_{five} \qquad \mathbf{40965}_{eleven}$$

	Level 8	Number 15

1.

	time	rate
H	4	$\dfrac{1}{4}$
S	(4)	$\dfrac{1}{4}$
tog	2	$\dfrac{1}{2}$

Hal can do a job alone in 4 hours. Find the time in hours for Sal to do the job alone if together they can complete it in 2 hours.

6. Draw the next figure in the pattern.

2. Square mentally.

35^2 1225	305^2 93,025
$3 \times 4 = 12$	$30 \times 31 = 930$
195^2 38,025	65^2 4225
$19 \times 20 = 380$	$6 \times 7 = 42$

7. Calculate the combinations.

$$C(5, 3) \ = \ \frac{5!}{3!\ 2!} \ = \ \frac{5 \cdot 4}{2 \cdot 1} \ = \ \boxed{10}$$

$$C(12, 9) \ = \ \frac{12!}{9!\ 3!} \ = \ \frac{12 \cdot 11 \cdot 10}{3 \cdot 2 \cdot 1} \ = \ \boxed{220}$$

$$C(6, 2) \ = \ \frac{6!}{2!\ 4!} \ = \ \frac{6 \cdot 5}{2 \cdot 1} \ = \ \boxed{15}$$

3. Complete for the geometric sets.

A • ———— B • ———————————— C •

$\overline{AB} \cup \overline{BC} = \underline{\ \overline{AC}\ }$ | $\overline{AC} \cup \overline{BC} = \underline{\ \overline{AC}\ }$

$\overline{AB} \cap \overline{BC} = \underline{\ B\ }$ | $\overline{AC} \cap \overline{BC} = \underline{\ \overline{BC}\ }$

8. Find the perimeter of an isosceles trapezoid with height 12, legs 15, and small base 10.

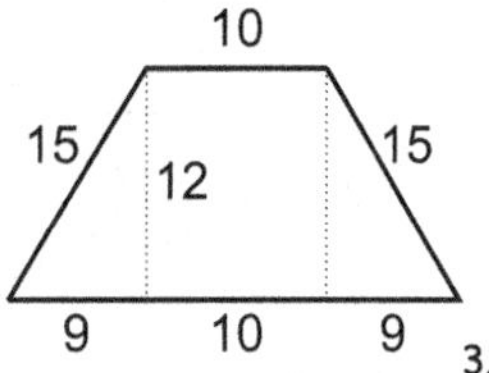

trapezoid labels: top 10, left leg 15, right leg 15, height 12, bottom 9, 10, 9

$$15 + 15 = 30$$
$$20 + 18 = 38$$
$$P = \boxed{68}$$

3, 4, 5 x3

4. How many positive integers less than 100 have an odd number of factors?

$1^2 \longrightarrow 9^2$ $\boxed{9}$

Only a perfect square has an odd number of factors. The square root is in the middle when listing factors in ascending order.

9. Find the slope given 2 points on a line.

Point #1	Point #2	Slope
(6, 10)	(8, 14)	2
(1, 2)	(3, 10)	4
(7, 7)	(9, 9)	1
(6, 4)	(7, 7)	3

5. Solve.

$(x + 4)(x + 7) = 0$	$(6x - 1)(9x + 4) = 0$
$x = \boxed{-4, -7}$	$x = \boxed{\dfrac{1}{6}, \dfrac{-4}{9}}$
$(x - 8)(x - 9) = 0$	$(8x - 3)(7x + 2) = 0$
$x = \boxed{8, 9}$	$x = \boxed{\dfrac{3}{8}, \dfrac{-2}{7}}$

10. Find the area of the triangle with $m\angle A = 30°$. (NTS)

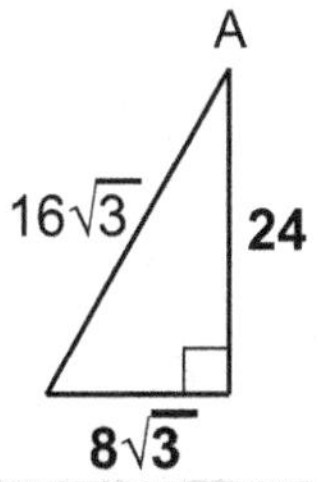

triangle with vertex A, left side $16\sqrt{3}$, right side 24, base $8\sqrt{3}$

$$A = \frac{(8\sqrt{3})(24)}{2}$$
$$A = \boxed{96\sqrt{3}}$$

Level 8 Number 16

1. Which fabric has a less costly unit rate per square foot: 3 yards of 48 inches wide for $15.84 or 4 yards of 36 inches wide for $15.84?

$\frac{15.84}{3 \times 3 \times 4}$ vs $\frac{15.84}{4 \times 3 \times 3}$ (same)

2. If n is an odd integer, label each even (E) or odd (O).

a.) $2n + 1$ O

b.) $n(n + 1)$ E

c.) n^2 O

3. A raw meatball with a 3-inch radius can be re-formed into how many meatballs with a 1-inch radius?

$$\frac{V \quad \frac{4\pi \cdot 3 \cdot 3 \cdot 3}{3}}{v \quad \frac{4\pi \cdot 1 \cdot 1 \cdot 1}{3}} = 3 \cdot 3 \cdot 3 = \boxed{27}$$

Divide the greater volume by the lesser volume.

4. Define operation ❁ as:

X ❁ Y = 2XY + X. For example,
3 ❁ 2 = 12 + 3 = 15.
Find the values.

4 ❁ X = 60 X = 7 8X + 4 = 60

A ❁ 8 = 51 A = 3 16A + A = 51

5. The measure of one angle of a parallelogram is 5 more than 4 times another. Find the 4 angles.

180 – A = 5 + 4A
175 = 5A
A = 35

(35, 145, 35, 145)

6. Find all digits d so that the 5-digit number 91,7d2 is divisible by 4.

ignore the 917 (1, 3, 5, 7, 9)

Find all digits x so that the 4-digit number 7x00 is divisible by 8. ignore the 7

100 has two factors of 2; need one more 2 (0, 2, 4, 6, 8)

7. S:U:N = 9:10:8 T:A:N = 12:11:6
Find T:S.

Common variable is N.
LCM(8,6) = 24
S:U:N = 27:30:24 (x3)
T:A:N = 48:44:24 (x4)
T:S = 48:27 = 16:9 (÷3)

(16:9)

8. Rolling a die twice, find the probability of at least one 5.

1 – P(not 5, not 5)

The greatest probability is 1.

$$1 - \frac{5}{6} \cdot \frac{5}{6} = 1 - \frac{25}{36} = \boxed{\frac{11}{36}}$$

9. Operate.

$(-5)^2 - 5^2 - (-5)^2 + 5^2$
25 – 25 – 25 + 25
(0)

$-5^2 - (-5^2) - (-5)^2$
–25 + 25 – 25
(–25)

$(-5)^3 - 5^3 - (-5)^3 + 5^3$
–125–125+125+125
(0)

$5^3 + (-5^3) - (-5)^3$
125 – 125 + 125
(125)

10. Find the volume in cubic units of a rectangular pyramid with length 14, width 15, and altitude 7 units.

$$\frac{14 \times 15 \times 7}{3} = 14 \times 5 \times 7 = 70 \times 7$$

$= \boxed{490}$

Level 8	Number 17

1. If c chickens lay e eggs in 1 day, how many eggs will k chickens lay in 30 days?

chickens	eggs	days	days?
c	e	1	
1	$\dfrac{e}{c}$	1	
k	$\dfrac{ek}{c}$	1	
k	$\left(\dfrac{30ek}{c}\right)$	30	

6. Find the sum of the:

first 4 primes. $\underline{17}$ 2 + 3 + 5 + 7

2 greatest 2-digit primes. $\underline{186}$ 97 + 89

2 least 3-digit primes. $\underline{204}$ 101 + 103

multiple of 2 or 5 primes. $\underline{7}$ 2 + 5

---

2. How many right triangles with whole number sides have one side of length 30 units?

3, 4, 5 $\longrightarrow$ 18, 24, 30
                        30, 40, 50   ④

8, 15, 17 $\longrightarrow$ 16, 30, 34

5, 12, 13 $\longrightarrow$ 30, 72, 78

7. Answer YES or NO as to whether the set is closed for the given operation.

rationals   squaring $\underline{YES}$

wholes   absolute value $\underline{YES}$

integers   absolute value $\underline{YES}$

---

3. Find the area of the rhombus in square units given perimeter P = 52 and diagonal d = 10.

[rhombus figure with sides 13, segments 5 and 12]

P = 52   A = 30 x 4
s = 13   or 12 x 10
d = 10
d/2 = 5   A = ⟨120⟩
5, 12, 13
D = 24

8. Find the angle in degrees formed by clock hands at 5:40.

[clock face showing hands]

360°/12 = 30° between neighboring numbers

60 + 10 = ⟨70⟩

Hour hand is 2/3 on way from 5 to 6. 1/3 is in angle.

---

4. Find the space diagonal of a cube with

edge = 7 $7\sqrt{3}$ e = 7

area of face = 81 $9\sqrt{3}$ e = 9

volume = 8 $2\sqrt{3}$ e = 2

face diagonal = $8\sqrt{2}$ $8\sqrt{3}$ e = 8

9. Solve. $|9x + 6| = 78$

9x + 6 = 78   OR   9x + 6 = –78
9x = 72   9x = –84
⟨x = 8⟩   ⟨x = $\dfrac{-28}{3}$⟩

---

5. Find the points of trisection of the line segment with the given endpoints.

(1, 5) & (34, 65) $\underline{(12, 25)}$ $\underline{(23, 45)}$

(–8, 12) & (–5, 21) $\underline{(-7, 15)}$ $\underline{(-6, 18)}$

(2, –8) & (41, 16) $\underline{(15, 0)}$ $\underline{(28, 8)}$

(1, –4) & (28, 23) $\underline{(10, 5)}$ $\underline{(19, 14)}$

10. Multiply as two binomials.

$4\dfrac{1}{6}$ x $9\dfrac{3}{5}$ $7\dfrac{9}{10}$ x $8\dfrac{1}{2}$

F = $\underline{36}$ F = $\underline{56}$
O = $\underline{2.4}$ O = $\underline{3.5}$
I = $\underline{1.5}$ ⟨40⟩   I = $\underline{7.2}$ ⟨67.15⟩
L = $\underline{0.1}$ L = $\underline{0.45}$

Level 8	Number 18

1. The more common name for arithmetic mean is _____ **average**.

Each answer or solution to an equation is called a _____ **root**.

A synonym for height is _____ **altitude**.

6. Two boats leave from the same dock at 11:00 AM. One travels north at 30 mph; the other travels east at 40 mph. When will they be 100 miles apart?

3, 4, 5 x 20
60, 80, 100

D = RT =
30 x 2 = 60

D = 100

D = RT = 40 x 2 = 80

11:00 + 2 hr
(1:00 PM)

---

2. Write the contrapositive of the conditional statement. Are they logically equivalent? (YES)

If a number is prime, then it is odd.

Both are FALSE.
Counterexample = 2

If a number is not odd, then it is not prime.

7. Simplify.

$$\frac{\overline{.18}}{\overline{.324}} \quad \frac{18}{99} \times \frac{999}{324} \quad \frac{2}{11} \times \frac{111}{36} \quad \left(\frac{111}{198}\right)$$

$$\frac{\overline{.185}}{\overline{.35}} \quad \frac{185}{999} \times \frac{99}{35} \quad \frac{37}{111} \times \frac{11}{7} \quad \left(\frac{11}{21}\right)$$

---

3. Find x degrees in the crook picture. (NTS)

152° / **28** by supp

28 by corr
x **71** by corr
71°

(99)

8. Find the area of a circle given its circumference is $10\pi^2$.

C = $10\pi^2$
C = $2r\pi$
r = 5π
A = πr^2
A = $\pi(5\pi)(5\pi)$
A = $\left(25\pi^3\right)$

---

4. The measures of the interior angles of a quadrilateral are 4x, 3x − 1, 2x + 21, and 5x − 10. Find the median of the angles.

4x + 3x − 1 + 2x + 21 + 5x − 10 = 360
14x + 10 = 360
14x = 350
x = 25
angles are 100, 74, 71, 115

Median = (74 + 100)/2

(87)

9. Find the value using fractions.

$66\frac{2}{3}$ % of $37\frac{1}{2}$ % of 60% of 760

$$\frac{2}{3} \times \frac{3}{8} \times \frac{3}{5} \times \cancel{760}^{\substack{38 \\ \cancel{190}}}$$

38 x 3 = 90 + 24 = (114)

---

5. Multiply mentally.

63 x 101	6363
71 x 1001	71,071
434 x 10,001	4,340,434
595 x 1001	595,595

Picture 101 as (100 + 1), 1001 as (1000 + 1), and 10,001 as (10,000 + 1).

10. Find the total value after compounding annually.

principal	rate	years	value	
$1500	10%	2	**$1815**	1650
$6000	5%	2	**$6615**	6300
$3500	10%	2	**$4235**	3850
$5000	2%	2	**$5202**	5100

Level 8	Number 19

1. Find the number of arrangements of the letters HOUSE if the last one must be S and the first one must be a vowel.

$$\underline{3} \cdot \underline{3} \cdot \underline{2} \cdot \underline{1} \cdot \underline{1} \quad \boxed{18}$$

↑ O,U,E 2nd   ↑ S 1st

6. Find the volume in cubic units of a rhomboidal prism with base diagonals 12 and 20 and altitude 12 units.

A base = 12 x 20 / 2 = 120

V = 120 x 12 = ⬭ 1,440

2. Is each the graph of a function?

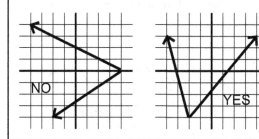

NO   YES

7. A garden 20 by 27 feet has a uniform sidewalk all around. Find the width and area of the sidewalk if the total area is 1248 square feet.

20 x 27 = 540
32 x 39 = 1248

A = 708
W = 6

Use T&E and ones digit instead of algebra.

3. The area of rectangle ABCD is 5 times the area of rectangle BCFE. Find the coordinates of E. (NTS)

30 + 15 = 45
45 x 9 = 5 x 81
EB = 9

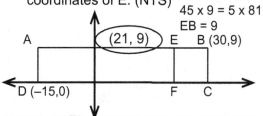

A   (21, 9) E B (30,9)

D (−15,0)   F   C

8. Find the angle in degrees formed by clock hands at 9:20.

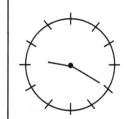

360°/12 = 30° between neighboring numbers

150 + 10 = 160

Hour hand is 1/3 of way from 9 to 10.

4. Evaluate the functions.

ABS(−125)	125	TRUNC(6.7)	6
SGN(−25)	−1	INT(−5.5)	−6
SQRT(25)	5	SIGMA(16)	5
SQR(−3)	9	INT(15.99)	15

9. Complete.

40 ≡ 6 (mod 17)	37 ≡ 1 (mod 18)
59 ≡ 29 (mod 30)	67 ≡ 7 (mod 12)
62 ≡ 12 (mod 50)	51 ≡ 3 (mod 16)
68 ≡ 2 (mod 22)	80 ≡ 10 (mod 35)

5. Find the mean and range of the 5 different integers 3x + 2, 2x − 10, −5, 24, and 4x − 1 if their median is 17.

−5	0	17	19	24
−5	2x−10	3x+2	4x−1	24

range = 24 − (−5) = ⬭ 29

mean = 55/5 = ⬭ 11

Draw 5 slots. Only 3x+2 can be 17. x = 5

10. Find the principal that yields $60 simple interest at 4.5% for 10 months.

I = PRT

$$60 = P \ \frac{4.5}{100} \ \frac{10}{12}$$

$$P = 60 \ \frac{80}{3}$$

$$60 = P \ \frac{45}{1200}$$

P = ⬭ $1600

MAVA Math: Middle Reviews Solutions Copyright © 2013 Marla Weiss

| Level 8 | Number 20 |

1.

Answer items below for the bar graph.

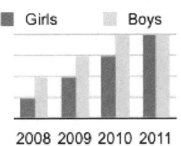

Girls ▪ Boys ▫

60
45
30
15
0

2008 2009 2010 2011

2. Find the percent increase: change/original

15 to 45
the number of girls in 2008 to
the number of girls in 2010 _200%_

30 to 60
the number of boys in 2008 to
the number of boys in 2011 _100%_

75 to 120
the number of children in 2009 to
the number of children in 2011 _60%_

3. Rewrite the statement using DeMorgan's Laws.

I am not a math major and I am not a cheerleader.

I am not a math major or a cheerleader.

4. Define operation ❋ as:

X ❋ Y = 3Y – 2X. For example,
5 ❋ 4 = 12 – 10 = 2.
Find the values.

(3 ❋ 2) ❋ 5 _15_ 3 ❋ 2 = 0

4 ❋ (6 ❋ 1) _–35_ 6 ❋ 1 = –9

5. Multiply using DPMA.

65 x 42 _2730_
(60 + 5)(40 + 2) = 2400 + 120 + 200 + 10

57 x 24 _1368_
(50 + 7)(20 + 4) = 1000 + 200 + 140 + 28

6. Find the perimeter of a rectangle given the area is 256 and the length is four times the width.

A = 256   L = 32
LW = 256   SP = 40
(4W)(W) = 256
(W)(W) = 64   P = ⬭ 80
W = 8

7. Find the area of the square inscribed in a circle of area 100π square units.

$A_{circ} = 100\pi$
R = 10
D = diag sq = 20
A_{sq} = 20 x 20 / 2 = ⬭ 200

A square is a rhombus. Use the area formula: half the product of the diagonals.

8. Six years from now, Kal's age will be twice Hal's age. Now Kal is 16 years older than Hal. How old was Hal 4 years ago?

	–4	now	+6
H	⬭ 6	H	H+6
K		2H+6	2H+12

H+16 = 2H+6
H = 10

9. Hal paid $1.50 each for pens. He sold all but 20 of them for $3.00 each. His profit was $15. How many pens did he buy?

$IN – $OUT = PROFIT
3(P – 20) – 1.5P = 15
3P – 60 – 1.5P = 15
1.5P = 75   P = ⬭ 50
15P = 750

10. Find the area of the triangle with m∠A = 30°. (NTS)

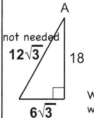

A

not needed

$12\sqrt{3}$ 18

$6\sqrt{3}$

$A = \dfrac{(6\sqrt{3})(18)}{2}$

A = ⬭ $54\sqrt{3}$

When the side across from 60° is whole, "divide by 3, tack on root 3."

Level 8	Number 21

1. Find the time in minutes needed for 2 people to do a job together given that: Person A can do the job in 2 hours; Person B can the same job in 4 hours.

Rate A = 1/2
Rate B = 1/4
Total Rate = 3/4
Total Time = 4/3

In work problems, rates and times are reciprocals.

⟨80⟩

6. Find the positive geometric mean between each pair. Show the proportion.

2 and 50	3 and 12	2 and 32
$\dfrac{2}{10} = \dfrac{10}{50}$	$\dfrac{3}{6} = \dfrac{6}{12}$	$\dfrac{2}{8} = \dfrac{8}{32}$
⟨10⟩	⟨6⟩	⟨8⟩

2. Find AC – BD.

A   B   C   D

–3 0 3

6 of length 3 – 6 of length 3

⟨0⟩

7. Estimate the specified root to the nearest whole number.

square root of 55   7   49 vs. 64

square root of 72   8   64 vs. 81

cube root of 155   5   125 vs. 216

cube root of 590   8   512 vs. 729

3. Multiply the matrices.

$$\begin{bmatrix} 6 & 1 & 5 \\ 4 & 9 & 2 \end{bmatrix} \times \begin{bmatrix} 11 & 4 & 2 \\ 5 & 0 & 1 \\ 3 & 6 & 10 \end{bmatrix} =$$

66 + 5 + 15 = 86
24 + 0 + 30 = 54
12 + 1 + 50 = 63
44 + 45 + 6 = 95
16 + 0 + 12 = 28
8 + 9 + 20 = 37

$$\begin{bmatrix} 86 & 54 & 63 \\ 95 & 28 & 37 \end{bmatrix}$$

8. Find the perimeter of an isosceles trapezoid with height 45, legs 51, and small base 22.

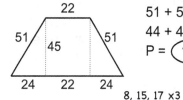

51 + 51 = 102
44 + 48 = 92
P = ⟨194⟩

8, 15, 17 x3

4. Find the endpoint of the line segment.

Endpoint	Midpoint	Endpoint
(2.5, 12)	(5, 14)	(7.5, 16)
(–9, 9)	(–4, 18)	**(1, 27)**
(3.3, –8)	(6.6, –1)	**(9.9, 6)**
(14, –6)	(19, 6)	(24, 18)

9. Find the angles of parallelogram ABCD given:

m∠A = 6x – 5   m∠B = 9x + 20

6x – 5 + 9x + 20 = 180
15x + 15 = 180
15x = 165
x = 11

⟨m∠A, C = 61°
m∠B, D = 119°⟩

Consecutive angles are supplementary.
Opposite angles are congruent.

5. Find the probability that 3 of the numbers 3, 5, 7, or 10 chosen randomly form the sides of a triangle.

3, 5, 7   YES
3, 5, 10   NO
3, 7, 10   NO
5, 7, 10   YES

10. Find x. (NTS)

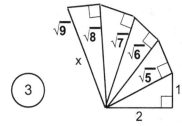

⟨3⟩

	Level 8	Number 22

1.

	time	rate
J	3	$\frac{1}{3}$
T	(1.5)	$\frac{2}{3}$
tog	1	1

Jim can do a job alone in 3 hours. Find the time in hours for Tim to do the job alone if together they can complete it in 1 hour.

6. Find the measure of one angle of each regular polygon.
Interior and exterior angles are supplements.

decagon 144   ext < = 360/10 = 36
int < = 180 - 36 = 144

icosagon 162   ext < = 360/20 = 18
int < = 180 - 18 = 162

60-gon 174   ext < = 360/60 = 6
int < = 180 - 6 = 174

Or 58 x 180 / 60 = 174

2. Find 2 consecutive even integers such that the triple the greater is 24 more than twice the lesser.
x, x + 2
3(x + 2) = 2x + 24
3x + 6 = 2x + 24
x = 18
(18, 20)

7. Simplify using DPMA.
513 x 405 − 513 x 5   513 x (405 − 5)
513 x 400
(205,200)
6.25% of 71 + 6.25% of 9 5/8 (71 + 9)
5/8 (80)
(50)
81² + 81 x 19   81(81 + 19)
81 x 100 (8100)

3. Graph on the number line, and write the solution to the quadratic inequality.
(x − 2)(x − 4) < 0
no   yes   no
0   2   4
(2 < x < 4)

8. Operate and simplify.
$8\frac{1}{4} \div 7\frac{7}{8}$ $\frac{33}{4} \times \frac{8}{63}$ $\frac{22}{21}$
$4\frac{2}{7} \times 4\frac{2}{5}$ $\frac{30}{7} \times \frac{22}{5}$ $\frac{6 \times 22}{7}$
$\frac{22}{21} \times \frac{7}{6 \times 22}$ $\left(\frac{1}{18}\right)$

4. Given parallelogram ABCD. Find x.
D C
155° x° (65)
155/25
A B

9. Answer YES or NO if the matrices may be multiplied. If yes, give the dimensions of the product.
2 by 3 times 4 by 5   NO
1 by 7 times 7 by 6   YES 1 by 6
3 by 6 times 4 by 6   NO

5. If f(x) = 2x + 3 and g(x) = 5x − 2, find g(f(4)) − f(g(3)).
f(4) = 8 + 3 = 11
g(11) = 55 − 2 = 53
g(3) = 15 − 2 = 13
f(13) = 26 + 3 = 29
53 − 29 = (24)

10. A regular hexagon has perimeter 54 units. Find its area in square units.
P = 54
S = 9
A = $\frac{6 \cdot 9^2 \sqrt{3}}{4}$ = $\left(\frac{243\sqrt{3}}{2}\right)$

Level 8	Number 23

1. The first two terms of a sequence are 4 and 10. Successive terms are found by taking the average of all preceding terms. What is the 50th term?

4, 10, 7, 7, 7, 7, . . .

$\boxed{7}$

6. Translate the cipher with each letter as its own digit.

$$\begin{array}{r} DS \\ DR\overline{)RRA} \\ DR \\ \hline DEA \\ DEA \\ \hline 0 \end{array}$$

$$\begin{array}{r} 19 \\ 12\overline{)228} \\ 12 \\ \hline 108 \\ 108 \\ \hline 0 \end{array}$$

DR x D = DR, so D=1. R–R = E, so E=0. R–1 = 1, so R=2. Divisor is 12. 12x8 is 96 (2 digits). So, S=9 and A=8.

2. Find the perimeter of the rhombus with area 1320 and shorter diagonal 22.

A = Dd/2
1320 = 22D/2
1320 = 11D
12 x 11 x 10 = 11D
D = 120
P = 61 x 4 = $\boxed{244}$

7. Find the area of a regular hexagon with edge 6.

$$\frac{6 \cdot s^2 \cdot \sqrt{3}}{4} = \frac{6 \cdot \overset{3}{\cancel{6}} \cdot \overset{3}{\cancel{6}} \cdot \sqrt{3}}{\cancel{4}} = \boxed{54\sqrt{3}}$$

Area of regular hexagon is 6 times area of equilateral triangle.

3. Point B divides segment $\overline{AC}$ such that BC:AB = 4:3. AC = 42. Find AB.

x 6

A   B   C

$\boxed{18}$ 24

8. Given the perimeter of a rectangle and the upper bound of the width. Find the range of values for the length.

P = 66; max W < 20	P = 90; max W < 36
SP = 33	SP = 45
0 < W < 20	0 < W < 36
$\boxed{13 < L < 33}$	$\boxed{9 < L < 45}$

4. Find the probability of drawing 2 blues successively without replacement from a bowl containing 10 green, 14 yellow, 16 red, and 25 blue marbles.

$$\frac{\overset{5}{\cancel{25}}}{\underset{13}{\cancel{65}}} \cdot \frac{\overset{3}{\cancel{24}}}{\underset{8}{\cancel{64}}} \quad \boxed{\frac{15}{104}}$$

9. Multiply.

$$\frac{ce}{ax} \cdot \frac{bx}{cd} \qquad \boxed{\frac{be}{ad}}$$

$$\frac{a}{x} \cdot \frac{y}{ac} \cdot \frac{x}{y} \qquad \boxed{\frac{1}{c}}$$

5. Find the mean of 15x – 7, 9x + 3, 4x – 5, 7x + 7, x – 1, and 6x + 9.

$$\frac{42x + 6}{6} = \boxed{7x + 1}$$

10. Find the distance between the points.

(6, –6) and (–2, 9)	(3, 11) and (–7, –13)
8, 15, $\boxed{17}$	10, 24, $\boxed{26}$
(10, 3) and (–14, –4)	(–1, 4) and (7, –2)
24, 7, $\boxed{25}$	8, 6, $\boxed{10}$

Level 8	Number 24

1. Find two consecutive positive integers with the given product.

210	14, 15	930	30, 31
132	11, 12	380	19, 20
2550	50, 51	600	24, 25

6. Of 130 girls, 64 wore sweaters, 72 wore coats, and 4 wore neither. How many wore both?

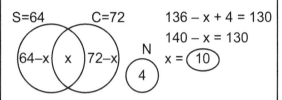

$136 - x + 4 = 130$
$140 - x = 130$
$x = \boxed{10}$

2. Find the length of the segment joining the midpoints of segments $\overline{AC}$ and $\overline{BC}$.

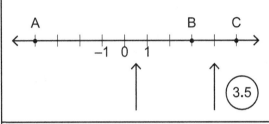

$\boxed{3.5}$

7. Label the lines perpendicular, parallel, or neither.

$2x + 3y = 7$
$-3x + 2y = 9$

$3x + 7y = 2$
$3x + 7y = 4$

(perpendicular)
negative reciprocal slopes

(parallel)
equal slopes

3. Find the volume of the cone in cubic units.

altitude = 12
slant height = 13

$V = \dfrac{5 \cdot 5 \cdot \pi \cdot 12}{3}$

$= 25 \cdot 4 \cdot \pi$

$= \boxed{100\pi}$

8. Calculate the combinations.

$C(11, 3) = \dfrac{11!}{3!\ 8!} = \dfrac{9 \cdot 10 \cdot 11}{3 \cdot 2 \cdot 1} = \boxed{165}$

$C(7, 4) = \dfrac{7!}{4!\ 3!} = \dfrac{5 \cdot 6 \cdot 7}{3 \cdot 2 \cdot 1} = \boxed{35}$

$C(12, 11) = \dfrac{12!}{11!\ 1!} = \boxed{12}$

4. Multiply mentally.

749 x 1001	749,749
65 x 1001	65,065
423 x 10,001	4,230,423
89 x 101	8989

Picture 101 as (100 + 1), 1001 as (1000 + 1), and 10,001 as (10,000 + 1).

9. Name a lattice point C in the picture that would yield △ABC with area 5.

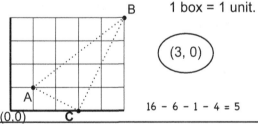

1 box = 1 unit.

$\boxed{(3, 0)}$

$16 - 6 - 1 - 4 = 5$

5. Find the space diagonal of the rectangular prism with the given dimensions.

2 by 3 by $\sqrt{5}$ $3\sqrt{2}$ $\sqrt{4 + 9 + 5}$

1 by 6 by 18   19 $\sqrt{1 + 36 + 324}$

10. How many numbers from 1 to 100 inclusive are multiples of 3 or 4?

$100 \div 3 = 33$
$100 \div 4 = 25$
$100 \div 12 = 8$
$33 + 25 - 8 = \boxed{50}$

Subtract the overlap: the multiples of 3 AND 4.

Level 8	Number 25

1. Of the apples on a tree, 20% fell off. Of those left, 15% were boxed. Of those left, 25% were discarded. 102 apples remained. How many were on the tree to start?

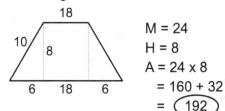

A

$\frac{A}{5}$ $\frac{4A}{5}$

$\frac{3A}{25}$ $\frac{17A}{25}$

$\frac{17A}{100}$ $\frac{51A}{100}$

$51A = 10,200$ **(200)**

6. Find the number of integer values that satisfy:

$|2x + 1| \le 7$ 8   −4 $\longrightarrow$ 3

$|x − 6| \le 11$ 23   −5 $\longrightarrow$ 17

$|2x + 3| \le 15$ 16   −9 $\longrightarrow$ 6

---

2. Solve.

$27^x = 3^6$ | $16^5 = 2^x$ | $8^5 = 32^{2x−3}$

$3^{3x} = 3^6$ | $2^{20} = 2^x$ | $2^{15} = 2^{10x−15}$

$3x = 6$ | $x = $ **(20)** | $10x = 30$

$x = $ **(2)** | | $x = $ **(3)**

7. 21 quarters, 19 dimes, and 37 nickels are what percent of $25?

$21 \times .25 = 5.25$ Use mental math for
$19 \times .10 = 1.90$ the multiplications.
$37 \times .05 = 1.85$
sum = 9.00

$\frac{9}{25}$ = **(36%)**

---

3. Find the area of an isosceles trapezoid with legs 10 and bases 18 and 30.

18

10 ╱ 8

6   18   6

M = 24
H = 8
A = 24 × 8
  = 160 + 32
  = **(192)**

8. Write in scientific notation.

$(7 \times 10^8) \div (8 \times 10^{−4})$ | $(1 \times 10^7) \div (4 \times 10^2)$

0.875×10^{12} 0.25×10^5

(8.75 × 10¹¹)   **(2.5 × 10⁴)**

8.75×10^{11} 2.5×10^4

---

4. Find the perimeter of parallelogram ABCD given AB = 15, BC = 12x − 3, and CD = 6x − 3.

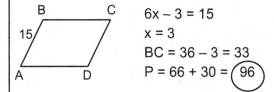

B   C

15

A   D

$6x − 3 = 15$
$x = 3$
$BC = 36 − 3 = 33$
$P = 66 + 30 = $ **(96)**

9. Complete the unit conversions.

Picture rectangles:

1 yard²   __9__   feet²   1 yd by 1 yd = 3 ft by 3 ft

3 yard²   __27__   feet²   3 yd by 1 yd = 9 ft by 3 ft

(3 yard)²   __81__   feet²   3 yd by 3 yd = 9 ft by 9 ft

---

5. Find the sum of all 2-digit numbers with exactly 5 factors.

16 ~~25~~ ~~36~~ ~~49~~ ~~64~~ 81

Only a perfect square has an odd number of factors. A prime to the 4th power has exactly 5 factors.

$16 + 81 = $ **(97)**

10. Answer YES or NO as to whether the 3 numbers form sides of a right triangle.

$\sqrt{5}$, 3, 4   __N__ | 5, 5, $5\sqrt{2}$ __Y__

2, 5, $\sqrt{29}$ __Y__ | 4, 5, $\sqrt{43}$ __N__

1, 2, $\sqrt{5}$ __Y__ | 3, $3\sqrt{3}$, 6   __Y__

Level 8	Number 26

1. A cube has a volume of 64 cubic units. Point A is in the center of one face, and point B is in the center of the opposite face. Find AB (the length of the segment connecting the 2 points.)

4 units   *congruent to an edge*

2. X varies inversely as Y. X = 75 when Y = 20. Find Y when X = 60.

XY = k

(75)(20) = k

k = 1500

60Y = 1500

Y = 25

3. Draw the reflection of each triangle over the line y = −x. *Reflect vertices first. (x,y) becomes (−y,−x).*

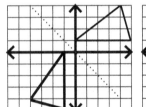 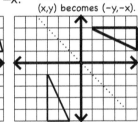

4. Find the surface area of the cone in square units.   *LA = "πrl"*

diameter = 40
slant height = 25

SA = 20 • 20 • π
  + 20 • 25 • π
= 400π + 500π
= 900π

5. The ratio of an angle to its complement is 1:8. Find the ratio of the angle to its supplement.

1 + 8 = 9

90 ÷ 9 = 10

angle = 10 (complement = 80)

supplement = 170   1:17

6. Calculate the sum of the arithmetic sequence.

12 + 15 + 18 + . . . + 72 + 75 + 78

4x3 5x3 6x3   26x3
1 to 26 is 26 numbers; omit 1, 2, 3; n = 23
F + L = 12 + 78 = 90
90 x 23 / 2 = 45 x 23 = 900 + 135 = 1035

7. Solve.

$9\sqrt{x} = x$

$81x = x^2$

$0 = x^2 - 81x$

$0 = x(x - 81)$

x = 0, 81

$5x^2 = 900$

$x^2 = 180$

$x = \pm\sqrt{180}$

$x = \pm 6\sqrt{5}$

8. Convert 4,594 in base ten to base eleven.

```
 4594
−3993
  601
− 484
  117
  110
    7
```

3	4	T	7
1331	121	11	1

34T7 eleven

9. Find the total value after compounding annually.

principal	rate	years	value	
$1200	5%	2	**$1323**	1260
$3000	10%	3	**$3993**	3300, 3630
$5000	8%	2	**$5832**	5400
$4000	10%	3	**$5324**	4400, 4840

10. Multiply as two binomials.

$2\frac{2}{5} \times 4\frac{7}{10}$

F = 8
O = 1.4
I = 1.6
L = 0.28   11.28

$6\frac{3}{4} \times 8\frac{5}{6}$

F = 48
O = 5
I = 6
L = .625   59.625

Level 8	Number 27

1. Find the area of each rectangle in square feet.

3 yards by 4 yards

9 feet by 12 feet → 90 + 18 (108)

60 inches by 84 inches

5 feet by 7 feet (35)

Converting to the unit of the answer first is easier.

6. Solve.

$X + Y = 9$
$2X + 5Y = 9$

$5X + 5Y = 45$
$2X + 5Y = 9$
_____
$3X = 36$ $12 + Y = 9$
$X = 12$ $Y = -3$

$3X + 4Y = 4$
$2X + Y = 6$

$8X + 4Y = 24$
$3X + 4Y = 4$
_____
$5X = 20$ $8 + Y = 6$
$X = 4$ $Y = -2$

2. In a set of 5 numbers, the mean, median, and mode are all equal. Four of the numbers are 50, 60, 10, and 40. Find the missing number.

10   (40)   40   50   60

Of the 4 given numbers, try 40 because it is their mean.

7. Is the fictitious operation ✹ defined by $x ✹ y = xy + 1$ associative?

$(x ✹ y) ✹ z = (xy + 1) ✹ z$
$= xyz + z + 1$

$x ✹ (y ✹ z) = x ✹ (yz + 1)$
$= xyz + x + 1$ (NO)

3. Solve for the variables in the matrices.

$$\begin{bmatrix} 3x + 6 & 89 \\ 65 & 4y - 3 \end{bmatrix} = \begin{bmatrix} 57 & 9z + 8 \\ 7w - 5 & 49 \end{bmatrix}$$

$3x + 6 = 57$ $7w - 5 = 65$
$x = 17$ $w = 10$
$9z + 8 = 89$ $4y - 3 = 49$
$z = 9$ $y = 13$

8. Determine if the triangle with the 3 sides is acute or obtuse.

6, 7, 8

36   85 is right
    64 is smaller
49   (acute)

3, 7, 8

9   58 is right
   64 is greater
49   (obtuse)

4. The radii of 2 spheres are in the ratio 2:5. Find the ratio of their volumes and surface areas.

$\dfrac{v}{V} = \dfrac{4/3 \cdot \pi \cdot 2 \cdot 2 \cdot 2}{4/3 \cdot \pi \cdot 5 \cdot 5 \cdot 5} = \left(\dfrac{8}{125}\right)$

$\dfrac{sa}{SA} = \dfrac{4 \cdot \pi \cdot 2 \cdot 2}{4 \cdot \pi \cdot 5 \cdot 5} = \left(\dfrac{4}{25}\right)$

9. Find the slope given 2 points on a line.

Point #1	Point #2	Slope
(7, 10)	(5, 4)	3
(3, 12)	(1, 2)	5
(−5, −5)	(0, 0)	1
(−6, 15)	(−7, 11)	4

5. Graph on the coordinate plane. One box equals one unit.

$x + 2y = 4$ $2x - y = 4$

Plot intercepts when a and b divide c.

10. A, B, C, D, and E are consecutive points on a line. AB:BC = 2:3, BC:CD = 3:7, and CD:DE = 1:5. Find AD:BE.

AB	2		2	12:45
BC	3	3	3	
CD		7	1	7	
DE			5	35	(4:15)

Level 8	Number 28

1. Walking 18 feet west, 4 feet south, 6 feet west, and 3 feet south in that order ends how many feet from the starting point? ⟨25⟩

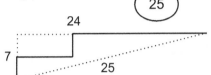

24

7

25

2. Find 4 consecutive odd integers such that the sum of the least and greatest is 100.

x, x + 2, x + 4, x + 6

x + (x + 6) = 100

2x = 94

x = 47

⟨47, 49, 51, 53⟩

3. Find the 45th term of each sequence.

12, 19, 26, 33, . . .   d = 7

12 + (44) (7) = 12 + 308 = ⟨320⟩

32, 40, 48, 56, . . .   d = 8

32 + (44) (8) = 32 + 352 = ⟨384⟩

4. Find the mean and range of the 5 different integers 3x + 6, 6x – 1, 22, 8, and 2x – 8 if their median is 15.

–2	8	15	17	22
2x–4	8	3x+6	6x–1	22

range = 22 – (–2) = ⟨24⟩   Draw 5 slots. Only 3x+6 can be 15. x = 3

mean = 60/5 = ⟨12⟩

5. Solve for n.

$$\frac{(n + 2)!}{(n - 2)!} = 840$$

(n + 2)(n + 1)(n)(n – 1) = 4 x 21 x 10

= 4 x 3 x 7 x 5 x 2

= 7 x 6 x 5 x 4

n = ⟨5⟩

6. Four foam blocks, each 2 m wide by 3 m long by 5 cm thick, contain how many cubic meters of foam?

4 x 2 x 3 x .05

12 x 0.1

⟨1.2⟩

7. Reflect each point over the line y = –x. Then translate as specified.

(9, 2)	(–2, –9)	up 10	(–2, 1)
(–7, 4)	(–4, 7)	right 10	(6, 7)
(–1, –6)	(6, 1)	left 10	(–4, 1)

8. Find the volume between 2 cylinders with concentric bases of diameters 18 and 8, both 10 high.

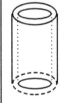

V1 = 9 • 9 • 10 • π

V1 = 81 • 10π

V2 = 4 • 4 • 10 • π

V2 = 16 • 10π

V_{band} = (81–16) • 10π

V_{band} = 65 • 10π = ⟨650π cu un⟩

9. Find the surface area in square units of a square pyramid with base edge 24 and lateral edge 13 units.

A_{sq} = 24 x 24 = 576

A_{tri} = 24 x 5 / 2 = 60

60 x 4 = 240

576
+ 240
⟨816⟩

13 5 12

10. Multiply.

2635_{seven} 1203_{four} 4785_{nine}

x 4_{seven} x 3_{four} x 5_{nine}

14506_{seven}   **10221_{four}**   **26367_{nine}**

| Level 8 | Number 29 |

1. The only even prime is the number ___two.___

Points on the coordinate plane with both x and y integers are called ___lattice points.___

The numbers 1, 8, 27, 64, 125, . . . are called ___perfect cubes.___

6.

A	B	real
12	12	12
11	2	1
10	4	2
9	6	3
8	8	4

4:00 PM

Broken clock A runs at the normal rate but backward. Broken clock B runs forward at twice the normal rate. If both clocks start at noon, find the correct time when both clocks next show the same time.

2. Find the measures of 2 supplementary angles, the lesser of which measures 54° less than the greater.

x + (x − 54) = 180
2x − 54 = 180
2x = 234
x = 117

(63 and 117)

7. A △ has sides 6, 8, 12. A similar △ has greatest side 15. Find the ratio of their:

sides ___4:5___

altitudes ___4:5___

perimeters ___4:5___

areas ___16:25___

3. Graph on the number line, and write the solution to the quadratic inequality.

$(x + 4)(x + 1) \le 0$

no   yes   no

−4   −1 0

($-4 \le x \le -1$)

8. Simplify. Answer in exponential form.

$2^5 \times 2^4$	$(5^5)^4$	$(8^3)^6$
(2^9)	(5^{20})	(8^{18})
$9^9 \div 9^3$	$12^6 \times 12^7$	$6^7 \div 6^5$
(9^6)	(12^{13})	(6^2)

4. Multiply using DPMA.

68 x 55   ___3740___
(60 + 8)(50 + 5) = 3000 + 300 + 400 + 40

52 x 36   ___1872___
(50 + 2)(30 + 6) = 1500 + 300 + 60 + 12

9. Find the number of subsets of each set.

{nor, and, the} 2^3 8

{3, 4, 5} U {4, 5, 6, 7, 8} 2^6 64

{10, 20, 30, 40, 50, 60, 70} 2^7 128

{H, E, L, L, O}   only 1 L 2^4 16

5. Write each repeating decimal as a simplified fraction.

.8̄4̄   8.8̄4̄ $8\frac{4}{9}$ $\frac{76}{9}$ $\frac{76}{90}$ ($\frac{38}{45}$)
x 10   ÷ 10

.0̄7̄8̄   7.8̄ $7\frac{8}{9}$ $\frac{71}{9}$ ($\frac{71}{900}$)
x 100   ÷ 100

10. Find the area and perimeter of the trapezoid with vertices (8, 15), (16, 15), (24, 0), and (0, 0).

A = (240) 16 x 15
P = (66) 34 + 32
B = 24   M = 16
b = 8   h = 15

(8,15)   8   (16,15)
NTS 15 17
8
(0,0)   (24,0)

MAVA Math: Middle Reviews Solutions Copyright © 2013 Marla Weiss

Level 8	Number 30

1. Which fabric has a less costly unit rate per square foot: 5 yards of 60 inches wide for $42.00 or 6 yards of 48 inches wide for $41.04?

$$\frac{42.00}{5 \times 3 \times 5} \quad \text{vs} \quad \frac{41.04}{6 \times 3 \times 4}$$

0.56   0.57

6. Find the circumference of a circle given its area is $36\pi^5$.

$$A = \pi r^2 = 36\pi^5$$
$$r^2 = 36\pi^4$$
$$r = 6\pi^2$$
$$D = 12\pi^2$$
$$C = D\pi$$
$$\boxed{C = 12\pi^3}$$

2. Find the area of the parallelogram bounded by the lines x = –1, x = 4, y = x + 3, and y = x – 3.

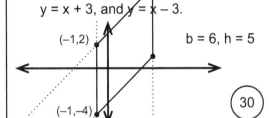

(–1,2)   b = 6, h = 5

(–1,–4)

$\boxed{30}$

7. Simplify.

$$\frac{\overline{.576}}{\overline{.72}} \quad \frac{576}{999} \times \frac{99}{72} \quad \frac{64}{111} \times \frac{11}{8} \quad \boxed{\frac{88}{111}}$$

$$\frac{\overline{.296}}{\overline{.64}} \quad \frac{296}{999} \times \frac{99}{64} \quad \frac{37}{111} \times \frac{11}{8} \quad \boxed{\frac{11}{24}}$$

3. A raw meatball with a 9-inch radius can be re-formed into how many meatballs with a 1-inch radius?

$$\frac{V \dfrac{4\pi \cdot 9 \cdot 9 \cdot 9}{3}}{V \dfrac{4\pi \cdot 1 \cdot 1 \cdot 1}{3}} = 9 \cdot 9 \cdot 9 = \boxed{729}$$

Divide the greater volume by the lesser volume.

8. Picking 2 cards from a standard deck without replacement, find the probability of at least one black card.

1 – P(not B, not B)

faster than P(B,R) + P(R,B) + P(B,B)

$$1 - \frac{\cancel{26}^{\,1}}{\cancel{52}_{\,2}} \cdot \frac{25}{51} = 1 - \frac{25}{102} = \boxed{\frac{77}{102}}$$

4. Find the area of a rectangle given the perimeter is 210 and the ratio of the sides is 5:2.

P = 210   A = 75 x 30
SP = 105
ratio sum = 7   A = $\boxed{2250}$
blow-up = 105/7 = 15   sq un
sides = 75 and 30

9. Solve. $\left| 6x + 5 \right| = 37$

6x + 5 = 37   OR   6x + 5 = –37
6x = 32   6x = –42
$\boxed{x = \dfrac{16}{3}}$ $\boxed{x = -7}$

5. Find the rate that yields $3250 simple interest on $6500 after 5 years.

I = PRT

$3250 = 6500 \cdot R \cdot 5$

$325 = 650 \cdot R \cdot 5$

$65 = 650 \cdot R$

R = .10 $\boxed{10\%}$

10. Find the next term in the sequence by successive differences.

7   7   11   20   37   69 $\boxed{129}$

   0   4   9   17   32   60

     4   5   8   15   28

       1   3   7   13

         2   4   6

| Level 8 | Number 31 |

1. Find the time in minutes needed for 2 people to do a job together given that: Person A can do the job in 2 hours; Person B can the same job in 3 hours.

Rate A = 1/2
Rate B = 1/3
Total Rate = 5/6
Total Time = 6/5

In work problems, rates and times are reciprocals.

$\left(72\right)$

6. Draw the next figure in the pattern.

---

2. Square mentally.

95^2 9025
9 x 10 = 90

405^2 164,025
40 x 41 = 1640

255^2 65,025
25 x 26 = 650
25x25 and add 25

125^2 15,625
12 x 13 = 156
12x12 and add 12

7. Find the shaded area in square units. The semicircles have diameters 8 and 6.

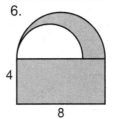

R = 4, r = 3
$32 + (8\pi - 4.5\pi)$

$\left(32 + 3.5\pi\right)$

---

3. Find the volume of the cone in cubic units.

altitude = 15
slant height = 17

$V = \dfrac{8 \cdot 8 \cdot \pi \cdot 15}{3}$

$= 64 \cdot 5 \cdot \pi$

$= \left(320\pi\right)$

8. Simplify.

$64^{\frac{7}{6}}$ $\left(128\right)$ $100^{\frac{1}{2}}$ $\left(10\right)$ $25^{\frac{3}{2}}$ $\left(125\right)$

$81^{\frac{3}{4}}$ $\left(27\right)$ $125^{\frac{2}{3}}$ $\left(25\right)$ $16^{\frac{3}{4}}$ $\left(8\right)$

---

4. Find the space diagonal of a cube with

edge = 8 $8\sqrt{3}$ e = 8

area of face = 49 $7\sqrt{3}$ e = 7

volume = 64 $4\sqrt{3}$ e = 4

face diagonal = $6\sqrt{2}$ $6\sqrt{3}$ e = 6

9. Find the perimeter of a rectangle given the area is 750 and the length is five more than the width.

A = 750   L = 30
LW = 750   SP = 55
(W + 5)W = 750
(30)(25) = 750   P = $\left(110\right)$
W = 25

Trial & error to find factors is easier/faster than solving a quadratic equation.

---

5. Solve.

x(x + 2) = 0
x = $\left(0, -2\right)$

(2x − 5)(9x + 1) = 0
x = $\left(\dfrac{5}{2}, \dfrac{-1}{9}\right)$

(x + 10)(x − 5) = 0
x = $\left(-10, 5\right)$

(3x − 8)(8x + 7) = 0
x = $\left(\dfrac{8}{3}, \dfrac{-7}{8}\right)$

10. How many triangles with whole sides have greatest side 4?

4, 4, 4   4, 3, 3
4, 4, 3   4, 3, 2
4, 4, 2
4, 4, 1

$\left(6\right)$

Other options do not form a triangle. The sum of 2 sides is greater than the 3rd.

| Level 8 | Number 32 |

1. The sum of the dates of all Fridays in a given month is 85. What is the date of the first Friday?

7 + 14 + 21 + 28 = 70 (4 Fridays)
must be 5 Fridays, thus < 7
85/5 = 17 = middle one
17 − 14 = 3   ⟨3rd⟩

6.

■ Bridges   ▨ Roads

Answer items below for the bar graph.

100 75 50 25 0
2005 2006 2007 2008

2. If n is an even integer, label each even (E) or odd (O).

a.) $n \div 2$ CBD   14 ÷ 2 is odd, but 16 ÷ 2 is even.

b.) $3n + 1$ O

c.) $n^2 + 5n$ E

7. Find the average increase as a fraction among the 3 consecutive pairs of years for roads.   change/original

05-06: 25/25 = 1 = 6/6
06-07: 25/50 = 1/2 = 3/6
07-08: 25/75 = 1/3 = 2/6
(11/6) ÷ 3 = 11/18 $\frac{11}{18}$

3. Find the area of an isosceles trapezoid with legs 17 and bases 12 and 28.

12
17 / 15
8   12   8

M = 20
H = 15
A = 20 x 15
= ⟨300⟩

8. Operate and simplify.

$$\frac{4\frac{2}{7} \times 5\frac{4}{9}}{6\frac{1}{9} \div 3\frac{2}{3}} \quad \frac{\frac{30}{7} \times \frac{49}{9}}{\frac{55}{9} \times \frac{3}{11}} \quad \frac{\frac{70}{3}}{\frac{5}{3}}$$

$$\frac{70}{3} \times \frac{3}{5} \quad \text{⟨14⟩}$$

4. Define operation ❋ as:

X ❋ Y = XY + Y − 3X. For example,
3 ❋ 2 = 6 + 2 − 9 = −1.
Find the values.

5 ❋ B = 21   B = 6   5B + B − 15 = 21

M ❋ 4 = 21   M = 17   4M + 4 − 3M = 21

9. If f(x) = 9x − 8 and g(x) = 2x + 5, find g(f(6)) − f(g(3)).

f(6) = 54 − 8 = 46
g(46) = 92 + 5 = 97
g(3) = 6 + 5 = 11
f(11) = 99 − 8 = 91

97 − 91 = ⟨6⟩

5. For the Fibonacci sequence, find the sum of the first 6 "n" such that Fib$_n$ is a prime number.

1, 1, 2, 3, 5, 8, 13, 21, 34, 55, 89, 144, 233

n = 3, 4, 5, 7, 11, 13

⟨43⟩

10. Find the remainder without dividing.

873,951 ÷ 50   873,950 div by 50   R 1

175,879 ÷ 75   175,875 div by 3, 25   R 4

555,159 ÷ 33   555,159 div by 3, 11   R 0

811,935 ÷ 66   811,932 div by 2, 3, 11   R 3

Level 8	Number 33

1.

	time	rate
B	4	$\frac{1}{4}$
R	(12)	$\frac{1}{12}$
tog	3	$\frac{1}{3}$

Bob can do a job alone in 4 hours. Find the time in hours for Rob to do the job alone if together they can complete it in 3 hours.

6. Solve.

$4X + 3Y = 40$
$X - 2Y = -1$

$4X + 3Y = 40$
$4X - 8Y = -4$
———————
$11Y = 44$
$(Y = 4)$

$X - 8 = -1$
$(X = 7)$

$-8Y = 40$
$2X - 3Y = 47$
$(Y = -5)$
$2X + 15 = 47$
$2X = 32$
$(X = 16)$

---

2. Find the formula that generates the pattern 5, 12, 21, 32, 45, ... Then find the 20th term using the formula.

1	5
2	12
3	21
4	32
5	45

$(n^2 + 4n)$

$20^2 + 4(20)$
$400 + 80$

(480)

7. Label the lines perpendicular, parallel, or neither.

$y = 6x - 5$
$2y = 12x + 3$

$(parallel)$

equal slopes

$4x + 9y = 3$
$-9x + 4y = 7$

$(perpendicular)$

negative reciprocal slopes

---

3. Complete for the geometric sets.

A   B   C   D

$\overline{AB} \cup \overline{BC} =$ $\underline{AC}$ | $\overline{AC} \cup \overline{CD} =$ $\underline{AD}$

$\overline{AC} \cap \overline{BD} =$ $\underline{BC}$ | $\overline{AD} \cap \overline{BC} =$ $\underline{BC}$

8. Write in scientific notation.

$(9 \times 10^8) - (7 \times 10^7)$

$(90 \times 10^7) - (7 \times 10^7)$

83×10^7

Can only add like terms.

(8.3×10^8)

$(2 \times 10^9) \div (5 \times 10^3)$

0.4×10^6

(4.0×10^5)

---

4. Define the binary operation ~ as:

~X~ = 4X − 3. For example,
~6~ = 24 − 3 = 21.
Find the values.

~(~5~)~ $\underline{65}$ Do ~5~ = 17 first.

~(~0.5~)~ $\underline{-7}$ Do ~0.5~ = −1 first.

9. Operate.

$\frac{5}{2a} \overset{5c}{\underset{}{-}} \frac{2}{5c} \overset{2a}{}$

$(\frac{25c - 4a}{10ac})$

$\frac{3a}{8} \cdot \frac{b}{6c} \cdot \frac{4c}{5d}$

$(\frac{ab}{20d})$

---

5. Find the points of trisection of the line segment with the given endpoints.

(3, 50) & (36, 20)	$(14, 40)$	$(25, 30)$
(−6, 5) & (18, 2)	$(2, 4)$	$(10, 3)$
(14, −9) & (29, 6)	$(19, -4)$	$(24, 1)$
(−12, 0) & (9, −9)	$(-5, -3)$	$(2, -6)$

10. The 3 outer triangles are equilateral. Two of their areas are given. Find the 3rd area. (NTS)

A = 20

9

A = 11

Solve for c^2 and b^2 using formula for area of equilateral Δ. Subsitute into $a^2 + b^2 = c^2$. Get 3rd A.

Level 8	Number 34

1. Draw a Venn Diagram to show the relationship among the sets.

A = {2, 4, 6, 8}
B = {2, 5, 6, 9}
C = {1, 3}

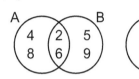

2. In a set of 5 whole numbers, the unique mode is 3 more than the mean. Four of the numbers are 34, 34, 36, and 39. Find the missing number.

(12)   34   34   36   39

mode = 34; mean = 31; sum = 31 x 5 = 155
34 + 34 + 36 + 39 = 143
155 − 143 = 12

3. Find x degrees in the crook picture. (NTS)

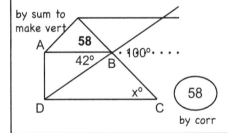

65° by corr
35 by alt int
(100)

4. Given trapezoid ABCD. Find x. (NTS)

by sum to make vert

58
42° B · 100° · · · ·
x°
(58)
by corr

5. The measure of one angle of a parallelogram is 4 less than 7 times another. Find the 4 angles.

180 − A = 7A − 4
184 = 8A
A = 23

(23, 157, 23, 157)

6. Find the next letter(s) in each pattern.

Z, e, i, y, F, j, X, g, __k__

Three sequences interleave. The 3rd is consecutive small letters starting with i.

c, Q, r, e, U, v, g, __Y__ __z__ __i__

In positions 1, 4, 7, 10 are small letters that skip one. Interleaving are 2 consecutive letters that skip two and start with a capital.

7. Find the area of the square inscribed in a circle of area 36π square units.

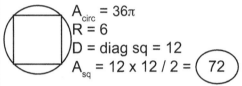

A_{circ} = 36π
R = 6
D = diag sq = 12
A_{sq} = 12 x 12 / 2 = (72)

A square is a rhombus. Use the area formula: half the product of the diagonals.

8. H:O:T = 7:8:20   T:I:P = 25:9:21
Find H:P.

Common variable is T.
LCM(20,25) = 100
H:O:T = 35:40:100 (x5)
T:I:P = 100:36:84 (x4)
H:P = 35:84 = 5:12 (÷7)

(5:12)

9. Max has 8 more dimes than quarters. The coins are worth $4.30. How many total coins does he have?

D = Q + 8
10D + 25Q = 430
10(Q + 8) + 25Q = 430
10Q + 80 + 25Q = 430
35Q = 350
Q = 10   D = 18

(28)

10. Find the area of the triangle with m∠A = 45°.

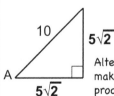

A = $\frac{(5\sqrt{2})(5\sqrt{2})}{2}$

A = (25)

Alternate method: Draw 2 lines to make a square. Use half-the-product-of-the-diagonals formula.
A square = 50; A △ = 50/2 = 25.

MAVA Math: Middle Reviews Solutions Copyright © 2013 Marla Weiss

Level 8	Number 35

1. If v movers can pack 1 room in d days, how many movers are needed to pack m rooms in 31 days?

movers	rooms	days
v	1	d
dv	1	1
$\dfrac{dv}{31}$	1	31
$\boxed{\dfrac{dmv}{31}}$	m	31

6. Find the sum of three primes whose product is:

935	2431
5 x 187	11 x 221
5 x 11 x 17	11 x 13 x 17
⭕ 33	⭕ 41

2. Find BD − AC.

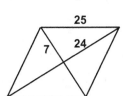

A   B   C   D

−4   0   4

7 of length 4 − 8 of length 4   ⭕ −4

7. Two similar triangles have sides 4, 8, 9 and 6, 12, 13.5. Find the ratio of their:

sides	2:3
altitudes	2:3
perimeters	2:3
areas	4:9

3. Find the area of the rhombus in square units given perimeter P = 100 and diagonal D = 48.

25, 24, 7

P = 100   A = 84 x 4
s = 25   or 48 x 7
D = 48
D/2 = 24   A = ⭕ 336
7, 24, 25
d = 14

8. Find the angle in degrees formed by clock hands at 1:20.

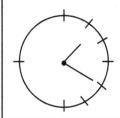

360°/12 = 30° between neighboring numbers

60 + 20 = ⭕ 80

Hour hand is 1/3 of way from 1 to 2. 2/3 is in angle.

4. Identify the net.

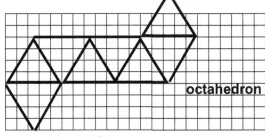

octahedron

9. Find the square root in the given mod.

square root of 9 mod 10	7
square root of 3 mod 4	none
square root of 1 mod 5	1, 4
square root of 2 mod 7	3, 4

3 x 3 = 9
9 − 7 = 2
and
4 x 4 = 16
16 − 14 = 2

5. Find the space diagonal of the rectangular prism with the given dimensions.

2 by 7 by 8 $3\sqrt{13}$ $\sqrt{4 + 49 + 64}$

3 by 6 by 9 $3\sqrt{14}$ $\sqrt{9 + 36 + 81}$

10. Multiply.

7328_{nine} 3042_{five} 6714_{eight}
x 4_{nine} x 3_{five} x 5_{eight}
32425_{nine}   **14231_{five}**   **42374_{eight}**

| Level 8 | Number 36 |

---

1. If the volume of a cube is increased by 700%, what is the percent increase in the edge of the cube?

Original cube: 1 by 1 by 1 V = 1
New cube: 2 by 2 by 2 V = 8

Volume change/original = 7/1 = 700%

Edge change/original = 1/1 = (100%)

---

6. Find all digits d such that the 5-digit number 1d,582 is divisible by 3.

1 + 5 + 8 + 2 = 16   (2, 5, 8)

Find all digits x such that the 4-digit number x564 is divisible by 4.

4 divides 64. A 4-digit   (1, 2, 3, 4, 5, 6, 7, 8, 9)
number cannot start with 0.

---

2. Write the contrapositive of the conditional statement. Are they logically equivalent?   (YES)

If a number is divisible by 10, then the number is not a prime.   Both are TRUE.

If a number is a prime, then the number is not divisible by 10.

---

7. Find the number of integer values that satisfy:

$|2x - 7| \le 19$ 20   −6 $\longrightarrow$ 13

$|x + 7| < 20$ 39   −26 $\longrightarrow$ 12

$|3x + 3| < 18$ 11   −6 $\longrightarrow$ 4

---

3. Multiply using the difference of 2 squares.

64 x 56	3584	$(60 + 4)(60 - 4)$ 3600−16
38 x 42	1596	$(40 - 2)(40 + 2)$ 1600−4
85 x 95	8075	$(90 - 5)(90 + 5)$ 8100−25
71 x 69	4899	$(70 + 1)(70 - 1)$ 4900−1

---

8. Show that primes of the form 4k+1 equal the sum of 2 perfect squares.

73 (k=18)	9 + 64	109 (k=27)	9 + 100
89 (k=22)	25 + 64	113 (k=28)	49 + 64
97 (k=24)	16 + 81	137 (k=34)	16 + 121
101 (k=25)	1 + 100	149 (k=37)	49 + 100

---

4. Randomly select two angles of a non-rectangular parallelogram.   C(4,2) = 6

Find the probability that they are congruent or supplementary.   1

Find the probability that they are congruent.   2/6 $\dfrac{1}{3}$

Find the probability that they are supplementary.   4/6 $\dfrac{2}{3}$

---

9. Find the value using fractions.

$83\dfrac{1}{3}$ % of $87\dfrac{1}{2}$ % of 40% of 792

$\dfrac{5}{6} \times \dfrac{7}{8} \times \dfrac{2}{5} \times$ ~~792~~

33 / ~~99~~ / 3

33 x 7 = 210 + 21 = (231)

---

5. The ratio of an angle to its supplement is 4:5. Find the ratio of the angle to its complement.

4 + 5 = 9
180 ÷ 9 = 20
angle = 80 (supplement = 100)
complement = 10   (8:1)

---

10. Find the volume in cubic units of a rectangular pyramid with length 21, width 13, and altitude 10 units.

$\dfrac{21 \times 13 \times 10}{3}$ = 7 x 13 x 10 = 91 x 10

= (910)

Level 8	Number 37

1. Find the volume of each rectangular prism in cubic feet.

4 yards by 5 yards by 2 yards

12 feet by 15 feet by 6 feet (1080)

12 x 15 = 180; 180 x 6 = 600 + 480

36 inches by 48 inches by 96 inches

3 feet by 4 feet by 8 feet (96)

6. Find the volume in cubic units of a triangular prism with base edges 7, 24, and 25 and altitude 11 units.

A base = 7 x 24 / 2 = 7 x 12 = 84

V = 84 x 11 = (924)

2. Point B divides segment $\overline{AC}$ such that AB:BC = 5:2. AC = 28. Find BC.

A   B   C

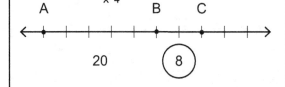

20   (8)

7. Answer YES or NO as to whether the set is closed for the given operation.

reals   square rooting   NO

The square root of a negative is imaginary.

naturals   division   NO

{−1, 0, 1}   addition   1 + 1 = 2   NO

3. Multiply the matrices.

$$\begin{bmatrix} 9 & 2 \\ 3 & 8 \end{bmatrix} \times \begin{bmatrix} 4 & 7 \\ 1 & 5 \end{bmatrix} = \begin{bmatrix} \mathbf{38} & \mathbf{73} \\ \mathbf{20} & \mathbf{61} \end{bmatrix}$$

36 + 2 = 38
63 + 10 = 73
12 + 8 = 20
21 + 40 = 61

8. Find the perimeter of an isosceles trapezoid with height 30, legs 34, and small base 25.

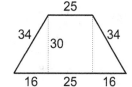

34 + 34 = 68
50 + 32 = 82
P = (150)

4. Find the probability of drawing yellow, green, yellow successively without replacement from a bowl containing 10 green, 14 yellow, 16 red, and 25 blue marbles.

$$\frac{\overset{1}{\cancel{14}}}{\underset{1}{65}} \cdot \frac{\overset{1}{\cancel{10}}}{\underset{16}{64}} \cdot \frac{\overset{1}{\cancel{13}}}{\underset{9}{63}} \quad \left(\frac{1}{144}\right)$$

9. Complete.

70 ≡ 10 (mod 15)	63 ≡ 3 (mod 10)
45 ≡ 20 (mod 25)	49 ≡ 9 (mod 20)
70 ≡ 0 (mod 35)	98 ≡ 8 (mod 30)
56 ≡ 11 (mod 45)	77 ≡ 37 (mod 40)

5. Find the probability that three of the numbers 4, 5, 7, or 8 chosen randomly form the sides of a triangle.

4, 5, 7   YES
4, 5, 8   YES
4, 7, 8   YES
5, 7, 8   YES   (1)

10. How many numbers from 1 to 100 inclusive are multiples of 3 or 5?

100 ÷ 3 = 33
100 ÷ 5 = 20
100 ÷ 15 = 6
33 + 20 − 6 = (47)

Subtract the overlap: the multiples of 3 AND 5.

Level 8	Number 38

1. Because dividing by zero is impossible, it is _undefined._

To prove a statement false, give only one _counterexample._

The only field property that uses two operations is _DPMA._

6. Find the measure of one angle of each regular polygon.

Interior and exterior angles are supplements.

15-gon	156	ext < = 360/15 = 24 int < = 180 - 24 = 156
36-gon	170	ext < = 360/36 = 10 int < = 180 - 10 = 170
octagon	135	ext < = 360/8 = 45 int < = 180 - 45 = 135

Or 13 x 180 / 15 = 156

2. Find four consecutive integers with the given average.

16	CBD	27.5	26, 27, 28, 29
30.5	29, 30, 31, 32	41.5	40, 41, 42, 43
18.5	17, 18, 19, 20	52.5	51, 52, 53, 54

7. Simplify using DPMA.

$251 \times 209 - 251 \times 9$ $251 \times (209 - 9)$
251×200
$\boxed{50,200}$

15% of $176 - 15\%$ of 36 $15\% (176 - 36)$
15% of 140
$\boxed{21}$

$35^2 - 35 \times 24$ $35(35 - 24)$
35×11 $\boxed{385}$

3. Find the perimeter of a rectangle given the area is 320 and the length is four more than the width.

$A = 320$ $L = 20$
$LW = 320$ $SP = 36$
$(W + 4)W = 320$
$(20)(16) = 320$ $P = \boxed{72}$
$W = 16$

Trial & error to find factors is easier/faster than solving a quadratic equation.

8. Four years from now the sum of Mia and Tia's ages is 95. Mia's age now is 9 less than Tia's age 6 years ago. Find Mia's age 6 years ago.

	−6	now	+4
M	⟨36⟩	M	M+4
T	M+9	M+3	M+7

$M+4+M+7 = 95$
$2M + 11 = 95$
$2M = 84$
$M = 42$
Mia now 42

4. The measures of the interior angles of a quadrilateral are x, 3x, 2x + 30, and 2x − 10. Find the sum of the least and greatest angles.

$x + 3x + 2x + 30 + 2x - 10 = 360$
$8x + 20 = 360$
$8x = 340$
$x = 42.5$ $\boxed{170}$
least = x greatest = 3x

9. Operate.

$(-2)^4 - (-2)^3 - (-2)^2$
$16 + 8 - 4$ $\boxed{20}$

$-2^5 - (-2)^5 - (-2^5)$
$-32 + 32 + 32$ $\boxed{32}$

$(-2)^3 + 2^6 - (-6)^2 - 2^3$
$-8 + 64 - 36 - 8$ $\boxed{12}$

$(-7)^2 + (-2^6) - (-2)^3$
$49 - 64 + 8$ $\boxed{-7}$

5. Multiply mentally.

374 x 1001	374,374
28 x 101	2828
56 x 10,001	560,056
93 x 1001	93,093

Picture 101 as (100 + 1), 1001 as (1000 + 1), and 10,001 as (10,000 + 1).

10. A regular hexagon has perimeter 48 units. Find its area in square units.

$P = 48$
$S = 8$
$A = \dfrac{6 \cdot 8^2 \sqrt{3}}{4} = \boxed{96\sqrt{3}}$

Level 8	Number 39

1. Find the number of ways to park 5 different colored cars in a row if the red one must be at one end and the blue one must be at the other end.

$$\underline{2} \cdot \underline{3} \cdot \underline{2} \cdot \underline{1} \cdot \underline{1} \quad \textcircled{12}$$

↑
R or B 1st
↑
R or B not used 2nd

6. Find the positive geometric mean between each pair. Show the proportion.

4 and 25	4 and 9	6 and 24
$\dfrac{4}{10} = \dfrac{10}{25}$	$\dfrac{4}{6} = \dfrac{6}{9}$	$\dfrac{6}{12} = \dfrac{12}{24}$
⑩	⑥	⑫

2. Find the length of the segment joining the trisection points of segment $\overline{AB}$.

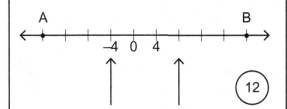

A   B
−4 0 4

↑   ↑

⑫

7. Translate the cipher with each letter as its own digit.

```
  SWAM      SW23      5823
x    4    x    4    x    4
-----     ------    ------
AMARA     232R2     23292
```

M≠0. M≠1. M≠2 or else R=M. Try M=3. Then A=2, R=9, W=8, and S=5.

3. The area of rectangle ABCD is 4 times the area of rectangle BCFE. Find the coordinates of E. (NTS) ⟨(21, 12)⟩

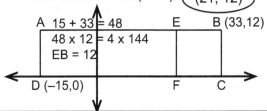

A 15 + 33 = 48   E   B (33,12)
48 x 12 = 4 x 144
EB = 12

D (−15,0)   F   C

8. Reflect each point over the line y = −x. Then translate as specified.

(1, 7)	(−7, −1)	down 5	(−7, −6)
(−4, 0)	(0, 4)	right 3	(3, 4)
(−9, −5)	(5, 9)	up 6	(5, 15)

4. Evaluate the functions.

SQRT(49)	7	TRUNC(4.3)	4
SQR(4)	16	INT(1.572)	1
ABS(4 − 5)	1	SIGMA(9)	3
SGN(13)	1	INT(−0.13)	−1

9. Find the angles of parallelogram ABCD given:

m∠A = 4x − 3   m∠B = 6x + 23

4x − 3 + 6x + 23 = 180
10x + 20 = 180
10x = 160
x = 16

⟨m∠A, C = 61°
m∠B, D = 119°⟩

Consecutive angles are supplementary. Opposite angles are congruent.

5. Find the mean and range of the 5 different integers 2x + 3, −1, x + 3, −3, and 2x − 2 if their median is 13.

−3	−1	13	18	23
−3	−1	x+3	2x−2	2x+3

range = 23 − (−3) = ㉖

mean = 50/5 = ⑩

Draw 5 slots. Only x+3 can be 13. x = 10

10. Find the area of the triangle with m∠A = 30°. (NTS)

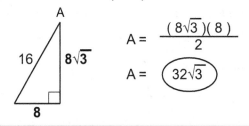

A
16 / 8√3
8

$$A = \frac{(8\sqrt{3})(8)}{2}$$

$$A = \boxed{32\sqrt{3}}$$

MAVA Math: Middle Reviews Solutions Copyright © 2013 Marla Weiss

Level 8	Number 40

1. The average of 4 tests is 88. What additional test would raise the average to 90?

$$\frac{352}{4} = 88 \qquad \frac{450}{5} = 90 \qquad \begin{array}{r} 450 \\ -\ 352 \\ \hline \textcircled{98} \end{array}$$

6. Two boats leave from the same dock at 11:00 AM. One travels north at 32 mph; the other travels east at 60 mph. When will they be 170 miles apart?

8, 15, 17 x 10
80, 150, 170

D = RT =
32 x 2.5 = 80
D = 170
D = RT = 60 x 2.5 = 150

11:00 + 2.5 hr
$\textcircled{1:30 PM}$

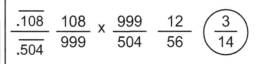

2. Draw the reflection of each trapezoid over the line y = x. Reflect vertices first. (x,y) becomes (y,x).

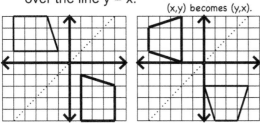

7. Simplify.

$$\frac{.\overline{108}}{.\overline{504}} \quad \frac{108}{999} \times \frac{999}{504} \quad \frac{12}{56} \quad \textcircled{\frac{3}{14}}$$

$$\frac{.\overline{315}}{.\overline{63}} \quad \frac{315}{999} \times \frac{99}{63} \quad \frac{35}{111} \times \frac{11}{7} \quad \textcircled{\frac{55}{111}}$$

3. Find x degrees in the crook picture. (NTS)

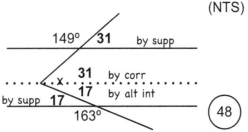

149° 31 by supp

31 by corr
x
17 by alt int
by supp 17
163° $\textcircled{48}$

8. Find the perimeter of an isosceles trapezoid with height 24, legs 26, and small base 16.

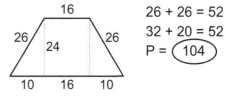

26 + 26 = 52
32 + 20 = 52
P = $\textcircled{104}$

16
26   26
24
10   16   10

4. Define operation ❀ as:

X ❀ Y = 4X – 2Y + 5. For example,
3 ❀ 4 = 12 – 8 + 5 = 9.
Find the values.

(5 ❀ 3) ❀ 6 ___69___   5 ❀ 3 = 19

1 ❀ (7 ❀ 2) ___–49___   7 ❀ 2 = 29

9. Answer YES or NO if the matrices may be multiplied. If yes, give the dimensions of the product.

2 by 4 times 2 by 5 _____ NO

5 by 3 times 7 by 3 _____ NO

8 by 1 times 1 by 2 ____ YES 8 by 2

5. If f(x) = 6x + 6 and g(x) = 3x – 7, find g(f(0)) • f(g(1)).

f(0) = 0 + 6 = 6
g(6) = 18 – 7 = 11
g(1) = 3 – 7 = –4
f(–4) = –24 + 6 = –18

11 • (–18) = $\textcircled{–198}$

10. Multiply as two binomials.

$$8\frac{2}{5} \times 3\frac{3}{4} \qquad \qquad 6\frac{3}{7} \times 7\frac{7}{10}$$

F =	24	F =	42
O =	6	O =	4.2
I =	1.2	I =	3
L =	0.3	L =	0.3

$\textcircled{31.5}$ $\textcircled{49.5}$

Level 8	Number 41

1. In a sequence, every odd-numbered term including the first is 2 and every even-numbered term including the second is –2. What is the sum of the first 99 terms?

2, –2, 2, –2, 2, –2, . . .
The first 98 terms sum to 0. (2)

6.

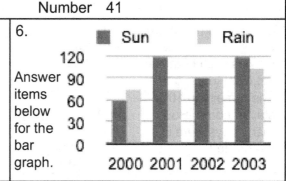

Answer items below for the bar graph.

7. Over the 4 years, find the sum of the values of Rain as a fraction of the sum of the values of Sun.

Rain: 75 + 75 + 90 + 105 = 345
Sun: 60 + 120 + 90 + 120 = 390

$\dfrac{345}{390}$ $\dfrac{69}{78}$ $\left(\dfrac{23}{26}\right)$

2. In a set of 5 numbers, the mean, median, and mode are all equal. Four of the numbers are 25, 50, 30, and 15. Find the missing number.

15   25   30   (30)   50

Of the 4 given numbers, try 30 because it is their mean.

3. Find the area of the trapezoid bounded by the lines y = –x, x = 3, y = 2 and the y-axis.

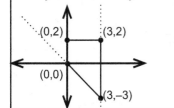

b = 2, B = 5,
M = 3.5, h = 3

(10.5)

8. Given the perimeter of a rectangle and the upper bound of the width. Find the range of values for the length.

P = 84; max W < 31	P = 58; max W < 16
SP = 42	SP = 29
0 < W < 31	0 < W < 16
(11 < L < 42)	(13 < L < 29)

4. Find the endpoint of the line segment.

Endpoint	Midpoint	Endpoint
(1.5, 13)	(4.5, 22)	**(7.5, 31)**
(–9, 3.2)	(–2, 5)	(5, 6.8)
(7.3, 16)	(7.4, 21)	**(7.5, 26)**
(–9, –3)	(–7, –8)	(–5, –13)

9. Name a lattice point C in the picture that would yield △ABC with area 8.

B   1 box = 1 unit

(5, 0)

A

(0,0)   C   4 x 4 / 2

5. Find the probability that three of the numbers 2, 4, 6, or 10 chosen randomly form the sides of a triangle.

2, 4, 6   NO
2, 4, 10   NO
2, 6, 10   NO
4, 6, 10   NO

(0)

10. Answer YES or NO as to whether the 3 numbers form sides of a right triangle.

2.5, 6, 6.5 x2 Y	4, 7.5, 9.5 N
1.6, 3, 3.4 x5 Y	1, 2.4, 2.6 x5 Y
3.1, 4.1, 5.1 N	2.2, 12, 12.2 x5 Y

174

Level 8	Number 42

1. One third of a group of boys went to the movies. Half of those left went to the game. The remaining 10 stayed home. How many boys were in the group?

B

$\dfrac{B}{3}$ $\dfrac{2B}{3}$

$\dfrac{B}{3}$ $\dfrac{B}{3}$ = 10

30

6. Calculate the sum of the arithmetic sequence.

14 + 16 + 18 + . . . + 92 + 94 + 96

7x2 8x2 9x2   48x2
1 to 48 is 48 numbers; omit 1–6; n = 42
F + L = 14 + 96 = 110
110 x 42 / 2 = 110 x 21 = 210 x 11

2310

2. X varies jointly as Y and Z. X = 450 when Y = 10 and Z = 15. Find Z when X = 540 and Y = 9.

X = kYZ

450 = (10)(15)k

k = 3

540 = (3)(9)Z

Z = 20

7. A △ has sides 6, 8, 12. A similar △ has greatest side 42. Find the ratio of their:

sides   2:7

altitudes   2:7

perimeters   2:7

areas   4:49

3. Rewrite the statement using DeMorgan's Laws.

I do not see a rainbow or a shadow.

I do not see a rainbow and I do not see a

shadow.

8. Calculate the combinations.

$C(10, 7)$ $= \dfrac{10!}{7! \ 3!}$ $= \dfrac{10 \cdot 9 \cdot 8}{3 \cdot 2 \cdot 1}$ $= 120$

$C(8, 2)$ $= \dfrac{8!}{2! \ 6!}$ $= \dfrac{8 \cdot 7}{2 \cdot 1}$ $= 28$

$C(9, 8)$ $= \dfrac{9!}{8! \ 1!}$ $= 9$

4. A raw meatball with a 6-inch radius can be re-formed into how many meatballs with a 1-inch radius?

$\dfrac{\dfrac{4\pi \cdot 6 \cdot 6 \cdot 6}{3}}{\dfrac{4\pi \cdot 1 \cdot 1 \cdot 1}{3}}$ = 6 · 6 · 6 = 216

Divide the greater volume by the lesser volume.

9. Find the total value after compounding annually.

principal	rate	years	value	
$5000	4%	2	**$5408**	5200
$6000	10%	3	**$7986**	6600, 7260
$5000	6%	2	**$5618**	5300
$8000	5%	3	**$9261**	8400, 8820

5. Find the surface area of the cone in square units.

diameter = 14   SA = 7 · 7 · π
slant height = 25   + 7 · 25 · π
                    = 49π + 175π
                    = 224π LA = "πrl"

10. Find the remainder without dividing.

562,179 ÷ 77 562,177 div by 7, 11   R 2

216,189 ÷ 60 216,180 div by 3, 4, 5 R 9

184,106 ÷ 50 184,100 div by 2, 25   R 6

432,135 ÷ 99 432,135 div by 9, 11   R 0

MAVA Math: Middle Reviews Solutions Copyright © 2013 Marla Weiss

Level 8	Number 43

1. Find the angle such that the sum of the measures of its complement and supplement is 230°.

$(90 - x) + (180 - x) = 230$
$270 - 2x = 230$
$40 = 2x$
$x = \boxed{20}$

6. Draw the next figure in the pattern.

Black moves clockwise. Gray moves counterclockwise.

2. Find the perimeter of the rhombus with diagonals 30 and 72.

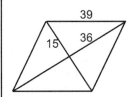

5, 12, 13 x 3
15, 36, 39
P = 4 x 39
P = $\boxed{156}$

7. B:I:G = 20:12:9   B:O:Y = 16:15:8
Find G:Y.

Common variable is B.
LCM(20,16) = 80
B:I:G = 80:48:36 (x4)
B:O:Y = 80:75:40 (x5)
G:Y = 36:40 = 9:10 (÷4)

$\boxed{9:10}$

3. Find the area of an isosceles trapezoid with legs 25 and bases 20 and 34.

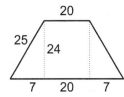

M = 27
H = 24
A = 24 x 27
  = 480 + 168
  = $\boxed{648}$

8. How many kg of pure acid must be added to 60 kg of a 70% acid solution to produce an 80% acid solution?

$(60) + (x) = (60 + x)$
$(.7)(60) + (1)(x) = (.8)(60 + x)$ $\boxed{30}$
$(7)(60) + (10)(x) = (8)(60 + x)$
$420 + 10x = 480 + 8x$
$2x = 60$
$x = 30$

4. Multiply using DPMA.

92 x 34   3128   Add 300 to 2700.
                   Add 128.
$(90 + 2)(30 + 4) = 2700 + 360 + 60 + 8$

54 x 38   2052   Add 400 and 100 to 1500.
                   Add 52.
$(50 + 4)(30 + 8) = 1500 + 400 + 120 + 32$

9. Find the slope given 2 points on a line.

Point #1	Point #2	Slope
(−1, 12)	(−3, 18)	−3
(−2, −13)	(2, −5)	2
(11, −7)	(13, −15)	−4
(−8, 5)	(−7, 4)	−1

5. Find the mean of $22x + 9$, $16x + 8$, $10x - 5$, and $8x - 4$.

$$\frac{56x + 8}{4} = \boxed{14x + 2}$$

10. Find x. (NTS)

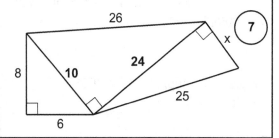

Level 8	Number 44

1. Find 2 consecutive even integers such that the twice the greater is 20 less than triple the lesser.

$x, x + 2$

$2(x + 2) = 3x - 20$

$2x + 4 = 3x - 20$

$24 = x$

$\boxed{24, 26}$

6. Of 90 students, 40 study French, 60 study Spanish, and 5 study neither. How many study both?

F=40   S=60

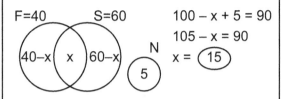

40–x   x   60–x   N

5

$100 - x + 5 = 90$

$105 - x = 90$

$x = \boxed{15}$

2. Multiply using the difference of 2 squares.

63 x 57   3591 $(60 + 3)(60 - 3)$ 3600-9

72 x 68   4896 $(70 + 2)(70 - 2)$ 4900-4

41 x 39   1599 $(40 + 1)(40 - 1)$ 1600-1

54 x 46   2484 $(50 - 4)(50 + 4)$ 2500-16

7. Label the lines perpendicular, parallel, or neither.

$y = 8x - 7$ $y = 6x + 3$
$y = 7x - 8$ $y = 6x + 2$

$\boxed{neither}$ $\boxed{parallel}$

equal slopes

3. Find the volume of the cone in cubic units.

radius = 7
slant height = 25

$V = \dfrac{7 \cdot 7 \cdot \pi \cdot 24}{3}$

$= 49 \cdot 8 \cdot \pi$

$= \boxed{392\pi}$

$50 \times 8 = 400$
$400 - 8 = 392$

8. Convert 698 in base ten to base sixteen.

	2	B	A
698	256	16	1
− 512
186
− 176
10

$\boxed{2BA_{sixteen}}$

4. Find the area of a rectangle given the perimeter is 192 and the ratio of the sides is 7:5.

P = 192   A = 56 x 40
SP = 96
ratio sum = 12   A = $\boxed{2240}$
blow-up = 96/12 = 8
sides = 56 and 40

9. Multiply and simplify.

$\sqrt{12} \cdot 2\sqrt{33} \cdot \sqrt{121} \cdot \sqrt{75}$

$2 \cdot \sqrt{3} \cdot 2 \cdot \sqrt{3} \cdot \sqrt{11} \cdot 11 \cdot 5 \cdot \sqrt{3}$

$2 \cdot 2 \cdot 3 \cdot 5 \cdot 11 \cdot \sqrt{33}$

$60 \cdot 11 \sqrt{33}$ $\boxed{660\sqrt{33}}$

5. The ratio of an angle to its supplement is 5:7. Find the ratio of the angle to its complement.

$5 + 7 = 12$
$180 \div 12 = 15$
angle = 75 (supplement = 105)
complement = 15 $\boxed{5:1}$

10. How many triangles with whole sides have greatest side 7?

7, 7, 7   7, 6, 6   7, 5, 5 $\boxed{16}$
7, 7, 6   7, 5, 4
                       7, 5, 3
          7, 6, 2   7, 4, 4
7, 7, 1

Other options do not form a triangle. The sum of 2 sides is greater than the 3rd.

| Level 8 | Number 45 |

1. Find the time in minutes needed for 2 people to do a job together given that: Person A can do the job in 1 hour; Person B can the same job in 3 hours.

Rate A = 1
Rate B = 1/3   *In work problems, rates and times are reciprocals.*
Total Rate = 4/3
Total Time = 3/4   ⟨45⟩

6. Find twin primes with sum equal to:

12	5, 7
60	29, 31
204	101, 103
360	179, 181

Because twin primes are 2 apart, halve the sum to get the median.

2. Is each the graph of a function?

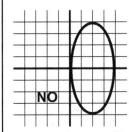 **NO**   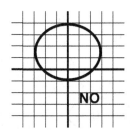 **NO**

7. 19 quarters, 17 dimes, and 155 pennies are what percent of $50?

19 x .25 = 4.75   *Use mental math for the multiplications.*
17 x .10 = 1.70
155 x .01 = 1.55
sum = 8.00

$\dfrac{8}{50}$ = ⟨16%⟩

3. Graph on the number line, and write the solution to the quadratic inequality.

$(x + 2)(x - 3) > 0$

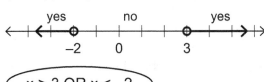

yes   no   yes
−2   0   3

⟨ x > 3 OR x < −2 ⟩

8. Determine if the triangle with the 3 sides is acute or obtuse.

5, 10, 12 | 5, 6, 7

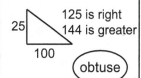 25 125 is right 144 is greater 100 ⟨obtuse⟩

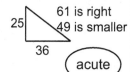 25 61 is right 49 is smaller 36 ⟨acute⟩

4. Find the space diagonal of a cube with

edge = 2 $2\sqrt{3}$ e = 2
area of face = 25 $5\sqrt{3}$ e = 5
volume = 27 $3\sqrt{3}$ e = 3
face diagonal = $7\sqrt{2}$ $7\sqrt{3}$ e = 7

9. Find the square root in the given mod.

square root of 4 mod 11   2, 9
square root of 9 mod 11   3, 8
square root of 1 mod 11   1, 10
square root of 2 mod 6   none

3 x 3 = 9 and 8 x 8 = 64 64 − 55 = 9

5. Find the sum of the greatest 2-digit number and the least 3-digit number, each with exactly 3 factors.

$7^2 + 11^2 = 49 + 121 =$ ⟨170⟩

Only the square of a prime has exactly 3 factors.

10. A, B, C, D, and E are consecutive points on a line. AB:BC = 1:1, BC:CD = 1:3, and CD:DE = 2:5. Find AB:CE.

AB	1		2
BC	1	1	2
CD		3	2	6
DE			5	15

⟨2:21⟩

Level 8	Number 46

1. Walking 6 feet east, 15 feet north, 5 feet east, and 45 feet north in that order ends how many feet from the starting point?

(61)

61 · 60 · 11

6. A concrete deck 24 m wide by 35 m long by 10 cm thick contains how many cubic meters of concrete?

10 cm = .1 m = 1/10 m

$24 \cdot 35 \cdot \dfrac{1}{10} = 12 \cdot 7 = $ (84)

2. Let n be an even integer. Half of n plus three-fourths of the next consecutive even integer equals 34. Find the odd integer between the 2 even integers.

26, 28
13 + 21 = 34   (27)

7. Solve.

$6\sqrt{x} = 3x$
$4x = x^2$
$0 = x^2 - 4x$
$0 = x(x-4)$
(x = 0, 4)

$3x^2 = 1125$
$x^2 = 375$
$x = \pm\sqrt{375}$
$x = \pm 5\sqrt{15}$

3. Define operation ❀ as:

X ❀ Y = 2Y + 3X. For example,
5 ❀ 4 = 8 + 15 = 23.
Find the values.

4 ❀ A = 80   A = 34   2A + 12 = 80

B ❀ 9 = 33   B = 5   18 + 3B = 33

8. Find the area of a circle given its circumference is $8\pi^4$.

$C = D\pi = 8\pi^4$
$D = 8\pi^3$
$r = 4\pi^3$
$A = \pi r^2$
$A = \pi(4\pi^3)(4\pi^3)$
$A = 16\pi^7$

4. Find the surface area of the cone in square units.   LA = "πrl"

radius = 8
slant height = 17

17, 15, 8

$SA = 8 \cdot 8 \cdot \pi$
$+ 8 \cdot 17 \cdot \pi$
$= 64\pi + 136\pi$
$= (200\pi)$

9. Solve. $|8x - 7| = 65$

8x − 7 = 65   OR   8x − 7 = −65
8x = 72   8x = −58
x = 9   x = −29/4

5. The measure of one angle of a parallelogram is 12 more than 3 times another. Find the 4 angles.

180 − A = 12 + 3A
168 = 4A
A = 42
(42, 138, 42, 138)

10. Find the area and perimeter of the trapezoid with vertices (3, 5), (8, 17), (22, 17), and (27, 5).

A = (228) 19 x 12
P = (64) 26 + 38

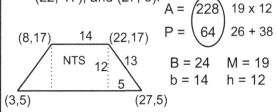
(8,17) 14 (22,17) NTS 12 13 5 (3,5) (27,5)

B = 24   M = 19
b = 14   h = 12

MAVA Math: Middle Reviews Solutions Copyright © 2013 Marla Weiss

Level 8 — Number 47

1.

	time	rate
M	2.5	$\frac{2}{5}$
T	(3.75)	$\frac{4}{15}$
tog	1.5	$\frac{2}{3}$

Mia can do a job alone in 2.5 hours. Find the time in hours for Tia to do the job alone if together they can complete it in 1.5 hours.

2. Square mentally.

115^2 — 13,225
$11 \times 12 = 132$

605^2 — 366,025
$60 \times 61 = 3660$

55^2 — 3025
$5 \times 6 = 30$

75^2 — 5625
$7 \times 8 = 56$

3. Find x degrees in the crook picture. (NTS)

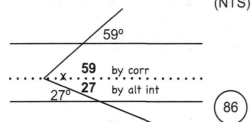

59°

59 by corr
x
27 by alt int
27°

(86)

4. The radii of 2 spheres are in the ratio 3:5. Find the ratio of their volumes and surface areas.

$$\frac{v}{V} = \frac{4/3 \cdot \pi \cdot 3 \cdot 3 \cdot 3}{4/3 \cdot \pi \cdot 5 \cdot 5 \cdot 5} = \left(\frac{27}{125}\right)$$

$$\frac{sa}{SA} = \frac{4 \cdot \pi \cdot 3 \cdot 3}{4 \cdot \pi \cdot 5 \cdot 5} = \left(\frac{9}{25}\right)$$

5. Graph on the coordinate plane. One box equals one unit.

$4x + 5y = 20$ $3x + 2y = 6$

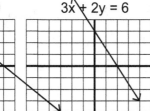

Plot intercepts when a and b divide c.

6. Find the number of integer values that satisfy:

$|x - 7| \le 12$ 25 $-5 \longrightarrow 19$

$|x + 4| \le 13$ 27 $-17 \longrightarrow 9$

$|x - 9| \le 9$ 19 $0 \longrightarrow 18$

7. Find the area of a regular hexagon with edge 10.

$$\frac{6 \cdot s^2 \cdot \sqrt{3}}{4} = \frac{6 \cdot \overset{5}{1\!\!/0} \cdot \overset{5}{1\!\!/0} \cdot \sqrt{3}}{\not{4}} = \left(150\sqrt{3}\right)$$

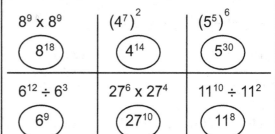

Area of regular hexagon is 6 times area of equilateral triangle.

8. Simplify. Answer in exponential form.

$8^9 \times 8^9$	$(4^7)^2$	$(5^5)^6$
(8^{18})	(4^{14})	(5^{30})
$6^{12} \div 6^3$	$27^6 \times 27^4$	$11^{10} \div 11^2$
(6^9)	(27^{10})	(11^8)

9. Find the next letter(s) in each pattern.

k, S, j, t, i, U, h, __v__ __g__

Consecutive small letters backwards interleave with alternating capital/small letters forwards.

r, s, j, t, u, h, v, w, __f__ __x__ __y__

Two consecutive letters forwards interleave with one letter backwards skipping one.

10. Find the distance between the points.

(1, 20) and (10, –20)

9, 40, (41)

(10, –16) and (–6, 14)

16, 30, (34)

(–9, 17) and (–2, –7)

7, 24, (25)

(–3, 3) and (6, –9)

9, 12, (15)

Level 8	Number 48

1. Find 3 consecutive integers with the given sum. *mean = median*

48	15, 16, 17	–21	–6, –7, –8
75	24, 25, 26	0	–1, 0, 1
36	11, 12, 13	60	19, 20, 21

6. Reflect each point over the line y = –x. Then translate as specified.

(0, 5)	(–5, 0)	down 9	(–5, –9)
(–6, 8)	(–8, 6)	right 4	(–4, 6)
(4, 9)	(–9, –4)	up 1	(–9, –3)

2. Find the area of the trapezoid bounded by the lines y = –x, x = 4, x = –2 and y = 4.

(–2,4) (4,4) (–2,2) (4,–4)

b = 2, B = 8, M = 5, h = 6

$\boxed{30}$

7. A:C:T = 5:6:9   T:W:O = 15:10:7 Find W:A.

Common variable is T.
LCM(9,15) = 45
A:C:T = 25:30:45 (x5)
T:W:O = 45:30:21 (x3)
W:A = 30:25 = 6:5 (÷5)

$\boxed{6:5}$

3. Find the 56th term of each sequence.

15, 19, 23, 27, . . .   d = 4

15 + (55)(4) = 15 + 220 = $\boxed{235}$

21, 30, 39, 48, . . .   d = 9

21 + (55)(9) = 21 + 495 = $\boxed{516}$

8. Find the volume between 2 cylinders with concentric bases of diameters 20 and 12, both 5 high.

$V_1 = 10 \cdot 10 \cdot 5 \cdot \pi$
$V_1 = 100 \cdot 5\pi$
$V_2 = 6 \cdot 6 \cdot 5 \cdot \pi$
$V_2 = 36 \cdot 5\pi$
$V_{band} = (100-36) \cdot 5\pi$
$V_{band} = 64 \cdot 5\pi = \boxed{320\pi \text{ cu un}}$

4. Define operation ♕ as:

X ♕ Y = 6X ÷ Y. For example, 3 ♕ 2 = 18 ÷ 2 = 9. Find the values. *Simplify first. Multiply last.*

(10 ♕ 5) ♕ 4   18   10 ♕ 5 = 12

12 ♕ (9 ♕ 6)   8   9 ♕ 6 = 9

9. Find the probability for any 2 people that at least one was born on Sunday.

1 – P(not Su, not Su)

The greatest probability is 1.

$1 - \dfrac{6}{7} \cdot \dfrac{6}{7} = 1 - \dfrac{36}{49} = \boxed{\dfrac{13}{49}}$

5. Find the principal that yields $27 simple interest at 4.5% for 9 months.

I = PRT

$27 = P \cdot \dfrac{4.5}{100} \cdot \dfrac{9}{12}$

$P = 27 \cdot \dfrac{100 \cdot 4}{4.5 \cdot 3}$

$P = 2 \cdot 100 \cdot 4$

$27 = P \cdot \dfrac{4.5 \cdot 3}{100 \cdot 4}$ $P = \boxed{\$800}$

10. Multiply.

1022_{three}	2543_{six}	3572_{twelve}
x 2_{three}	x 3_{six}	x T_{twelve}
2121_{three}	**12513_{six}**	**$2T7E8_{twelve}$**

| Level 8 | Number 49 |

1. Is each the graph of a function?

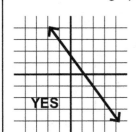 

YES NO

6. Solve.

$4X - Y = 2$
$10X - 3Y = 4$

$3X = 45$
$2X + 5Y = 60$

$12X - 3Y = 6$
$10X - 3Y = 4$
_____
$2X = 2$ $4 - Y = 2$
$X = 1$ $Y = 2$

$X = 15$
$30 + 5Y = 60$
$5Y = 30$
$Y = 6$

---

2. Point B divides segment $\overline{AC}$ such that AB:BC = 5:3. AC = 48. Find BC.

x 6

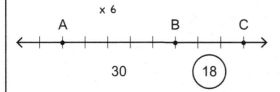

A   B   C

30   ⟨18⟩

7. A picture 12 by 25 feet has a uniform mat all around. Find the width and area of the mat if the total area is 338 square feet.

12 x 25 = 300
13 x 26 = 338

⟨A = 38
W = .5⟩

Use T&E and ones digit instead of algebra.

---

3. Solve for the variables in the matrices.

$$\begin{bmatrix} 6a - 9 & 21 \\ 41 & 5c + 7 \end{bmatrix} = \begin{bmatrix} 57 & 7b - 7 \\ 7d - 8 & 47 \end{bmatrix}$$

$6a - 9 = 57$ $7b - 7 = 21$
a = 11   **b = 4**
$7d - 8 = 41$ $5c + 7 = 47$
d = 7   **c = 8**

8. Simplify.

| $64^{\frac{3}{2}}$ ⟨512⟩ | $8^{\frac{4}{3}}$ ⟨16⟩ | $16^{\frac{1}{2}}$ ⟨4⟩ |
| $64^{\frac{5}{6}}$ ⟨32⟩ | $125^{\frac{4}{3}}$ ⟨625⟩ | $27^{\frac{2}{3}}$ ⟨9⟩ |

---

4. AB = BC = BD. Find x. (NTS)

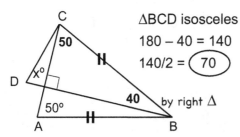

△BCD isosceles
180 − 40 = 140
140/2 = ⟨70⟩

by right △

9. Find the number of subsets of each set.

{C, O, U} ∪ {N, T} 2^5 32

{P, E, A} ∪ {P, O, D}   only 1 P 2^5 32

{1, 2, 3, 4, 5, 6} 2^6 64

{1, 2, 3, . . . , 7, 8, 9} 2^9 512

---

5. Solve.

$(x - 2)(x + 8) = 0$ $(7x - 6)(4x + 3) = 0$

$x = $ ⟨2, −8⟩ $x = $ ⟨$\frac{6}{7}$, $\frac{-3}{4}$⟩

$(x + 3)(x - 11) = 0$ $(5x - 7)(6x + 5) = 0$

$x = $ ⟨−3, 11⟩ $x = $ ⟨$\frac{7}{5}$, $\frac{-5}{6}$⟩

10. How many numbers from 1 to 200 inclusive are multiples of 2 or 5?

200 ÷ 2 = 100
200 ÷ 5 = 40
200 ÷ 10 = 20
100 + 40 − 20 = ⟨120⟩

Subtract the overlap: the multiples of 2 AND 5.

MAVA Math: Middle Reviews Solutions Copyright © 2013 Marla Weiss

Level 8	Number 50

1. Which fabric has a less costly unit rate per square foot: 7 yards of 60 inches wide for $73.50 or 9 yards of 36 inches wide for $55.08?

$$\frac{73.50}{7 \times 3 \times 5} \quad \text{vs} \quad \frac{55.08}{9 \times 3 \times 3}$$

0.70   0.68

6. Find the positive geometric mean between each pair. Show the proportion.

4 and 36	16 and 25	5 and 45

$$\frac{4}{12} = \frac{12}{36} \quad \frac{16}{20} = \frac{20}{25} \quad \frac{5}{15} = \frac{15}{45}$$

12   20   15

2. Sam has 12 more quarters than dimes with value $12.45. How many total quarters and dimes does he have?

Q = D + 12
10D + 25Q = 1245
2D + 5Q = 249
2D + 5(D + 12) = 249
7D + 60 = 249   D = 27
7D = 189   Q = 39

66

7. Find the area of the square inscribed in a circle of area 16π square units.

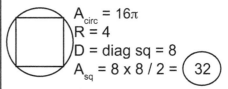

$A_{circ} = 16\pi$
R = 4
D = diag sq = 8
A_{sq} = 8 x 8 / 2 = 32

A square is a rhombus. Use the area formula: half the product of the diagonals.

3. Find the area of a rectangle given the perimeter is 200 and the ratio of the sides is 7:3.

P = 200   A = 70 x 30
SP = 100
ratio sum = 10   A = 2100
blow-up = 100/10 = 10   sq un
sides = 70 and 30

8. Seven years from now, Bob's age will be twice Pam's age then but triple Pam's age 3 years ago. How old was Bob 3 years ago?

	–3	now	+7	3P = 2(P+10)
P	P	P+3	P+10	3P = 2P+20
B	50	53	3P	P = 20

4. Find the mean of 11x – 8, 17x + 15, 12 – 5x, 9x + 8, x – 10, and 3x + 7.

$$\frac{36x + 24}{6} = 6x + 4$$

9. Find the value using fractions.

$55\frac{5}{9}$ % of $33\frac{1}{3}$ % of 80% of 1890

$$\frac{5}{9} \times \frac{1}{3} \times \frac{4}{5} \times \overset{70}{\underset{\cancel{210}}{\cancel{1890}}}$$ 280

5. Solve for n.

$$\frac{(n + 2)!}{(n - 2)!} = 7920$$

(n + 2)(n + 1)(n)(n – 1) = 72 x 11 x 10
          = 11 x 10 x 9 x 8
n = 9

10. Multiply as two binomials.

$7\frac{1}{3}$ x $9\frac{3}{4}$ $6\frac{1}{2}$ x $5\frac{3}{5}$

F =	63		F =	30
O =	5.25		O =	3.6
I =	3		I =	2.5
L =	0.25	71.5	L =	0.3	36.4

Level 8	Number 51

1. Draw the discards on the left.

Dana ate 1/4 of her candies and froze 1/3 of those left. She then packed 1/3 of those remaining, leaving 20. How many did Dana have to start?

C

$\dfrac{C}{4}$ $\dfrac{3C}{4}$

$\dfrac{C}{4}$ $\dfrac{C}{2}$

60 $\dfrac{C}{6}$ $\dfrac{C}{3} = 20$

6. A clock that is correctly set at 2:00 AM loses 5 minutes every hour. What is the correct time when the clock reads 9:00 PM the next day?

2:00 AM to 9:00 PM is 19 hours
real time is extra 5 minutes per hour
5 x 19 = 95 = 60 + 35 minutes

$10{:}35\ PM$

---

2. Solve.

$9^3 = 27^{2x-1}$	$4^x = 8^8$	$16^{x+5} = 32^{x-2}$
$3^6 = 3^{6x-3}$	$2^{2x} = 2^{24}$	$2^{4x+20} = 2^{5x-10}$
$6x = 9$	$2x = 24$	$5x = 4x + 30$
$x = \boxed{3/2}$	$x = \boxed{12}$	$x = \boxed{30}$

7. Answer YES or NO as to whether the set is closed for the given operation.

$\{0, 2, 4\}$ square rooting   NO
root 2 not in set

$\{-2, 0, 2\}$ subtraction   NO
$2 - (-2) = 4$

reals   cube rooting   YES

---

3. A raw meatball with a 4-inch radius can be re-formed into how many meatballs with a 2-inch radius?

$V = \dfrac{4\pi \cdot 4 \cdot 4 \cdot 4}{3}$

$V = \dfrac{4\pi \cdot 2 \cdot 2 \cdot 2}{3}$

$\dfrac{4 \cdot 4 \cdot 4}{2 \cdot 2 \cdot 2} = \boxed{8}$

The 4π and 3 simplify when dividing the 2 volumes.

8. Find the angle in degrees formed by clock hands at 12:40.

$360°/12 = 30°$ between neighboring numbers

$120 + 20 = \boxed{140}$

Hour hand is 2/3 from 12 to 1. Use acute angle always.

---

4. Find the perimeter of parallelogram ABCD given AD = 22, CD = 11x + 4, and BC = 5x – 8.

$5x - 8 = 22$
$x = 6$
$CD = 66 + 4 = 70$
$P = 44 + 140 = \boxed{184}$

9. Complete.

$35 \equiv \underline{5}$	(mod 10)	$50 \equiv \underline{6}$	(mod 11)
$35 \equiv \underline{11}$	(mod 12)	$50 \equiv \underline{2}$	(mod 24)
$35 \equiv \underline{6}$	(mod 29)	$50 \equiv \underline{19}$	(mod 31)
$35 \equiv \underline{1}$	(mod 17)	$50 \equiv \underline{2}$	(mod 16)

---

5. Find the points of trisection of the line segment with the given endpoints.

(–2, 20) & (1, 29)	(–1, 23)	(0, 26)
(–7, 15) & (5, 45)	(–3, 25)	(1, 35)
(0, –8) & (36, –2)	(12, –6)	(24, –4)
(–5, 1) & (10, 22)	(0, 8)	(5, 15)

10. Define operation ❋ as:

X ❋ Y = XY – X – 7. For example,
4 ❋ 2 = 8 – 4 – 7 = –3.

Find the values.   Not commutative

9 ❋ A = 56 $A = 8$ 9A – 9 – 7 = 56

B ❋ 9 = 57 $B = 8$ 9B – B – 7 = 57

Level 8	Number 52

1. The average of 6 tests is 73. If the worst grade is dropped, the new average is 77. What was the worst grade?

$\dfrac{438}{6} = 73$ $\dfrac{385}{5} = 77$ $\begin{array}{r} 438 \\ -\ 385 \\ \hline \boxed{53} \end{array}$

6.

■ School A   ░ School B

Answer items below for the bar graph.

100 75 50 25 0

2005 2006 2007 2008

2. X varies inversely as the square of Y. X = 100 when Y = 3. Find Y when X = 36.

$XY^2 = k$

$100 \cdot 9 = k$

$k = 900$

$36Y^2 = 900$

$Y^2 = 25$ $Y = \boxed{5}$

7. Express all consecutive increases in values as percents.

	School A	School B
'05 - '06	50%	100%
'06 - '07	$33.\overline{3}$%	50%
'07 - '08	25%	$33.\overline{3}$%

3. Rewrite the statement using DeMorgan's Laws.

I do not eat meat or fish.
_____

I do not eat meat and I do not eat fish.
_____

_____

8. Given the perimeter of a rectangle and the upper bound of the width. Find the range of values for the length.

P = 80; max W < 38 | P = 96; max W < 43

SP = 40 | SP = 48

0 < W < 38 | 0 < W < 43

$\boxed{2 < L < 40}$ | $\boxed{5 < L < 48}$

4. Find the mean and range of the 5 different integers 4x − 9, 6x + 9, 30, −20, and 2x if their median is 11.

−20	10	11	30	39
−20	2x	4x−9	30	6x+9

range = 39 − (−20) = $\boxed{59}$ Draw 5 slots.

mean = 70/5 = $\boxed{14}$ Only 4x−9 can be 11. x = 5

9. Find the surface area in square units of a square pyramid with base edge 16 and lateral edge 17 units.

$A_{sq} = 16 \times 16 = 256$

$A_{tri} = 16 \times 15 / 2 = 120$

120 x 4 = 480

$\begin{array}{r} 256 \\ +\ 480 \\ \hline \boxed{736} \end{array}$

17 15 8

5. Find the space diagonal of the rectangular prism with the given dimensions.

3 by 4 by 5 $5\sqrt{2}$ $\sqrt{9 + 16 + 25}$

4 by 5 by 6 $\sqrt{77}$ $\sqrt{16 + 25 + 36}$

10. Find the area of the triangle with m∠A = 45°.

$A = \dfrac{(9\sqrt{2})(9\sqrt{2})}{2}$

$A = \boxed{81}$

18    $9\sqrt{2}$

A $9\sqrt{2}$

Alternate method: Draw 2 lines to make a square. Use half-the-product-of-the-diagonals formula. A square = 162; A △ = 162/2 = 81.

Level 8	Number 53

1. If e chefs can prepare m meals in 1 week, f chefs can prepare how many meals in 1 hour?

chefs	meals	hours
e	m	7 · 24
1	$\frac{m}{e}$	7 · 24
f	$\frac{mf}{e}$	7 · 24
f	$\boxed{\frac{mf}{168e}}$	1

6. Draw the next figure in the pattern.

Clockwise, delete 1, 2, 1, 2 lines.

2. Find the time in minutes needed for 2 people to do a job together given that: Person A can do the job in 2 hours; Person B can the same job in 6 hours.

Rate A = 1/2
Rate B = 1/6
Total Rate = 4/6 = 2/3
Total Time = 3/2

In work problems, rates and times are reciprocals.

$\boxed{90}$

7. Find the area of a regular hexagon with edge 12.

$$\frac{6 \cdot s^2 \cdot \sqrt{3}}{4} = \frac{6 \cdot \cancel{12} \cdot 12 \cdot \sqrt{3}}{\cancel{4}} = \boxed{216\sqrt{3}}$$

Area of regular hexagon is 6 times area of equilateral triangle.

3. Multiply the matrices.

$$\begin{bmatrix} 2 & 11 & 5 \\ 2 & 9 & 4 \end{bmatrix} \times \begin{bmatrix} 10 & 8 \\ 0 & 6 \\ 5 & 1 \end{bmatrix} =$$

20 + 0 + 25 = 45
16 + 66 + 5 = 87
20 + 0 + 20 = 40
16 + 54 + 4 = 74

$$\begin{bmatrix} \mathbf{45} & \mathbf{87} \\ \mathbf{40} & \mathbf{74} \end{bmatrix}$$

8. Simplify. Answer in exponential form.

$9^9 \times 9^{11}$	$(8^9)^2$	$(2^7)^3$
$\boxed{9^{20}}$	$\boxed{8^{18}}$	$\boxed{2^{21}}$
$13^{11} \div 13^3$	$23^7 \times 23^{10}$	$3^{13} \div 3^4$
$\boxed{13^8}$	$\boxed{23^{17}}$	$\boxed{3^9}$

4. Find the probability of drawing green, red, then blue or yellow successively without replacement from a bowl containing 10 green, 14 yellow, 16 red, and 25 blue marbles.

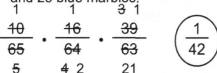

$\boxed{\frac{1}{42}}$

9. Operate.

$$\frac{w}{x} \div \frac{y}{2x} \qquad \frac{6x}{5a} \cdot \frac{ab}{9x} \cdot \frac{3a}{bc}$$

$$\boxed{\frac{2w}{y}} \qquad \boxed{\frac{2a}{5c}}$$

5. Write each repeating decimal as a simplified fraction.

$.6\overline{5}$ $6.\overline{5}$ $6\frac{5}{9}$ $\frac{59}{9}$ $\boxed{\frac{59}{90}}$

×10   ÷10

$.05\overline{7}$ $5.\overline{7}$ $5\frac{7}{9}$ $\frac{52}{9}$ $\frac{52}{900}$ $\boxed{\frac{13}{225}}$

×100   ÷100

10. BC = 12. Find AB + BD.

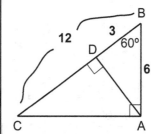

AB = 6 (opposite 30° is 1/2 hypotenuse)

AB is hypotenuse of smallest triangle
BD = 3 $\boxed{9}$

| Level 8 | Number 54 |

1. Find the number of 5-digit even numbers less than 60,000 with the middle digit 3, the leftmost digit odd, and no repeated digits.

$$\underline{2} \cdot \underline{7} \cdot \underline{1} \cdot \underline{6} \cdot \underline{5} \quad \boxed{420}$$

↑ 1 or 5 2nd ↑ 3 1st ↑ even 3rd

6. Find the area of a circle given its circumference is 100.

$$C = \pi D = 100$$
$$D = \dfrac{100}{\pi}$$
$$r = \dfrac{50}{\pi}$$

$$A = \dfrac{50}{\pi} \cdot \dfrac{50}{\pi} \cdot \pi$$

$$A = \boxed{\dfrac{2500}{\pi}}$$

2. Find 3 consecutive integers with the given product.

60	3, 4, 5	−720	−8, −9, −10
6	1, 2, 3	120	4, 5, 6
990	9, 10, 11	−210	−5, −6, −7

7. Simplify.

$$\dfrac{.\overline{518}}{.\overline{81}} \quad \dfrac{518}{999} \times \dfrac{99}{81} \quad \dfrac{14}{27} \times \dfrac{11}{9} \quad \boxed{\dfrac{154}{243}}$$

$$\dfrac{.\overline{207}}{.\overline{69}} \quad \dfrac{207}{999} \times \dfrac{99}{69} \quad \dfrac{3}{111} \times \dfrac{11}{1} \quad \boxed{\dfrac{11}{37}}$$

3. Solve for the variables in the matrices.

$$\begin{bmatrix} 8a - 9 & 55 \\ 87 & 4c + 7 \end{bmatrix} = \begin{bmatrix} 63 & 2b - 11 \\ 9d + 6 & 79 \end{bmatrix}$$

$$8a - 9 = 63 \qquad 2b - 11 = 55$$
$$\mathbf{a = 9} \qquad\qquad \mathbf{b = 33}$$
$$9d + 6 = 87 \qquad 4c + 7 = 79$$
$$\mathbf{d = 9} \qquad\qquad \mathbf{c = 18}$$

8. Calculate the combinations.

$$C(9, 6) = \dfrac{9!}{6!\,3!} = \dfrac{9 \cdot 8 \cdot 7}{3 \cdot 2 \cdot 1} = \boxed{84}$$

$$C(7, 2) = \dfrac{7!}{2!\,5!} = \dfrac{7 \cdot 6}{2 \cdot 1} = \boxed{21}$$

$$C(8, 5) = \dfrac{8!}{5!\,3!} = \dfrac{8 \cdot 7 \cdot 6}{3 \cdot 2} = \boxed{56}$$

4. Find the first 3 abundant numbers: the sum of the proper divisors is greater than the number.

$$1 + 2 + 3 + 4 + 6 \qquad \boxed{12}$$

$$1 + 2 + 3 + 6 + 9 \qquad \boxed{18}$$

$$1 + 2 + 4 + 5 + 10 \qquad \boxed{20}$$

The number itself is not a proper divisor.

9. How many kg of water must be added to 160 kg of a 5% salt solution to make a 4% acid solution?

$$(160) + (x) = (160 + x)$$
$$(.05)(160) + (0)(x) = (.04)(160 + x)$$
$$(5)(160) + (0)(x) = (4)(160 + x)$$
$$800 = 640 + 4x$$
$$160 = 4x$$
$$x = \boxed{40}$$

5. Multiply mentally.

47 x 101	4747
52 x 1001	52,052
829 x 10,001	8,290,829
608 x 1001	608,608

Picture 101 as (100 + 1), 1001 as (1000 + 1), and 10,001 as (10,000 + 1).

10. How many right triangles with whole number sides have one side of length 20 units?

$$3, 4, 5 \longrightarrow \begin{matrix} 12, 16, 20 \\ 15, 20, 25 \end{matrix} \qquad \boxed{4}$$

$$5, 12, 13 \longrightarrow 20, 48, 52$$

$$20, 21, 29$$

187

Level 8	Number 55

1.

	time	rate
B	5	$\frac{1}{5}$
J	(7.5)	$\frac{2}{15}$
tog	3	$\frac{1}{3}$

Bo can do a job alone in 5 hours. Find the time in hours for Jo to do the job alone if together they can complete it in 3 hours.

2. Solve.

$64^x = 2^{x+5}$	$81^x = 3^{2x-2}$	$9^9 = 27^{4x+2}$
$2^{6x} = 2^{x+5}$	$3^{4x} = 3^{2x-2}$	$3^{18} = 3^{12x+6}$
$6x = x + 5$	$4x = 2x - 2$	$18 = 12x + 6$
$x = \boxed{1}$	$x = \boxed{-1}$	$x = \boxed{1}$

3. Complete for the geometric sets.

A —— B ———————— C —— D

$\overrightarrow{AB} \cup \overrightarrow{BC} = \underline{\overrightarrow{AC}}$ $\overleftarrow{CA} \cup \overline{CD} = \underline{\overrightarrow{DA}}$

$\overleftrightarrow{AB} \cap \overline{CD} = \underline{\overline{CD}}$ $\overrightarrow{AC} \cap \overline{AC} = \underline{\overline{AC}}$

4. The sum of the first 10 positive perfect squares is 385. Find the sum of the first:

9 positive perfect squares   285   - 100

11 positive perfect squares   506   + 121

5. Find the probability that three of the numbers 3, 5, 7, or 9 chosen randomly form the sides of a triangle.

3, 5, 7   YES
3, 5, 9   NO
3, 7, 9   YES $\boxed{\frac{3}{4}}$
5, 7, 9   YES

6. Find the volume in cubic units of a trapezoidal prism with parallel base edges 4 and 20, base height 5, and altitude 15 units.

b = 4, B = 20, M = 12
A base = 12 x 5 = 60

V = 15 x 60 = $\boxed{900}$

7. Translate the cipher with each letter as its own digit.

```
     EA              31
AS) BED         17) 534
    BA              51
    CD              24
    AS              17
     S               7
```

AS x A = AS, so A=1. C=2 for last minus to be 1-digit. E−1=2, so E=3. Sx3 ends in 1, so S=7. D=4. B=5.

8. Write in scientific notation.

$(3 \times 10^6) \div (8 \times 10^{-3})$ $(3 \times 10^9) \div (4 \times 10^4)$

0.375×10^9 0.75×10^5

$\boxed{3.75 \times 10^8}$ $\boxed{7.5 \times 10^4}$

9. Complete the unit conversions.

Picture cubes:

1 yard³	27	feet³	1 yd by 1 yd by 1 yd = 3 ft by 3 ft by 3 ft
2 yard³	54	feet³	2 yd by 1 yd by 1 yd = 6 ft by 3 ft by 3 ft
(2 yard)³	216	feet³	2 yd by 2 yd by 2 yd = 6 ft by 6 ft by 6 ft

10. Find the slope given 2 points on a line.

Points	Points	Points
(−3, 12)	(7, 12)	(−7, −5)
(−11, 6)	(9, 11)	(−13, −1)
Slope	Slope	Slope
$\frac{3}{4}$	$\frac{-1}{2}$	$\frac{-2}{3}$

MAVA Math: Middle Reviews Solutions Copyright © 2013 Marla Weiss

| Level 8 | Number 56 |

1. Draw a Venn Diagram to show the relationship among the sets.

A = {2, 3, 4, 7, 9}
B = {1, 3, 5, 7}
C = {1, 3, 8, 9}

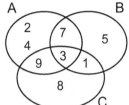

6. Find all digits d such that the 5-digit number 55,d35 is divisible by 9.

15 + 3 = 18   (0, 9)

___

Find all digits x such that the 4-digit number 5x14 is divisible by 6.

5 + 1 + 4 = 10   already divisible   (2, 5, 8)
by 2

2. Write the contrapositive of the conditional statement. Are they logically equivalent?   (YES)

If a shape is a polygon, then it has sides.

Both are TRUE.

If a shape does not have sides, then it is

not a polygon.

7. Two similar triangles have sides 6, 6, 9 and 8, 8, 12. Find the ratio of their:

sides   3:4

altitudes   3:4

perimeters   3:4

areas   9:16

3. Draw the reflection of each trapezoid over the line y = −x. Reflect vertices first. (x,y) becomes (−y,−x).

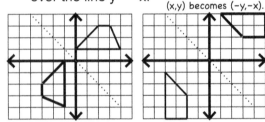

8. Operate and simplify.

$$\frac{6\frac{3}{4} \times 7\frac{1}{9}}{7\frac{2}{5} \div 4\frac{5}{8}} \quad \frac{\frac{27}{4} \times \frac{64}{9}}{\frac{37}{5} \times \frac{8}{37}} \quad \frac{48}{\frac{8}{5}}$$

$$48 \times \frac{5}{8} \quad (30)$$

4. The measures of the interior angles of a pentagon are 2x, x + 30, 3x + 10, 2x + 18, and 4x − 10. Find the angle measure that is a perfect square.

2x + x + 30 + 3x + 10 + 2x + 18 + 4x − 10 = 540
12x + 48 = 540
12x = 492
x = 41
2(41) + 18 = 82 + 18 = 100   (100)

9. Find the value using fractions.

$44\frac{4}{9}$ % of $12\frac{1}{2}$ % of 25% of 720

$$\frac{4}{9} \times \frac{1}{8} \times \frac{1}{4} \times \overset{10}{\cancel{720}}$$

(10)

5. If f(x) = 5x − 5 and g(x) = 4x + 4, find g(f(4)) + f(g(5)).

f(4) = 20 − 5 = 15
g(15) = 60 + 4 = 64
g(5) = 20 + 4 = 24
f(24) = 120 − 5 = 115

64 + 115 = (179)

10. A regular hexagon has perimeter 42 units. Find its area in square units.

P = 42
S = 7
$$A = \frac{6 \cdot 7^2 \sqrt{3}}{4} = \left(\frac{147\sqrt{3}}{2} \right)$$

Level 8	Number 57

1. Find the volume of each rectangular prism in cubic yards.

24 feet by 18 feet by 33 feet

8 yards by 6 yards by 11 yards $\quad$ (528)

36 inches by 72 inches by 54 inches

1 yard by 2 yards by 1.5 yards $\quad$ (3)

2. Square mentally.

45^2 $\quad$ 2025	395^2 $\quad$ 156,025
4 x 5 = 20	39 x 40 = 1560
205^2 $\quad$ 42,025	85^2 $\quad$ 7225
20 x 21 = 420	8 x 9 = 72

3. Find the area of the rhombus in square units given perimeter P = 40 and diagonal D = 16.

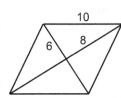

P = 40 $\qquad$ A = 24 x 4
s = 10 $\qquad$ or 8 x 12
D = 16
D/2 = 8 $\qquad$ A = (96)
6, 8, 10
d = 12

4. Find the endpoint of the line segment.

Endpoint	Midpoint	Endpoint
(2.5, 11)	(3, 11.5)	**(3.5, 12)**
(−10, 3.1)	(0, 10.5)	(10, 17.9)
(1.3, 14)	(7.3, 25)	**(13.3, 36)**
(4.7, −18)	(8.8, −33)	(12.9, −48)

5. The radii of 2 spheres are in the ratio 4:5. Find the ratio of their volumes and surface areas.

$$\frac{v}{V} = \frac{4/3 \cdot \pi \cdot 4 \cdot 4 \cdot 4}{4/3 \cdot \pi \cdot 5 \cdot 5 \cdot 5} = \left(\frac{64}{125}\right)$$

$$\frac{sa}{SA} = \frac{4 \cdot \pi \cdot 4 \cdot 4}{4 \cdot \pi \cdot 5 \cdot 5} = \left(\frac{16}{25}\right)$$

6. Find the sum of 4 consecutive primes whose product is:

1155	5005
3 x 5 x 7 x 11	5 x 7 x 11 x 13
(26)	(36)
See factors of 3, 5, and 11.	See factors of 5 and 11 but not of 3.

7. Find the number of integer values that satisfy:

|x + 8| < 14 $\quad$ 27 $\quad$ −21 $\longrightarrow$ 5

|x − 2| < 16 $\quad$ 31 $\quad$ −13 $\longrightarrow$ 17

|2x + 5| ≤ 17 $\quad$ 18 $\quad$ −11 $\longrightarrow$ 6

8. Convert 926 in base ten to base fifteen.

$$\begin{array}{r} 926 \\ -\ 900 \\ \hline 26 \\ -\ 15 \\ \hline 11 \end{array}$$

4	1	B
225	15	1

$\left(41B_{\text{fifteen}}\right)$

9. Find the angles of parallelogram ABCD given:

m∠A = 9x + 7 $\qquad$ m∠B = 2x + 8

9x + 7 + 2x + 8 = 180
11x + 15 = 180
11x = 165
x = 15

m∠A, C = 142°
m∠B, D = 38°

Consecutive angles are supplementary.
Opposite angles are congruent.

10. Find the perimeter of the trapezoid with vertices (2, 1), (8, 9), (5, 9), and (2, 5).

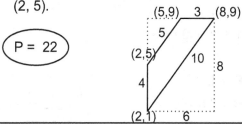

P = 22

| Level 8 | Number 58 |

1. Walking 18 feet west, 8 feet south, 18 feet west, and 7 feet south in that order ends how many feet from the starting point?

(39)

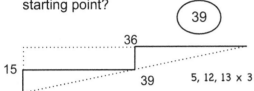

36

15

39   5, 12, 13 x 3

6. Find the measure of one angle of each regular polygon.

Interior and exterior angles are supplements.

triangle   60   180/3 = 60

hexagon   120   ext < = 360/6 = 60
                    int < = 180 − 60 = 120

30-gon   168   ext < = 360/30 = 12
                   int < = 180 − 12 = 168

Or 28 x 180 / 30 = 168

2. Let n be an even integer. Half of n plus two-thirds of the next consecutive even integer equals 41. Find the odd integer between the 2 even integers.

34, 36

17 + 24 = 41   (35)

7. Simplify using DPMA.

371 x 173 + 371 x 27   371 x (173 + 27)
                             371 x 200
                             (74,200)

40% of 359 + 40% of 241   40% (359 + 241)
                           40% of 600
                           (240)

12 x 13 + 12 x 45 + 12²   12(13 + 45 + 12)
                          12 x 70 (840)

3. Find the area of a rectangle given the perimeter is 80 and the length is triple the width.

P = 80   W = 10
SP = 40   L = 30
L + W = 40   A = 30 x 10
3W + W = 40
4W = 40   A = (300)

8. Flipping 4 coins, find the probability of at least one tail landing face up.

1 − P(no tail) = 1 − P(all heads)

faster than summing all of the options

$1 - \dfrac{1}{16} = \left(\dfrac{15}{16}\right)$

4. Define operation ✴ as:

X ✴ Y = X² + Y³ − 9. For example,
3 ✴ 2 = 9 + 8 − 9 = 8.
Find the values.

5 ✴ B = 16   B = 0   25 + B³ − 9 = 16

N ✴ 4 = 104   N = ±7   N² + 64 − 9 = 104

9. Name a lattice point C in the picture that would yield △ABC with area 9.

B   1 box = 1 unit

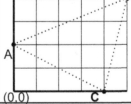

A

(0,0)   C

(4, 0)

20 − 5 − 2 − 4 = 9

5. Find the rate that yields $100 simple interest on $500 after 8 years.

I = PRT

100 = 500 • R • 8

1 = 5 • R • 8

R = 1/40

R = 2.5/100   (2.5%)

10. Find the volume in cubic units of a rectangular pyramid with length 18, width 7, and altitude 5 units.

$\dfrac{18 \times 7 \times 5}{3}$ = 6 x 7 x 5 = 30 x 7

= (210)

| Level 8 | Number 59 |

---

1. Find the 6-digit mystery number.
Clue 1: The number is divisible by 36.
Clue 2: The digits are in ascending order.
Clue 3: The 6 digits are distinct.
Clue 4: The number is less than 200,000.

Starts with 1. Ends in 6 or 8 by #1
and #2. Not 123456.
Wrong sums: 123458; 123468;
123478; 134578; 145678.　　$\boxed{134,568}$

6. Find the positive geometric mean between each pair. Show the proportion.

2 and 8	8 and 50	2 and 18
$\dfrac{2}{4} = \dfrac{4}{8}$	$\dfrac{8}{20} = \dfrac{20}{50}$	$\dfrac{2}{6} = \dfrac{6}{18}$
$\boxed{4}$	$\boxed{20}$	$\boxed{6}$

---

2. Find AC − BD.

A　　B　　　　　C　　　　D

−5　0　5

6 of length 5 − 7 of length 5　　$\boxed{-5}$

7. Answer YES or NO as to whether the set is closed for the given operation.

integers　　　　　　subtraction　　　YES

positive integers　subtraction　　　NO
　　　　　　　　2 − 3 = −1

negative integers　subtraction　　　NO
　　　　　　　　−2 − (−3) = 1

---

3. The area of rectangle BEFC is 1/2 the area of rectangle AEFD. Find the coordinates of E. (NTS)　$\boxed{(30, 20)}$

A　　　　　　E　　　　B (60,20)

D (−30,0)　　　　F　　　　C

8. Find the perimeter of an isosceles trapezoid with height 20, legs 25, and small base 30.

$25 + 25 = 50$
$60 + 30 = 90$
P = $\boxed{140}$

30
25　20　25
15　30　15

---

4. Multiply using DPMA.

67 x 29　　1943
$(60 + 7)(20 + 9) = 1200 + 540 + 140 + 63$

56 x 43　　2408
$(50 + 6)(40 + 3) = 2000 + 150 + 240 + 18$

9. Complete.

$60 \equiv 5$ (mod 11)	$60 \equiv 4$ (mod 14)
$60 \equiv 28$ (mod 32)	$60 \equiv 12$ (mod 24)
$60 \equiv 2$ (mod 29)	$67 \equiv 33$ (mod 34)
$60 \equiv 9$ (mod 51)	$89 \equiv 1$ (mod 44)

---

5. Graph on the coordinate plane. One box equals one unit.

$3x + 4y = 12$　　　$2x − 3y = 6$

Plot intercepts when a and b divide c.

10. Multiply.

3256_{seven}　　　3312_{four}　　　$8T47_{eleven}$
x 3_{seven}　　x 2_{four}　　x 3_{eleven}
13134_{seven}　**13230_{four}**　**$2492T_{eleven}$**

Level 8	Number 60

1. The sum of the dates of all Fridays in a given month is 75. What is the date of the first Friday?

7 + 14 + 21 + 28 = 70 (4 Fridays) < 75
must be 5 Fridays, thus < 7
75/5 = 15 = middle one
15 – 14 = 1   (1st)
1, 8, 15, 22, 29

6. Calculate the sum of the arithmetic sequence.

300 + 301 + 302 + . . . + 499 + 500

Because consecutive, use "subtract, add 1" to get the number of numbers. n = 201
F + L = 300 + 500 = 800
800 x 201 / 2 = 400 x 201 = (80,400)
Do 201 x 4. Tack on two 0s.

2. Find 4 consecutive integers such that 4 times the 3rd decreased by 3 times the 2nd is 39.
$x, x + 1, x + 2, x + 3$
$4(x + 2) - 3(x + 1) = 39$
$4x + 8 - 3x - 3 = 39$
$x = 34$
(34, 35, 36, 37)

7. Label the lines perpendicular, parallel, or neither.

$3x + 8y = 2$ $3y + 8x = 2$
$-3x + 8y = 1$ $8y - 3x = 1$

(neither)   (perpendicular)

negative reciprocal slopes

3. The complement of an angle is 36° greater than twice the angle. Find half the compement.
$90 - x = 36 + 2x$
$54 = 3x$
$x = 18$
comp = 72   (36)

8. Find the volume between 2 cylinders with concentric bases of diameters 16 and 6, both 4 high.

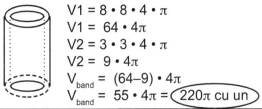

$V1 = 8 \cdot 8 \cdot 4 \cdot \pi$
$V1 = 64 \cdot 4\pi$
$V2 = 3 \cdot 3 \cdot 4 \cdot \pi$
$V2 = 9 \cdot 4\pi$
$V_{band} = (64-9) \cdot 4\pi$
$V_{band} = 55 \cdot 4\pi =$ (220π cu un)

4. Given parallelogram ABCD. Find x.

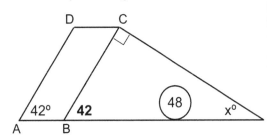

9. Randomly select two angles of a non-rectangular parallelogram.   C(4,2) = 6

Find the probability that they are both right.   0

Find the probability that they are both obtuse. $\frac{1}{6}$

Find the probability that they are both acute. $\frac{1}{6}$

5. The ratio of an angle to its complement is 7:3. Find the ratio of the angle to its supplement.

7 + 3 = 10
90 ÷ 10 = 9
angle = 63 (complement = 27)
supplement = 117   (7:13)
A:S = 63:117

10. Find the space diagonal of the rectangular prism with the given dimensions.

2 by 5 by 6 $\sqrt{65}$ $\sqrt{4 + 25 + 36}$

1 by 4 by 8   9 $\sqrt{1 + 16 + 64}$

Level 8	**Number 61**

1. Find the number of 5 character codes from left to right with no repetitions: vowel (not Y), even digit, odd digit, vowel (not Y), and any digit.

$$\underline{\ 5\ } \cdot \underline{\ 5\ } \cdot \underline{\ 5\ } \cdot \underline{\ 4\ } \cdot \underline{\ 8\ } \quad \boxed{4000}$$

↑   ↑   ↑   ↑   ↑
1st   2nd   3rd   4th   5th

6. Of 140 houses, 84 need new roofs, 99 need new shrubs, and 9 need neither. How many need both?

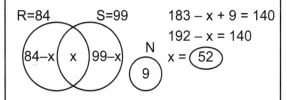

R=84   S=99   183 − x + 9 = 140
                                     192 − x = 140
                                     x = ⬭52

2. Find the perimeter of the rhombus with area 840 and longer diagonal 42.

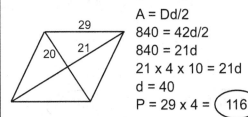

A = Dd/2
840 = 42d/2
840 = 21d
21 x 4 x 10 = 21d
d = 40
P = 29 x 4 = ⬭116

7. A lawn 16 by 16 feet has a uniform walkway all around. Find the area of the walkway in square feet if the total area is 576 square feet.

16 x 16 = 256
24 x 24 = 576
A =   320
W =   4

Use T&E and ones digit instead of algebra.

3. Graph on the number line, and write the solution to the quadratic inequality.

(x + 1)(x − 5) < 0

        no   yes   no
←┼──┼──┼──●──┼──┼──┼──┼──●──┼→
         −1 0   5

⬭−1 < x < 5

8. Translate the cipher, with each letter as its own digit.

SELFS	5EL05	**59705**
x _____ 3	x _____ 3	x _____ **3**
PLEPPS	1LE115	**179115**

In 1s, S=5. Try P=1 in 100,000s. Then F=0, L=7, and E=9.

4. Find the space diagonal of a cube with

edge = 6 $6\sqrt{3}$ e = 6

area of face = 64 $8\sqrt{3}$ e = 8

volume = 729 $9\sqrt{3}$ e = 9

face diagonal = $2\sqrt{2}$ $2\sqrt{3}$ e = 2

9. Find the probability of drawing not blue, not blue successively without replacement from a bowl containing 10 green, 14 yellow, 16 red, and 25 blue marbles.

$$\frac{\cancel{40}}{\cancel{65}} \cdot \frac{\cancel{39}}{\cancel{64}} \quad \boxed{\frac{3}{8}}$$

8 ... 3 ... 13 ... 8

5. Find the mean of 14x − 10, 18x + 14, 4 − 5x, 3 − 9x, and 7x − 6.

$$\frac{25x + 5}{5} = \boxed{5x + 1}$$

10. Multiply and simplify.

$$\sqrt{15} \cdot \sqrt{45} \cdot \sqrt{21} \cdot \sqrt{14}$$

$$\sqrt{5} \cdot \sqrt{3} \cdot 3 \cdot \sqrt{5} \cdot \sqrt{7} \cdot \sqrt{3} \cdot \sqrt{7} \cdot \sqrt{2}$$

$$5 \cdot 3 \cdot 3 \cdot 7 \cdot \sqrt{2}$$

$$5 \cdot 63 \sqrt{2} \qquad \boxed{315\sqrt{2}}$$

Level 8	**Number 62**

1. Original money earning interest is called _____ principal.

A fraction with a fraction as its numerator or denominator is called _____ complex.

"Positive, negative, zero" is an example of a _____ trichotomy.

2. Y varies directly as the square of X. Y = 800 when X = 5. Find Y when X = 2.
$Y = kX^2$
$800 = 25k$
$k = 32$
$Y = 32 \times 4$
Y = (128)

3. Find the 81st term of each sequence.

19, 22, 25, 28, . . . d = 3

19 + (80)(3) = 19 + 240 = (259)

3, 14, 25, 36, . . . d = 11

3 + (80)(11) = 3 + 880 = (883)

4. Find the surface area of the cone in square units. LA = "πrl"
diameter = 10 SA = 5 · 5 · π
slant height = 13 + 5 · 13 · π
= 25π + 65π
= (90π)

5. Find the points of trisection of the line segment with the given endpoints.

(5, 2) & (17, 17) (9, 7) (13, 12)

(−1, 3) & (8, 24) (2, 10) (5, 17)

(−6, 10) & (0, 13) (−4, 11) (−2, 12)

(4, −4) & (28, 5) (12, −1) (20, 2)

6. Two boats leave from the same dock at 9:30 AM. One travels north at 25 mph; the other travels east at 60 mph. When will they be 195 miles apart?
5, 12, 13 x 15
75, 180, 195
D = RT = 25 x 3 = 75 D =195
9:30 + 3 hr
D = RT = 60 x 3 = 180 (12:30 PM)

7. Solve.
$3\sqrt{x} = 2x$
$9x = 4x^2$
$0 = 4x^2 − 9x$
$0 = x(4x − 9)$
(x = 0, 9/4)

$5x^2 = 875$
$x^2 = 175$
$x = \pm\sqrt{175}$
(x = ± 5√7)

8. Convert 1153 base ten to base thirteen.
1153 − 1014 = 139 − 130 = 9
6 A 9
169 13 1
(6A9 thirteen)

9. Answer YES or NO if the matrices may be multiplied. If yes, give the dimensions of the product.

3 by 9 times 9 by 6 YES 3 by 6

6 by 9 times 9 by 3 YES 6 by 3

5 by 7 times 5 by 5 NO

10. Find the area of a rectangle given the perimeter is 54 and the ratio of the sides is 5:4.
P = 54 A = 15 x 12
SP = 27
ratio sum = 9 A = (180) sq un
blow-up = 27/9 = 3
sides = 15 and 12

Level 8	Number 63

1. Two broken clocks show the correct time at 2:00. One runs backward and the other forward, both at twice the normal rate. When will both clocks next show the same time?

F	B
2	2
4	12
6	10
8	8

$\boxed{8:00}$

6. Solve.

$4X - Y = 9$
$X - 3Y = 16$

$12X - 3Y = 27$
$X - 3Y = 16$
___
$11X = 11$ $4 - Y = 9$
$X = 1$ $Y = -5$

$9X = 63$
$3X - 2Y = 71$

$X = 7$
$21 - 2Y = 71$
$-2Y = 50$
$Y = -25$

2. Is each the graph of a function?

YES

NO

7. Find the area of a regular hexagon with edge 14.

$$\frac{6 \cdot s^2 \cdot \sqrt{3}}{4} = \frac{6 \cdot \overset{7}{\cancel{14}} \cdot \overset{7}{\cancel{14}} \cdot \sqrt{3}}{\cancel{4}} = \boxed{294\sqrt{3}}$$

Area of regular hexagon is 6 times area of equilateral triangle.

3. Complete for the geometric sets.

A • B • C

$\overrightarrow{AB} \cup \overrightarrow{BC} = \underline{\overrightarrow{AB}}$ $\overrightarrow{BA} \cup \overrightarrow{BC} = \underline{\overleftrightarrow{AC}}$

$\overrightarrow{AB} \cap \overrightarrow{BC} = \underline{\overrightarrow{BC}}$ $\overline{AC} \cap \overline{AC} = \underline{\overline{AC}}$

8. Simplify.

$8^{\frac{2}{3}}$ $\boxed{4}$ $9^{\frac{5}{2}}$ $\boxed{243}$ $64^{\frac{2}{3}}$ $\boxed{16}$

$4^{\frac{3}{2}}$ $\boxed{8}$ $32^{\frac{2}{5}}$ $\boxed{4}$ $81^{\frac{3}{2}}$ $\boxed{729}$

4. Find the perimeter of parallelogram ABCD given AD = 17, CD = 13x + 1, and BC = 8x − 7.

$8x - 7 = 17$
$x = 3$
$CD = 39 + 1 = 40$
$P = 34 + 80 = \boxed{114}$

9. Find the square root in the given mod.

square root of 5 mod 11 $\underline{4, 7}$

square root of 4 mod 7 $\underline{2, 5}$

square root of 5 mod 6 $\underline{none}$

square root of 6 mod 10 $\underline{4, 6}$

4 x 4 = 16
16 − 10 = 6
and
6 x 6 = 36
36 − 30 = 6

5. Find the sum of all 2-digit numbers with an odd number of factors.

$16 + 25 + 36 + 49 + 64 + 81 = \boxed{271}$

Only a perfect square has an odd number a factors.

10. Find x. (NTS)

$180 - 51 - 33 = 96$
$x = 96/2 = \boxed{48}$

$x°$ $x°$ $51°$ $33°$

Level 8	Number 64

1. Which fabric has a less costly unit rate per square foot: 9 yards of 54 inches wide for $97.20 or 10 yards of 60 inches wide for $120.00?

$$\frac{97.20}{9 \times 3 \times 4.5} \quad vs \quad \frac{120.00}{10 \times 3 \times 5}$$

0.80   0.80   (same)

6. Find the measure of one angle of each regular polygon.

Interior and exterior angles are supplements.

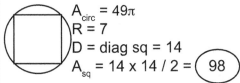

square   90   360/4 = 90

45-gon   172   ext < = 360/45 = 8
                             int < = 180 – 8 = 172

90-gon   176   ext < = 360/90 = 4
                             int < = 180 – 4 = 176

Or 43 × 180 / 45 = 172

2. If x is even, what is the sum of the next two odd numbers greater than 3x + 4?

x even
3x even
3x + 4 even
3x + 5 + 3x + 7   (6x + 12)

7. Find the area of the square inscribed in a circle of area 49π square units.

A_{circ} = 49π
R = 7
D = diag sq = 14
A_{sq} = 14 × 14 / 2 = (98)

A square is a rhombus. Use the area formula: half the product of the diagonals.

3. Find the volume of the cone in cubic units.

diameter = 18
slant height = 15

$$V = \frac{9 \cdot 9 \cdot \pi \cdot 12}{3}$$

= 81 · 4 · π

= (324π)

8. Reflect each point over the line y = –x. Then translate as specified.

(–3, 1)	(–1, 3)	right 6	(5, 3)
(8, 0)	(0, –8)	left 7	(–7, –8)
(7, 2)	(–2, –7)	up 2	(–2, –5)

4. The measure of one angle of a parallelogram is 9 more than 8 times another. Find the 4 angles.

180 – A = 9 + 8A
171 = 9A
A = 19

(19, 161, 19, 161)

9. Solve. | 5x + 4 | = 59

5x + 4 = 59   OR   5x + 4 = –59
5x = 55   5x = –63
(x = 11)   (x = –63/5)

5. Solve for n.

$$\frac{n!}{4!(n-3)!} = \frac{n!}{6!(n-4)!}$$

If 2 equal fractions have equal numerators, they have equal denominators.

4!(n – 3)! = 6!(n – 4)!
(n – 3)! = 30(n – 4)!
n – 3 = 30
n = (33)

10. How many numbers from 1 to 200 inclusive are multiples of 5 or 7?

200 ÷ 5 = 40
200 ÷ 7 = 28
200 ÷ 35 = 5
40 + 28 – 5 = (63)

Subtract the overlap: the multiples of 5 AND 7.

Level 8	Number 65

1. A machine knits a sweater by making 5 knit stitches followed by 5 purl stitches and repeating the pattern thereafter. What are the 68th, 69th, 70th, 71st, and 72nd stitches in a row?

KKKKKPPPPP
cycle length 10
60th and 70th are full patterns.   (**PPPKK**)

6. Find the volume in cubic units of a triangular prism with base edges 10, 13, and 13, and altitude 16 units.

h of isosceles $\triangle$ = 12
A base = 10 x 12 / 2 = 60
V = 16 x 60 = (**960**)

2. In a set of 5 numbers, the unique mode is greater than the median. Four of the numbers are 13, 27, 35, and 42. Find the missing number.

13   27   35   42   (**42**)

7. 11 quarters, 13 dimes, and 39 nickels are what percent of $20?

11 x .25 = 2.75   Use mental math for
13 x .10 = 1.30   the multiplications.
39 x .05 = 1.95
sum = 6.00
$\frac{6}{20}$ = (**30%**)

3. Find the area of an isosceles trapezoid with legs 61 and bases 50 and 72.

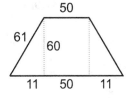

M = 61
H = 60
A = 61 x 60
= 3600 + 60
= (**3660**)

8. Given the perimeter of a rectangle and the upper bound of the width. Find the range of values for the length.

P = 56; max W < 19   P = 72; max W < 18
SP = 28   SP = 36
0 < W < 19   0 < W < 18
(9 < L < 28)   (18 < L < 36)

4. Evaluate the functions.

SQR(0)	0	SIGMA(10)	4
SQRT(0)	0	INT(0)	0
SGN(0)	0	TRUNC(2.9)	2
ABS(0)	0	INT(–5.7)	–6

9. Find the slope given 2 points on a line.

Point #1	Point #2	Slope
(6, 5)	(6, 8)	**undefined**
(8, 5)	(2, 5)	**0**
(7, –9)	(–7, –9)	**0**
(–5, –4)	(–5, 16)	**undefined**

5. Solve.

(x + 9)(2x – 9) = 0   (x – 6)(10x + 1) = 0
x = (-9 , $\frac{9}{2}$)   x = (6 , $\frac{-1}{10}$)

x(x + 12) = 0   (11x – 4)(5x + 6) = 0
x = (0, –12)   x = ($\frac{4}{11}$, $\frac{-6}{5}$)

10. Answer YES or NO as to whether the 3 numbers form sides of a right triangle.

$\sqrt{8}$, 3, 4	N	7, 7, $7\sqrt{2}$	Y
2, 3, $\sqrt{13}$	Y	3, 7, $\sqrt{59}$	N
1, 3, $\sqrt{10}$	Y	4, $4\sqrt{3}$, 8	Y

MAVA Math: Middle Reviews Solutions Copyright © 2013 Marla Weiss

198

Level 8	Number 66

1. The average of 5 tests is 84. What additional test would raise the average to 86?

$\dfrac{420}{5} = 84$ $\dfrac{516}{6} = 86$ $\begin{array}{r}516\\-420\\\hline \boxed{96}\end{array}$

6. Find the volume in cubic meters of each rectangular prism.

30 cm x 40 cm x 200 cm .3 x .4 x 2   .24

50 cm x 60 cm x 2 m   .5 x .6 x 2   .6

20 cm x 25 cm x 50 cm $\dfrac{1 \times 1 \times 1}{5 \times 4 \times 2}$ .025

2. Find the measures of two complementary angles such that one is 6 less than triple the other.
x + (3x – 6) = 90
4x = 96
x = 24

(24 and 66)

7. F:R:Y = 14:12:7   F:I:G = 10:11:8
Find I:Y.

Common variable is F.
LCM(14,10) = 70
F:R:Y = 70:60:35 (x5)
F:I:G = 70:77:56 (x7)
I:Y = 77:35 = 11:5 (÷7)

(11:5)

3. Multiply using the difference of 2 squares.

25 x 35   875   (30 – 5)(30 + 5) 900-25

43 x 37   1591   (40 + 3)(40 – 3) 1600-9

21 x 19   399   (20 + 1)(20 – 1) 400-1

76 x 84   6384   (80 – 4)(80 + 4) 6400-16

8. Five years ago the sum of the ages of Sue and Kim was 80. Kim's age now is 35 more than Sue's age 5 years ago. How old will Sue be in 8 years?

	–5	now	+8
S	S	30	(38)
K	S+30	S+35	

S+S+30 = 80
2S = 50
S = 25

4. The measures of the interior angles of a hexagon are 2x, 3x, 4x – 7, 5x – 13, 6x – 15, and 7x – 55. Find the greatest angle.

27x – 90 = 720
27x = 810
x = 30
6x – 15 = 165   (165)
7x – 55 = 155

9. A bowl holds 2 white, 3 red, and 4 blue marbles. Selecting 2 randomly in succession without replacement, find the probability of at least one red.

1 – P(not R, not R)

$1 - \dfrac{\overset{2}{\cancel6}}{\underset{3}{\cancel9}} \cdot \dfrac{5}{\underset{4}{\cancel8}} = 1 - \dfrac{5}{12} = \boxed{\dfrac{7}{12}}$

5. Write each repeating decimal as a simplified fraction.

$.5\overline{2}$ $5.\overline{2}$ $5\dfrac{2}{9}$ $\dfrac{47}{9}$ $\boxed{\dfrac{47}{90}}$
x 10   ÷ 10

$.02\overline{6}$ $2.\overline{6}$ $2\dfrac{2}{3}$ $\dfrac{8}{3}$ $\dfrac{8}{300}$ $\boxed{\dfrac{2}{75}}$
x 100   ÷ 100

10. Find the remainder without dividing.

607,642 ÷ 44 607,640 div by 4, 11   R 2

435,548 ÷ 55 435,545 div by 5, 11   R 3

714,078 ÷ 70 714,070 div by 7, 10   R 8

992,201 ÷ 88 992,200 div by 8, 11   R 1

Middle School Math Vocabulary

absolute value	calculator
abundant number	calendar problem
acute angle	center
acute triangle	centimeter
add	central angle
addition	chart
Addition Property of Equality	chord
additive inverse	cipher
adjacent	circle
adjacent angles	circle graph
adjacent sides	circumference
age problem	circumscribed
algebra	Closure Property of Addition
algebraic expression	Closure Property of Multiplication
algorithm	coefficient
alternate exterior angles	coin problem
alternate interior angles	collinear
altitude	combination
AND	common denominator
angle	common difference
apothem	common element
approximation	common factor
arc	common fraction
area	common multiple
arithmetic	common ratio
arithmetic mean	Commutative Property of Addition
arithmetic sequence	Commutative Property of Multiplication
array	compass
ascending order	complement
Associative Property of Addition	complementary angles
Associative Property of Multiplication	complementation
at least	complex fraction
at most	complex number
average	composite number
average rate	composition
axis (axes)	compound interest
bar graph	concatenation
base	concave
billion	concentric
binary	cone
bisect	congruent
bisection	consecutive
boundary problem	consecutive even

consecutive odd	divisor
constant	dodecagon
convex	dodecahedron
coordinate plane	domain
coordinates	double
coplanar	edge
corresponding angles	element
corresponding sides	empty set
counterexample	endpoint
counting number	equal
cross multiply	equality
cube	equation
cube root	equiangular
cylinder	equidistant
data	equilateral
decagon	equilateral triangle
decimal	equivalent
decimeter	equivalent fractions
decrease	estimation
decrement	Euler Graph
definition	evaluate
degree	even number
denominator	event
dependent events	expanded notation
descending order	exponent
diagonal	exponential form
diameter	exponentiation
dichotomy	expression
difference	exterior
digit	exterior angle
dimension	externally tangent
direct variation	face
directed graph	factor
discount	factorial
discrete sets	fence post problem
disjoint sets	Fibonacci Sequence
distance	fictitious operation
distance-rate-time problem	field
distinct	First-Plus-Last Method
Distributive Property	flip
dividend	foot (feet)
divisibility rule	formula
divisible	fraction
division	fractional part

frequency distribution	inscribed
frequency table	insufficient information
function	integer
geometric mean	intercept
geometric sequence	interest
geometry	interior
Goldbach's Conjecture	interior angle
graph	internally tangent
greater	interpolate
greater than	intersect
greater than or equal to	intersecting lines
greatest	intersection
greatest common divisor	inverse
greatest common factor	inverse operation
half	Inverse Property of Addition
halve	Inverse Property of Multiplication
height	inverse variation
hemisphere	irrational number
heptagon	isosceles right triangle
hexagon	isosceles trapezoid
hexagonal prism	isosceles triangle
hexagonal pyramid	kilometer
histogram	kite
How Many? problem	lateral
hundred	lateral area
hundredth	lattice point
hypotenuse	least
i	least common denominator
icosahedron	least common multiple
identity element	leg
Identity Property of Addition	length
Identity Property of Multiplication	less
imaginary number	less than
improper fraction	less than or equal to
in terms of	like terms
inch	line
increase	line of symmetry
increment	line segment
independent events	linear equation
inequality	linear function
infinite	long division
infinite sequence	lowest common denominator
infinite series	lowest terms
infinity	magic square

mathematical maturity	number line
mathematician	number theory
mathematics	numeral
matrix (matrices)	numerator
matrix addition	obtuse angle
matrix multiplication	obtuse triangle
maximum	octagon
mean	octahedron
measure	odd number
measurement	odds
median	ones digit
member	ones place
mental math	operation
meter	opposite
metric system	opposite angles
midline	opposite sides
midpoint	OR
millimeter	order
million	ordered pair
minimum	organized list
mixed number	origin
mixture problem	original cost
mode	outcome
modulo	outlier
multiple	palindrome
multiplication	parabola
Multiplication Principle	parallel
Multiplication Property of Equality	parallelepiped
multiplicative inverse	parallelogram
mutually exclusive	parenthesis (parentheses)
natural number	pattern
negate	pentagon
negative number	pentagonal prism
negative sign	pentagonal pyramid
net	percent
nonagon	percent change
noncollinear	perfect cube
nonoverlapping	perfect number
nonzero	perfect square
NOT	perimeter
not to scale	permutation
nth term	perpendicular
number	perpendicular bisector
number base	perpendicular lines

perpendicular planes	ratio
pi	rational
place	ray
place value	real number
plane	reciprocal
point	rectangle
polygon	rectangular prism
polyhedron	rectangular pyramid
positive number	rectangular solid
power	reflection
power of ten	Reflexive Property of Equality
power of two	region
preceding term	region bounded by
prime factorization	regular
prime number	regular polygon
principal	relation
prism	relatively prime
probability	remainder
problem solving	repeating decimal
product	rhombus
profit problem	right angle
proof	right cylinder
proper fraction	right prism
proper subset	right pyramid
property	right triangle
proportion	rigid motion
protractor	root
prove	rotation
pyramid	sale price
Pythagorean primitive or triplet	sales tax
Pythagorean Theorem	scalar multiplication
Quadrant I, II, III, and IV	scalene trapezoid
quadratic equation	scalene triangle
quadratic formula	scientific notation
quadrilateral	sector
quadruple	segment
quantity	semicircle
quarter	semiperimeter
quotient	sequence
radical	series
radius	set
random number	shaded area
range	short division
rate	side

similar	term
similar polygons	terminating decimal
similar triangles	tetrahedron
simple interest	thousand
simplest form	thousandth
simplify	transformation
skew lines	Transitive Property of Equality
slant height	transitivity
slide	translation
slope	transversal
solid	trapezoid
solution	triangle
solution set	trichotomy
solve	trillion
space	trillions place
special right triangle	triple
sphere	triplet problem
square	trisect
square root	trisection
standard form	twin primes
standard notation	undefined
statistics	uniform border problem
stem-and-leaf plot	union
straight angle	unit
straight edge	unit fraction
subset	units digit
substitution	units place
subtraction	universal set
succeeding term	value
successive terms	variable
sum	Venn Diagram
sum of the digits	vertex (vertices)
supplement	vertical angles
supplementary angles	volume
surface area	whole number
symbol	width
symmetry	work problem
symmetric about	x-axis
Symmetric Property of Equality	x-coordinate
system of equations	yard
tangent	y-axis
ten billions place	y-coordinate
ten millions place	zero
tens place	Zero Property of Multiplication

Pre-algebra Sample Test A: 90 minutes, No calculator	
1. Find the median of the data. 9.1, 5.3, 4.6, 23.8, 7.8, 6.4 4.6 5.3 6.4 7.8 9.1 23.8 $\boxed{7.1}$ With an even number of numbers, average the 2 middlemost. The spread is 1.4. Add 0.7 to the lesser.	6. Find the surface area in square inches of a rectangular prism with edges 4, 5, and 9 inches. 4 5 9 (20 + 45 + 36) • 2 101 • 2 Think of the groupings as a happy face. $\boxed{202}$
2. In which quadrant or on which axis does (−19, −13) lie? $\boxed{\text{III}}$ Start in (pos, pos) and count counter-clockwise.	7. Find the balance when $7000 compounds at 2% for 2 years. 7000 + 140 = 7140 7140 + 71.40 + 71.40 = $\boxed{\$7282.80}$
3. Find the midpoint of the segment with endpoints (−2, 9) and (−7, −2). $\boxed{(-4.5, 3.5)}$ Average the xs; average the ys.	8. A segment has endpoints (2, 7) and (4, 8). Name its endpoints after a reflection over the line y = 1. 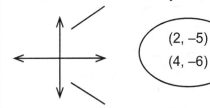 $\boxed{\begin{array}{c}(2, -5)\\(4, -6)\end{array}}$
4. Find the slope of the line through (4, −7) and (12, −3). $\dfrac{-3 - -7}{12 - 4} = \dfrac{4}{8} = \boxed{\dfrac{1}{2}}$ Subtract the ys over subtract the xs.	9. Express in scientific notation. $(70.5 \times 10^3)(1.01 \times 10^4)$ $(705)(10^2)(101 \times 10^2)$ $(705)(101)(10^2)(10^2)$ $(70,500 + 705)(10^4)$ $(71,205)(10^4)$ $\boxed{(7.1205)(10^8)}$
5. Simplify. $-6 - 3 \cdot |5 - 12| - 12 \div 4 \cdot 3$ $-6 - 3 \cdot 7 - 3 \cdot 3$ $-6 - 21 - 9$ $\boxed{-36}$	10. Two angles of a triangle are 40.7° and 100.1°. Find the third. $\boxed{39.2°}$

Pre-algebra Sample Test A	
11. Find the volume of a sphere with diameter 12 feet. $V = \dfrac{4}{3}\pi r^3 = \dfrac{4}{3}\pi \cdot 6 \cdot 6 \cdot 6$ $= 4 \cdot \pi \cdot 72$ 288π cu ft	16. Solve for a. $x = c\,(b + a)$ Distribute to release the desired variable from parentheses. $x = cb + ca$ Then isolate it. $x - cb = ca$ $\dfrac{x - bc}{c} = a$
12. Name the property illustrated. $13\left(\dfrac{1}{13}\right) = 1$ Inverse Property of Multiplication or InPM A number times its multiplicative inverse (reciprocal) is 1.	17. Find the distance between the points $(7, -6)$ and $(11, -2)$ as a simplified radical. $\sqrt{16 + 16}$ $\sqrt{32}$ $4\sqrt{2}$ Subtract the xs and square. Subtract the ys and square. The distance formula is equivalent to the Pythagorean Theorem.
13. Write as an algebraic expression. half the sum of two numbers, increased by 6 $\dfrac{x + y}{2} + 6$ By PEMDAS, "half" (division) occurs before "increased" (addition).	18. Simplify. $6\sqrt{32} + 5\sqrt{8}$ $6\sqrt{16}\sqrt{2} + 5\sqrt{4}\sqrt{2}$ $24\sqrt{2} + 10\sqrt{2}$ $34\sqrt{2}$
14. Find. GCF $(15x^2y^4z^9,\ 18xy^2z^6)$ $3xy^2z^6$ Do constants and each variable separately.	19. Complete. 7.5 pints = $\underline{\ \ 15\ \ }$ cups 3.5 gallons = $\underline{\ \ 28\ \ }$ pints 2.5 quarts = $\underline{\ \ 5\ \ }$ pints Decide the conversion for one unit. Then multiply proportionally.
15. List all factors of 48. 1, 2, 3, 4, 6, 8, 12, 16, 24, 48 Write factor pairs from left and right concurrently, working toward the center. Using 2 lines and leaving center white space is fine.	20. Simplify. $\dfrac{22x^3y^2z^5}{33xy^6z^2}$ $\dfrac{2x^2z^3}{3y^4}$ Do constants and each variable separately.

Pre-algebra Sample Test A	
21. Solve. Answer as a fraction. $\dfrac{11}{x} = \dfrac{6}{7}$ $\quad$ $\dfrac{x}{13} = \dfrac{4}{5}$ $6x = 77$ $\qquad$ $5x = 52$ $\boxed{x = \dfrac{77}{6}}$ $\quad$ $\boxed{x = \dfrac{52}{5}}$	**26.** Find the percent change when 80 becomes 15. $\dfrac{80 - 15}{80}$ $\quad$ $\dfrac{65}{80}$ $\quad$ $\dfrac{16.25}{20}$ $\quad$ $\dfrac{81.25}{100}$ Using mental math, change $\quad$ $\boxed{81.25\% \text{ D}}$ $65 \div 4$ to $64 \div 4$ plus $1 \div 4$.
22. Solve. $|x + 7| = 12$ $\boxed{x = 5 \text{ or } x = -19}$ $x + 7$ can equal 12 or -12.	**27.** Evaluate. $GCF(19, 57) + GCF(49,125)$ $\qquad$ $19 + 1$ $\qquad$ 49 has only 7s $\qquad\qquad\qquad\qquad$ while 125 has $\qquad\qquad\qquad\qquad$ only 5s. $\qquad\qquad$ $\boxed{20}$ 57 is a multiple of 19.
23. How many terms does the polynomial have? $6a^2 - 12ab + 4b^2 + 7$ $\qquad\qquad\qquad$ $\boxed{4}$ Terms are separated by plus or minus.	**28.** Simplify. $(10x^4)\,(7x^3)\,(5x^7) + (3x^7)^2$ $350x^{14} + 9x^{14}$ $\boxed{359x^{14}}$
24. Find the original cost when a selling price of \$91 represents a 30% loss. $\dfrac{7}{10}\,P = 91$ $P = (91)\,\dfrac{10}{7}$ $P = (13)\,(10)$ $\quad$ $\boxed{\$130}$	**29.** Find: (5 hrs 13 mins 12 secs) $\div$ 3 $\boxed{1 \text{ hr} \quad 44 \text{ mins} \quad 24 \text{ secs}}$ $3\;|\;\overline{5 \text{ hrs} \quad 13 \text{ mins} \quad 12 \text{ secs}}$ $\dfrac{3}{2}$ $\qquad$ $\begin{matrix}120\\133\\132\\\hline 1\end{matrix}$ $\qquad$ $\begin{matrix}60\\72\end{matrix}$
25. Find the simple interest when \$350 is invested at 3.6% for 8 months. $\dfrac{(350)\,(3.6)\,(8)}{(100)\,(12)}$ $\quad$ $\dfrac{(35)\,(1.2)}{5}$ $\dfrac{(35)\,(3.6)\,(2)}{(10)\,(3)}$ $\quad$ $(7)\,(1.2)$ $\qquad\qquad\qquad$ $\boxed{\$8.40}$	**30.** Find the range of the data. 3.5, 2.1, 20.9, 1.9, 50.8, 6.7, 20.9 $\qquad$ $50.8 - 1.9$ $\qquad\qquad$ $\boxed{48.9}$

Pre-algebra Sample Test A	
31. Find the probability of rolling an odd prime with one roll of a standard 1 through 6 number cube. 3, 5 $\dfrac{2}{6}$ $\left(\dfrac{1}{3}\right)$	**36.** Write in scientific notation. 375.43×10^6 $\left(3.7543 \times 10^8\right)$
32. Express as a simplified fraction. $7.\overline{27}$ $7\dfrac{27}{99}$ $7\dfrac{3}{11}$ $\left(\dfrac{80}{11}\right)$	**37.** Name the space figure depicted by the net. 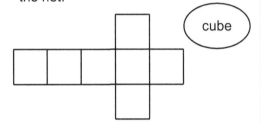 $\left(\text{cube}\right)$
33. Solve. Answer as a fraction. $\dfrac{1}{8}(8x-3)=10$ $8x-3=80$ $8x=83$ $\left(x=\dfrac{83}{8}\right)$	**38.** Find the LCM. $\text{LCM}(20x^3y^2z,\ 25x^2yz^4)$ $\left(100x^3y^2z^4\right)$ *Do constants and each variable separately.*
34. Operate. $^{(y)}\ \ y \qquad\quad 4x\ ^{(2x)}$ $\dfrac{1}{6x} + \dfrac{2}{3y}$ $\left(\dfrac{y+4x}{6xy}\right)$ $6xy \qquad 6xy$ *Multiply each fraction by 1: the left by y/y; the right by 2x/2x.*	**39.** What percent of 11 is 5? $\dfrac{5}{11}$ $\dfrac{45}{99}$ $\left(45.\overline{45}\%\right)$ *Write the "is over of" numbers as a fraction. A fraction with all digits of 9 converts directly to a decimal.*
35. Write the prime factorization of 720. 72 x 10 8 x 9 x 2 x 5 $\left(2^4 \times 3^2 \times 5\right)$	**40.** Divide. $\begin{array}{c}{}^{1}\\[-2pt]\dfrac{11}{8xy}\\[-2pt]{}_{y}\end{array} \div \begin{array}{c}{}^{2z}\\[-2pt]\dfrac{16xz}{33}\\[-2pt]\dfrac{16xz}{33}\\[-2pt]{}_{3}\end{array}$ $\left(\dfrac{2z}{3y}\right)$

Pre-algebra Sample Test A	
41. Solve. Answer as a fraction. $6x - 11 = 2(x + 3) - x$ $7x - 11 = 2x + 6$ $5x = 17$ $$x = \frac{17}{5}$$	46. Find the mode of the data. 5, 9, 6, 2, 3, 9, 2, 4, 1, 9, 4, 7, 8, 5 ↑ ↑ ↑ 9
42. Find the volume in cubic cm of a cone with radius 6 cm and height 10 cm. $$\frac{6 \cdot 6 \cdot \pi \cdot 10}{3}$$ 120 π The volume of a cone is 1/3 the volume of the cylinder in which it just fits.	47. Find the diagonal in feet of a square with side 12 feet. x x $x\sqrt{2}$ 12 12 $12\sqrt{2}$
43. Operate. $\sqrt{72} \cdot \sqrt{25} \cdot \sqrt{48}$ $\sqrt{36} \cdot \sqrt{2} \cdot 5 \cdot \sqrt{16} \cdot \sqrt{3}$ $6 \cdot 5 \cdot 4 \cdot \sqrt{6}$ $120\sqrt{6}$	48. Express as a decimal. $$\frac{2}{11} + \frac{4}{9} \qquad \frac{18}{99} + \frac{44}{99} \qquad \frac{62}{99}$$ $.\overline{62}$
44. In how many ways may 8 different CDs be stacked vertically? $8 \cdot 7 \cdot 6 \cdot 5 \cdot 4 \cdot 3 \cdot 2 \cdot 1$ $7 \cdot 24 \cdot 24 \cdot 10$ $7 \cdot 576 \cdot 10$ $4032 \cdot 10$ 40,320 Use Multiplication Principle.	49. Find the surface area in square units of a sphere with diameter 10 units. $4 \pi r^2 = (4\pi)(25) =$ 100π
45. Round 6.1285 to each place. ones 6 tenths 6.1 hundredths 6.13 thousandths 6.129	50. Find LCM (34, 85). 2 x 17 5 x 17 2 x 5 x 17 10 x 17 170

Pre-algebra Sample Test A	
51. Find the perimeter in cm of a rectangle with area 72 sq cm and width 6 cm. A = 72 6L = 72 L = 12 W = 6 P = 18 + 18 = (36)	56. Solve: x (x + 5) (x – 7) (x + 1) = 0 (x = 0, –5, 7, –1) Solve: (9x – 4) (6x + 7) = 0 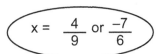 (x = $\frac{4}{9}$ or $\frac{-7}{6}$)
52. A book costs $16.00. The sales tax is 6%. You pay with a twenty-dollar bill. How much change do you receive? 0.06 x 16 is 0.96 20 – 16.96 = 3.04 ($3.04)	57. 2, 3, and 7 are factors of n. Name five other factors of n. (1, 6, 14, 21, 42) 1 is a factor of every natural number. If 2 and 3 (primes) are factors, then 2x3 is a factor, and so on.
53. Label each number rational (Q) or irrational (Irr). $\sqrt{36}$ (Q) $\frac{1}{7}$ (Q) $\sqrt{30}$ (Irr) $0.5\overline{06}$ (Q)	58. Find the probability of doubles when tossing 2 dice. 1, 1 ⟶ 6, 6 $\frac{6}{6 \times 6}$ ($\frac{1}{6}$)
54. Operate. 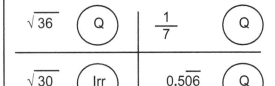 ($\frac{1}{4}$)	59. Name the property illustrated. 3x = 3x (RPE) Reflexive Property of Equality
55. Write as a percent: 2.0043 (200.43%) The percent symbol means "parts of 100" or divide by 100. To compensate, multiply by 100.	60. A square has area 8100 sq un. Find its perimeter in units. S = 90 P = (360)

Pre-algebra Sample Test A	
61. Find the slope and y-intercept of the line $3x - 12y = 5$. $-12y = -3x + 5$ $12y = 3x - 5$ slope $\dfrac{1}{4}$ y-intercept $\dfrac{-5}{12}$	**66.** Graph on the number line. $x > -3$  $x \le 4$ OR $x = 5$ 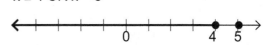
62. The area of a circle is 121π sq un. Find its radius in un. $\boxed{11}$ ――――――――― The circumference of a circle is 528π un. Find its radius in un. $\boxed{264}$	**67.** Find the greatest 2-digit number with an odd number of factors plus the least 3-digit number with an odd number of factors. $81 + 100 = \boxed{181}$ A number must be a perfect square to have an odd number of factors.
63. Solve for M. $21AB^2 = 35B^3AMC$ $\dfrac{21AB^2}{35B^3AC} = \dfrac{35B^3AMC}{35B^3AC}$ $\boxed{\dfrac{3}{5BC} = M}$	**68.** Evaluate. $\dfrac{10!}{3!\,6!}$ $\dfrac{7 \cdot \overset{4}{8} \cdot \overset{3}{9} \cdot 10}{3 \cdot 2}$ $7 \cdot 12 \cdot 10$ $\boxed{840}$
64. A man left 1/2 of his estate to his wife, \$40,000 to his son, 1/2 of what remained to a friend, and the final \$6000 to charity. What was the value of his estate? $\boxed{\$104{,}000}$ $\dfrac{x}{2} + 40{,}000 + \dfrac{\dfrac{x}{2} - 40{,}000 + 6000}{2} = x$ $x + 80{,}000 + .5x - 40{,}000 + 12{,}000 = 2x$	**69.** Find the surface area in sq un of a cylinder with radius 5 un and height 13 un. A circ = 25π A rec = Ch = $\pi dh = 130\pi$ $25\pi + 25\pi + 130\pi = \boxed{180\pi}$
65. Find the surface area in sq un of a cube with volume 1728 cu un. V = 1728 e = 12 A face = 144 SA = 6 x 144 = $\boxed{864}$	**70.** Name all of the line segments in the figure. 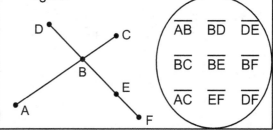 $\overline{AB}$ $\overline{BD}$ $\overline{DE}$ $\overline{BC}$ $\overline{BE}$ $\overline{BF}$ $\overline{AC}$ $\overline{EF}$ $\overline{DF}$

Pre-algebra Sample Test A	
71. How many 3-letter permutations may be made using the letters PENCIL? 6 x 5 x 4　　(120) Multiplication Principle	76. Calculate the combination. $C(6, 3) = \dfrac{6!}{3!\ 3!} = \dfrac{4 \cdot 5 \cdot 6}{3 \cdot 2 \cdot 1} = (20)$
72. In △ABC, m∠A = 90, AB = 3, and AC = 3. Classify the triangle by angle and side. By angle　　right By side　　isosceles	77. Simplify. 5a – 3b + 2(a + b) – 3(3a – 2b) 5a – 3b + 2a + 2b – 9a + 6b (–2a + 5b)
73. Find 8.4% of 250. 1% of 250 = 2.5 8% of 250 = 20 1/5 % of 250 = 0.5 2/5 % of 250 = 1 20 + 1 = (21)	78. A triangle has endpoints (0, 0), (0, –8), and (4, 0). Name its endpoints after a reflection over the x-axis followed by a reflection over the y-axis. 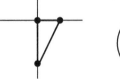　　(0, 0) 　　(–4, 0) 　　(0, 8)
74. Name the property illustrated. If x = y, then y = x. (SPE) Symmetric Property of Equality	79. Find the volume in cu un of a square pyramid with base edge 6 units and altitude 11 units. $\dfrac{6 \times 6 \times 11}{3}$　　2 x 6 x 11 　　　　　　12 x 11 (132)
75. Find $1\frac{2}{9}$ minus its reciprocal. $\dfrac{11}{9} - \dfrac{9}{11} = \dfrac{121}{99} - \dfrac{81}{99} = \left(\dfrac{40}{99}\right)$	80. Simplify. $2\sqrt{25} - |\,10 - (-9)\,| + 2 \cdot (3)^2$ 10 – 19 + 18 (9)

Pre-algebra Sample Test A

81. Write an equation of the line through (2, 5) and (0, 4).

SL $\dfrac{5-4}{2-0}$ $\dfrac{1}{2}$

$\boxed{y = \dfrac{1}{2} x + 4}$

86. Evaluate for x = –2 and y = – 4.

$2xy - 3y^2 - 5x - 5$

$16 - 48 + 10 - 5$

$\boxed{-27}$

82. Write as a unit rate.

39 boxes in 6 minutes

$\dfrac{39}{6}$ boxes in 1 minute

$\dfrac{13}{2}$ boxes in 1 minute

$\boxed{\text{6.5 boxes in 1 minute}}$

87. Complete each with the correct vocabulary word.

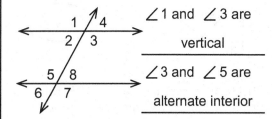

∠1 and ∠3 are __vertical__

∠3 and ∠5 are __alternate interior__

83. Find the volume in cu ft of a rectangular prism that measures 24 in by 6 in by 8.35 feet.

(2) (0.5) (8.35)

(1) (8.35)

$\boxed{8.35}$

88. Find the rate in mph of a vehicle that travels 270 miles in 4.5 hours.

D = RT

$270 = R \cdot \dfrac{9}{2}$

$\dfrac{2}{9} \; 270 = R \cdot \dfrac{9}{2} \; \dfrac{2}{9}$

$\boxed{60 = R}$

84. Write the next two terms and the 41st term of the arithmetic sequence.

4, 7.5, 11, 14.5, 18, __21.5__   __25__

41st term = 4 + (40)(3.5)
        = 4 + (120 + 20)

= $\boxed{144}$

89. An equilateral triangle has perimeter 24.6 units. Find the measure of a side in units.

24.6 ÷ 3 = $\boxed{8.2}$

85. Complete.

13.02 mm = __1.302__ cm

13 L = __13,000__ mL

Use 2 steps: determine the number of places away and whether more or fewer are needed to balance the equation.

90. Express the English as math: Twice the product of two numbers is greater than the sum of the numbers increased by two.

$\boxed{2xy > x + y + 2}$

Pre-algebra Sample Test A	
91. Find x if 30% of x is 111. $\dfrac{30}{100} \cdot x = 111$ $3x = 1110$ $x = \boxed{370}$	96. A bowl has 11 yellow, 10 red, 20 blue, and 14 orange marbles. Find, with replacement, P(BLUE, RED) after 2 draws. $\dfrac{20}{55} \cdot \dfrac{10}{55} \quad \dfrac{4}{11} \cdot \dfrac{2}{11} \quad \boxed{\dfrac{8}{121}}$
92. Find the 2 missing congruent lengths. $\boxed{8\sqrt{2}}$ "Divide by 2, tack on the root 2" is faster than doing out the Pythagorean Theorem.	97. Given $\triangle CAB \cong \triangle DEB$, m$\angle$B = 90, m$\angle$C = 53.1, BA = 8, and CB = 6. Find: m$\angle$E = 36.9 | EB = 8 m$\angle$D = 53.1 | BD = 6 m$\angle$A = 36.9 | DE = 10
93. Solve. $\dfrac{5}{9} \cdot x - 1 = 29$ $\dfrac{5}{9} \cdot x = 30$ $\dfrac{9}{5}\,\dfrac{5}{9} \cdot x = 30\,\dfrac{9}{5} \qquad x = \boxed{54}$	98. A diameter of a circle has endpoints (3, 11) and (7, 3). Find the circle's center. $\boxed{(5,\ 7)}$ The circle's center is the midpoint of the diameter.
94. Write the equation of the line through the origin with slope –3. $\boxed{y = -3x}$ Lines through the origin are of the form y = mx.	99. Complete the missing numbers in the Pythagorean triplets. 8, 8, $8\sqrt{2}$ 5, 12 13 4, 5, $\sqrt{41}$
95. Find the hypotenuse in simplest radical form. 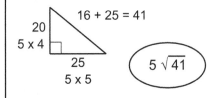 20 5 x 4 16 + 25 = 41 25 5 x 5 $\boxed{5\sqrt{41}}$	100. Express as math: the sum of 3 consecutive odd numbers. $\boxed{x + (x + 2) + (x + 4)}$ An even number is 2x; an odd number is 2x + 1. However, in an algebra word problem, both may be expressed as x. If impossible, the equation will lead to no solution.

Pre-algebra Sample Test A	
101. A trapezoid has A = 180, h = 12, and B = 21. Find b. 180 = 12M M = 15 (9) 15 21	**106.** Label the measures of the 7 angles. (NTS) 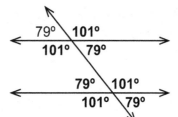
102. Find the final sales price on a $280 item after a 15% markup and a 10% discount. 280 280 + 28 + 14 = 322 322 − 32.20 (289.80)	**107.** What is the effect on the area of a rectangle if both the length and width are tripled? (L) (W) (3L) (3W) (9) (LW) (times 9)
103. Label each sequence A (arithmetic), G (geometric), or N (neither). 1, 4, 9, 16, 25, . . . __N__ perfect squares 1, −2, 4, −8, 16, . . . __G__ x −2	**108.** Find the 2 missing sides. 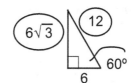
104. Find the number of codes that may be formed of the type: LETTER–DIGIT–EVEN DIGIT 26 x 10 x 5 (1300)	**109.** Given a square with perimeter 120 units and an inscribed circle. Find the area in sq un between the circle and square. P sq = 120 s = 30 d = 30 r = 15 A sq = 900 A circ = 225π 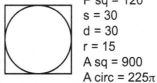 $(900 - 225\pi)$
105. Solve: $-7x - 10 > -8$ $-7x > 2$ $x < \dfrac{-2}{7}$ When multiplying or dividing both sides of an inequality by a negative number, the sign flips.	**110.** Write the equation of the line shown. 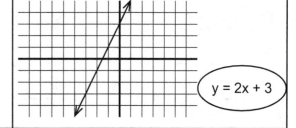 $(y = 2x + 3)$

Pre-algebra Sample Test A	
111. Solve. $3(2x + 4) - x + 1 = 3(3x + 2)$ $6x + 12 - x + 1 = 9x + 6$ $5x + 13 = 9x + 6$ $7 = 4x$ $x = \boxed{\dfrac{7}{4}}$	116. Graph: $y = -x + 2$ 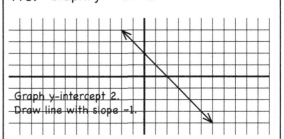 Graph y-intercept 2. Draw line with slope −1.
112. Write the function rule. | x	f(x)
2	−5
0	3
−1	7 | $\boxed{f(x) = -4x + 3}$ Note linear relationship: As x increases by 1, y decreases by 4. $y = mx + b$ Use (0, 3) $y = mx + 3$ Use (−1, 7) $y = -4x + 3$	117. Evaluate for $x = \dfrac{1}{2}$, $y = \dfrac{2}{3}$. $4x^2 - 6xy - 9y^2 - 6$ $1 - 2 - 4 - 6$ $\boxed{-11}$
113. Solve the system by substitution. $2x + 3y = 4$ $x - 2y = -5 \quad x = 2y - 5$ $2(2y - 5) + 3y = 4$ $4y - 10 + 3y = 4$ $7y = 14$ $y = 2$ $\boxed{\begin{array}{l} x = -1 \\ y = 2 \end{array}}$	118. Find the area in sq un of the figure: a semi-circle attached to a parallelogram. $d = 10$ $r = 5$ A circ $= 25\pi$ A para $= 10 \times 6$ 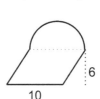 6 10 $\boxed{60 + 12.5\pi}$
114. Multiply. $(5x - 2)(3x - 5)$ $15x^2 - 25x - 6x + 10$ $\boxed{15x^2 - 31x + 10}$	119. Operate. $\dfrac{5}{14} + \dfrac{1}{7} + \dfrac{7}{10} \quad \dfrac{25 + 10 + 49}{2 \cdot 5 \cdot 7} \quad \boxed{\dfrac{6}{5}}$ $7 \cdot 2 \quad 7 \quad 5 \cdot 2 \qquad 84$ Do not multiply denominator. Simplify first.
115. Complete. $1^{36} \quad \underline{1}$ $36^0 \quad \underline{1}$ $0^{36} \quad \underline{0}$ $5^{-2} \quad \underline{\dfrac{1}{25}}$ $(-2)^6 \quad \underline{64}$ $-2^6 \quad \underline{-64}$	120. Circle the numbers that are divisors of the 15-digit number 436,951,785,053,124. ① ② ③ ④ 5 ⑥ 8 ⑨ 10 ⑫ 15 ⑱ Divisibility rule for 18: by 2 and by 9.

Pre-algebra Sample Test B: 90 minutes, No calculator	
1. Find the median of the data. 6.7, 8.4, 2.3, 5.9, 4.1, 11.8 2.3 4.1 5.9 6.7 8.4 11.8 6.3 With an even number of numbers, average the 2 middlemost. The spread is 0.8. Add 0.4 to the lesser.	**6.** List all factors of 60. 1, 2, 3, 4, 5, 6, 10, 12, 15, 20, 30, 60 Write factor pairs from left and right concurrently, working toward the center. Using 2 lines and leaving center white space is fine.
2. Simplify. $3\sqrt{75} + 4\sqrt{12}$ $3\sqrt{25}\sqrt{3} + 4\sqrt{4}\sqrt{3}$ $15\sqrt{3} + 8\sqrt{3}$ $23\sqrt{3}$	**7.** A segment has endpoints $(4, 6)$ and $(5, 9)$. Name its endpoints after a reflection over the line $x = 1$. 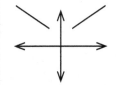 $(-2, 6)$ $(-3, 9)$
3. Find the volume of a sphere in cubic inches with diameter 18 inches. $V = \dfrac{4}{3}\pi r^3 = \dfrac{4}{3}\pi \cdot 9 \cdot 9 \cdot 9$ $= 4 \cdot \pi \cdot 243$ 972π	**8.** Solve for b. $xy = 5 + 3(a + b)$ Distribute to release $xy = 5 + 3a + 3b$ the desired variable from parentheses. Then $xy - 5 - 3a = 3b$ isolate it. $\dfrac{xy - 3a - 5}{3} = b$
4. Find. GCF($20a^2b^2c^5$, $30ab^4c^8$) $10ab^2c^5$ Do constants and each variable separately.	**9.** Name the property illustrated. $ab + 1 = ba + 1$ CPM Commutative Property of Multiplication
5. Write as an algebraic expression: the product of three numbers decreased by five. 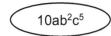 xyz − 5	**10.** Two angles of a triangle are 31.2° and 104.5°. Find the third. 44.3°

Pre-algebra Sample Test B	
11. Find the midpoint of the segment with endpoints (13, –3) and (–2, –11). (5.5, –7) Average the xs; average the ys.	16. Solve. Answer as a fraction. $\dfrac{5}{9} = \dfrac{x}{17}$ $\dfrac{8}{25} = \dfrac{7}{x}$ $9x = 85$ $8x = 175$ $x = \dfrac{85}{9}$ $x = \dfrac{175}{8}$
12. Simplify. $-9 - (-7) - 2 \cdot |7 - 9| + 18 \div 3 \cdot 6$ $-9 + 7 - 2 \cdot 2 + 6 \cdot 6$ $-2 - 4 + 36$ 30	17. Find the distance between the points (–3, 5) and (8, –3) as a simplified radical. $\sqrt{121 + 64}$ $\sqrt{185}$ Subtract the xs and square. Subtract the ys and square. The distance formula is equivalent to the Pythagorean Theorem.
13. Find the slope of the line through (–3, 6) and (8, –16). $\dfrac{6 - -16}{-3 - 8} = \dfrac{22}{-11} =$ –2 Subtract the ys over subtract the xs.	18. In which quadrant or on which axis does (16, –12) lie? IV Start in (pos, pos) and count counterclockwise.
14. Write in scientific notation. $(2.3 \times 10^3)(11.24 \times 10^2)$ $(2.3)(11.24) \times 10^5$ $(22 + 0.48 + 3.3 + 0.072) \times 10^5$ 25.852×10^5 2.5852×10^6	19. Find the range of the data. 8.2, 2.8, 21.7, 8.6, 44.7, 3.4, 26.7 $44.7 - 2.8$ 41.9
15. Find the surface area in sq cm of a rectangular prism with edges 6, 8, and 10 cm. 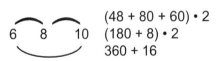 6 8 10 $(48 + 80 + 60) \cdot 2$ $(180 + 8) \cdot 2$ $360 + 16$ Think of the groupings as a happy face. 376	20. Complete. 42 cups = 10.5 quarts 34 pints = 8.5 half-gallons 13 cups = 6.5 pints Decide the conversion for one unit. Then multiply proportionally.

Pre-algebra Sample Test B	
21. Express as a simplified fraction. $5.\overline{18}$ $5\frac{18}{99}$ $5\frac{2}{11}$ $\boxed{\frac{57}{11}}$	26. Operate. (2y) 10y 9x (3x) $\frac{5}{6x}$ + $\frac{3}{4y}$ $\boxed{\frac{9x + 10y}{12xy}}$ 12xy 12xy
22. Find the probability of a number less than 6 landing face up with one roll of a standard 1 through 6 number cube. 1, 2, 3, 4, 5 $\boxed{\frac{5}{6}}$	27. Find the balance when $8000 compounds at 3% for 2 years. 8000 + 240 = 8240 8240 + 82.40 + 82.40 + 82.40 = 8240 + 246 + 1.20 = $\boxed{\$8487.20}$
23. Find the original cost when a selling price of $133 represents a 30% loss. $\frac{7}{10}P = 133$ $P = (133)\frac{10}{7}$ $P = (19)(10)$ $\boxed{\$190}$	28. How many terms does the polynomial have? $13x^2 - 4xy + 16y^3$ $\boxed{3}$ Terms are separated by plus or minus.
24. Find the percent change when 80 becomes 35. $\frac{80 - 35}{80}$ $\frac{45}{80}$ $\frac{11.25}{20}$ $\frac{56.25}{100}$ Using mental math, change 45 ÷ 4 to 44 ÷ 4 plus 1 ÷ 4. $\boxed{56.25\% \text{ D}}$	29. Solve. $\lvert x + 5 \rvert = 13$ $\boxed{x = 8 \text{ or } x = -18}$ x + 5 can equal 13 or –13.
25. Write the prime factorization of 990. 99 x 10 9 x 11 x 2 x 5 $\boxed{2 \times 3^2 \times 5 \times 11}$	30. Write in scientific notation. $(50.2 \times 10^4)(10.1 \times 10^5)$ $(502)(10^3)(101 \times 10^4)$ $(502)(101)(10^3)(10^4)$ $(50{,}200 + 502)(10^7)$ $(50{,}702)(10^7)$ $\boxed{(5.0702)(10^{11})}$

Pre-algebra Sample Test B	
31. Simplify. $$\frac{36wx^2y^7z^3}{60wx^3y^5z^5}$$ $\left(\dfrac{3y^2}{5xz^2}\right)$ Do constants and each variable separately.	**36.** Simplify. $(3x)\,(4x^2)\,(2x^7) + (4x^5)^2$ $24x^{10} + 16x^{10}$ $\left(40x^{10}\right)$

32. Evaluate.
 	
$\text{GCF}(14, 77) + \text{GCF}(49,100)$
 	
$\qquad 7 + 1$ 49 has only 7s
           while 100 has
           no 7s.
$\quad\left(8\right)$
	37. Solve. Answer as a fraction.
	
	$7x - 27 = 5\,(x - 4) - 2x$
	$9x - 27 = 5x - 20$
	$4x = 7$
	
	$\left(x = \dfrac{7}{4}\right)$

33. Find the greatest 4-digit number having 2, 3, and 8 as three of its digits and 6 as a factor. $\left(8832\right)$ 9832 is not divisible by 3.	**38.** Divide. $\qquad\qquad\qquad\;\; 2b$ $\quad 6b \quad\;\; 14bc$ $\dfrac{30ab}{49c} \div \dfrac{25a}{14bc} \quad \left(\dfrac{12b^2}{35}\right)$ $\quad\; 7 \qquad 25a$ $\qquad\qquad\;\; 5$
34. Solve. Answer as a fraction. $\dfrac{1}{5}\,(5x - 1) = 11$ $5x - 1 = 55$ $5x = 56$ $\left(x = \dfrac{56}{5}\right)$	**39.** What percent of 33 is 16? $\dfrac{16}{33} \quad \dfrac{48}{99} \quad \left(48.\overline{48}\%\right)$ Write the "is over of" numbers as a fraction. A fraction with all digits of 9 converts directly to a decimal.
35. Find the simple interest when \$480 is invested at 4.5% for 10 months. $\dfrac{(480)\,(4.5)\,(10)}{(100)\,(12)} \qquad (4)\,(4.5)$ $\dfrac{(48)\,(4.5)}{(12)} \qquad \left(\$18.00\right)$	**40.** Name the space figure depicted by the net. $\left(\begin{array}{c}\text{triangular}\\\text{pyramid}\end{array}\right)$

Pre-algebra Sample Test B

41. Nine years ago Hal was as old as Jen will be in 2 years. If Hal is now 23 years old, how old was Jen 7 years ago?

	−9	now	+2
H	14	23	25
J		12	14

12 − 7 =

(5)

46. Fnd the mode of the data.

15, 11, 6, 6, 15, 14, 31, 6, 7, 15, 1

(6 and 15)

42. Label each number rational (Q) or irrational (Irr).

$\sqrt{27}$ (Irr) | $\dfrac{1}{11}$ (Q)

$\sqrt{11}$ (Irr) | $0.\overline{127}$ (Q)

47. Find the surface area in sq cm of a sphere with diameter 20 cm.

$4 \pi r^2 = (4\pi)(100) =$ (400π)

43. Find the volume in cu cm of a cone with height 15 cm and radius 9 cm.

$$\dfrac{\overset{3}{\cancel{9}} \cdot 9 \cdot \pi \cdot 15}{\cancel{3}} \quad 3 \cdot 135 \cdot \pi \quad (405\,\pi)$$

The volume of a cone is 1/3 the volume of the cylinder in which it just fits.

48. Operate.

$$\dfrac{\overset{2}{\cancel{22}}}{\underset{2}{\cancel{18}}} \cdot \dfrac{\overset{3}{\cancel{27}}}{\underset{2}{\cancel{30}}} \cdot \dfrac{\overset{1}{\cancel{15}}}{\underset{3}{\cancel{33}}} \quad \left(\dfrac{1}{2} \right)$$

44. Find the perimeter in ft of a rectangle with area 90 sq ft and length 5 ft.

A = 90
5W = 90
W = 18
L = 5
P = 23 + 23 = (46)

49. Find the LCM.

LCM($15ab^2c^4$, $25a^3bc^5$)

($75a^3b^2c^5$)

Do constants and each variable separately.

45. Find the diagonal in cm of a square with side 10 cm.

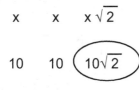

x   x   x$\sqrt{2}$

10   10   ($10\sqrt{2}$)

50. Round 7.6354 to each place.

ones	8
tenths	7.6
hundredths	7.64
thousandths	7.635

Pre-algebra Sample Test B	
51. Operate. $\sqrt{75} \cdot \sqrt{36} \cdot \sqrt{98}$ $\sqrt{25} \cdot \sqrt{3} \cdot 6 \cdot \sqrt{49} \cdot \sqrt{2}$ $5 \cdot 6 \cdot 7 \cdot \sqrt{6}$ $\boxed{210\sqrt{6}}$	56. Name the property illustrated. $(15 + x) + c = 15 + (x + c)$ $\boxed{\text{APA}}$ Associative Property of Addition
52. In how many ways may 7 children line up at a water fountain? $7 \cdot 6 \cdot 5 \cdot 4 \cdot 3 \cdot 2 \cdot 1$ $7 \cdot 6 \cdot 12 \cdot 10$ $7 \cdot 72 \cdot 10$ $504 \cdot 10$ $\boxed{5040}$ Use Multiplication Principle.	57. Given 2 sides of a triangle 3 and 17 units, write 2 inequalities: one for the range of the 3rd side (s) and one for the range of the perimeter (p). $\boxed{\begin{array}{c} 14 < s < 20 \\ 34 < p < 40 \end{array}}$
53. A CD costs \$17.00. The sales tax is 5%. You pay with a \$20 bill. How much change do you receive? 0.05 x 17 is 0.85 20 − 17.85 = 2.15 $\boxed{\$2.15}$	58. Evaluate. $\dfrac{15!}{10! \, 6!} \qquad \dfrac{11 \cdot \overset{7}{\cancel{12}} \cdot 13 \cdot \cancel{14} \cdot \cancel{15}}{\underset{2}{\cancel{6} \cdot \cancel{5} \cdot 4 \cdot 3 \cdot \cancel{2}}} \qquad \dfrac{11 \cdot 91}{2}$ $\boxed{\dfrac{1001}{2}}$
54. Find LCM (55, 95). 5 x 11 5 x 19 5 x 11 x 19 11 x 95 $\boxed{1045}$	59. When tossing 2 dice, find the probability of a sum of 2 or 3. 1, 1 1, 2 $\dfrac{3}{6 \times 6}$ $\boxed{\dfrac{1}{12}}$ 2, 1
55. Write as a percent: 3.0056 $\boxed{300.56\%}$ The percent symbol means "parts of 100" or divide by 100. To compensate, multiply by 100.	60. Name all of the line segments in the figure. $\boxed{\begin{array}{ccc} \overline{AB} & \overline{CD} & \overline{AE} \\ \overline{BC} & \overline{AD} & \\ \overline{AC} & \overline{DE} & \end{array}}$

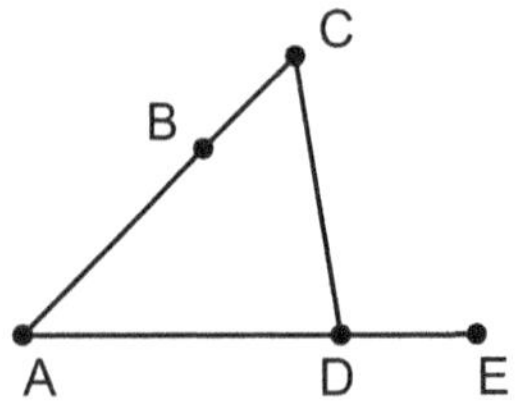

Pre-algebra Sample Test B

61. Solve for A.

$$30AB^3CD = 42BC^4D$$

$$\frac{30AB^3CD}{30B^3CD} = \frac{42BC^4D}{30B^3CD}$$

$$A = \frac{7C^3}{5B^2}$$

66. Solve: $x(x+3)(x-9)(x+4) = 0$

$$x = 0, -3, 9, -4$$

Solve: $(5x-2)(4x+3) = 0$

$$x = \frac{2}{5} \text{ or } \frac{-3}{4}$$

62. How many 4-letter permutations may be made using the letters of the word ORANGE ?

$6 \times 5 \times 4 \times 3$ $\boxed{360}$

Multiplication Principle

67. If 8 workers load 12 trucks of hay in 1 day, how many similar workers are needed to load 21 of the same trucks in 1 day?

$$\frac{8}{12} = \frac{w}{21} \qquad \frac{2}{3} = \frac{14}{21}$$

$$\frac{2}{3} = \frac{w}{21} \qquad \boxed{14}$$

63. A square has area 2500 sq un. Find its perimeter in units.

A = 2500

S = 50

P = $\boxed{200}$

68. Graph on the number line.
$x = -3$ OR $x > 0$

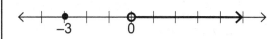

$x \le 2$

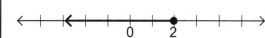

64. Find the product of the complement of 40° and the supplement of 150°.

C(40) = 50
S(150) = 30
50 × 30 = $\boxed{1500}$

69. In △ABC, m∠A = 39 and m∠C = 51. Classify the triangle by angle and side.

By angle   right

By side   scalene

39 + 51 + 90 = 180. A triangle with 3 different angle measures has 3 different side measures.

65. The area of a circle is 3600π sq un. Find its radius in un.

$\boxed{60}$

The circumference of a circle is 120π un. Find its radius in un.

$\boxed{60}$

70. Find the surface area of a cylinder in sq un with diameter 12 un and height 8 un.

A circ = 36π
A rec = Ch = $\pi dh = 96\pi$
$36\pi + 36\pi + 96\pi =$ $\boxed{168\pi}$

Pre-algebra Sample Test B	
71. The number n has factors 3, 5, and 11. Name 5 other factors of n. $\boxed{1,\ 15,\ 33,\ 55,\ 165}$ 1 is a factor of every natural number. If 3 and 5 (primes) are factors, then 3x5 is a factor, and so on.	76. An equilateral triangle has perimeter 51.15 units. What is the measure of a side in units? $51.15 \div 3 =$ $\boxed{17.05}$
72. Find the slope and y-intercept of the line $5x - 10y = 4$. $-10y = -5x + 4$ $10y = 5x - 4$ slope $\dfrac{1}{2}$ y-intercept $\dfrac{-2}{5}$	77. Simplify. $4\sqrt{81} - 2\,|\,6 - 14\,| + 5 \cdot (-3)^2$ $36 - 16 + 45$ $\boxed{65}$
73. Find 4.75% of 400. 1% of 400 = 4 4% of 400 = 16 1/4 % of 400 = 1 3/4 % of 400 = 3 $16 + 3 =$ $\boxed{19}$	78. Write an equation of the line through (4,6) and (0,3). SL $\dfrac{6-3}{4-0}$ $\dfrac{3}{4}$ $\boxed{y = \dfrac{3}{4}\,x + 3}$
74. Evaluate for $a = -10$ and $b = -5$. $-2ab - 3b^2 - 4a + 3b - 2$ $-100 - 75 + 40 - 15 - 2$ $\boxed{-152}$	79. Write the correct vocabulary word. 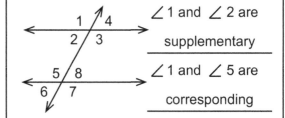 $\angle\,1$ and $\angle\,2$ are supplementary $\angle\,1$ and $\angle\,5$ are corresponding
75. Name the property illustrated. If $x = y$, then $x + 2 = y + 2$ $\boxed{\text{APE}}$ Addition Property of Equality	80. Calculate the combination. $C(7,4) = \dfrac{7!}{4!\ 3!} = \dfrac{5 \cdot 6 \cdot 7}{3 \cdot 2 \cdot 1} = \boxed{35}$

Pre-algebra Sample Test B	
81. Find the volume of a square pyramid in cubic inches with base edge 9 inches and height 8 inches. $\dfrac{9 \times 9 \times 8}{3}$ $\qquad$ $3 \times 9 \times 8$ $\qquad\qquad$ 3×72 216	86. Find the volume in cubic feet of a rectangular prism that measures 18 inches by 3.1 feet by 9 feet. (1.5) (3.1) (9) (13.5) (3.1) 40.5 + 1.35 41.85
82. Write as a unit rate. 44 pages in 8 minutes $\dfrac{44}{8}$ pages in 1 minute $\dfrac{11}{2}$ pages in 1 minute 5.5 pages in 1 minute	87. Write as an algebraic expression: Half the product of two numbers is greater than the absolute value of the difference of the numbers. 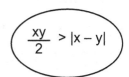 $\dfrac{xy}{2} > |x - y|$
83. A rectangular block 35 by 63 by 70 units is cut into cubes measuring 7 units on a side. How many such cubes are formed? 5 x 9 x 10 450	88. Find the rate in mph of a vehicle that traveled 280 miles in 3.5 hours. D = RT $280 = R \cdot \dfrac{7}{2}$ $\dfrac{2}{7}\, 280 = R \cdot \dfrac{7}{2}\, \dfrac{2}{7}$ 80 = R
84. Simplify. $5(2x - y) + 4x - 2y - 4(2x + 3y)$ $10x - 5y + 4x - 2y - 8x - 12y$ 6x – 19y	89. Complete. 86.1 mm = $\underline{\quad 8.61 \quad}$ cm 72 kg = $\underline{\quad 72{,}000 \quad}$ g Use 2 steps: determine the number of places away and whether more or fewer are needed to balance the equation.
85. Write the next two terms and the 51st term of the arithmetic sequence. 2.5, 7, 11.5, 16, 20.5, $\underline{\quad 25 \quad}$ $\underline{\quad 29.5 \quad}$ 51st term = 2.5 + (50)(4.5) $\qquad\quad$ = 2.5 + (200 + 25) = 227.5	90. A triangle has endpoints (0, 0), (0, –6), and (–8, 0). Name its endpoints after a reflection over the y-axis followed by a reflection over the x-axis. 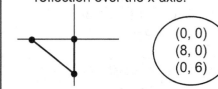 (0, 0) (8, 0) (0, 6)

Pre-algebra Sample Test B	
91. Write the equation of the line through the origin with slope –6. $\boxed{y = -6x}$ Lines through the origin are of the form y = mx.	96. Solve: –3x – 1 < 15 –3x < 16 x > $\dfrac{-16}{3}$ When multiplying or dividing both sides of an inequality by a negative number, the sign flips.
92. Label each sequence A (arithmetic), G (geometric), or N (neither). 5, –15, 45, –135, . . . <u> G </u> × –3 2, 3, 5, 7, 11, 13, . . . <u> N </u> primes	97. Given $\triangle CAB \cong \triangle DEF$, m $\angle$D = 90, m $\angle$A = 67.5, AC = 5, and FE = 13. Find: m $\angle$E = <u> 67.5 </u> | BC = <u> 12 </u> m $\angle$C = <u> 90 </u> | AB = <u> 13 </u> m $\angle$F = <u> 22.5 </u> | DE = <u> 5 </u>
93. Solve. $\dfrac{4}{5} \cdot x + 2 = 74$ $\dfrac{4}{5} \cdot x = 72$ $\dfrac{5}{4} \ \dfrac{4}{5} \cdot x = 72 \ \dfrac{5}{4}$ x = $\boxed{90}$	98. Find the number of codes that may be formed of the type: DIGIT–EVEN DIGIT–ODD DIGIT–LETTER 10 x 5 x 5 x 26 10 x 25 x (25 + 1) 10 x 650 $\boxed{6500}$
94. Find the hypotenuse in simplest radical form. 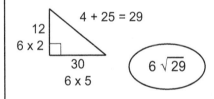 $\boxed{6\sqrt{29}}$	99. Label the measures of the other 7 angles. (NTS) 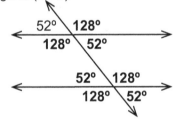
95. Find the final sales price on a $300 item after a 15% markup and a 10% discount. 300 300 + 30 + 15 = 345 345 – 34.50 $\boxed{\$310.50}$	100. Express algebraically: the sum of 4 consecutive even numbers. $\boxed{x + (x + 2) + (x + 4) + (x + 6)}$ An even number is 2x; an odd number is 2x + 1. However, in an algebra word problem, both may be expressed as x. If impossible, the equation will lead to no solution.

MAVA Math: Middle Reviews Solutions Copyright © 2013 Marla Weiss

Pre-algebra Sample Test B	

101. What is the effect on the area of a rectangle if both the length and width are quadrupled?

(L) (W)
(4L) (4W)
(16) (LW)

times 16

106. Find the two missing sides.

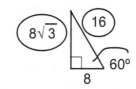

102. A diameter of a circle has endpoints (5,15) and (11,3). Find the circle's center.

(8, 9)

The circle's center is the midpoint of the diameter.

107. Find x if 60% of x is 75.

60% is 75
20% is 25
100% is *125*

103. Find the 2 missing congruent lengths.

6 √2

"Divide by 2, tack on the root 2" is faster than doing the Pythagorean Theorem.

108. A bowl has 13 striped, 5 speckled, and 8 solid marbles. Draw 3 in succession without replacement. Find the probability.
P(STRIPED, SPECKLED, SOLID)

$$\frac{13}{26} \cdot \frac{5}{25} \cdot \frac{8}{24} \qquad \frac{1}{2} \cdot \frac{1}{5} \cdot \frac{1}{3} \quad \left(\frac{1}{30}\right)$$

104. A trapezoid has A = 120, h = 12, and b = 6. Find B.

120 = 12M
M = 10

6   10   *14*

109. Circle the numbers that are divisors of the 15-digit number 275,096,251,836,015.

1   2   *3*   4   *5*   6

8   9   10   12   *15*   18

Simplified digit sum = 15 after all 9s are eliminated.

105. Complete the missing numbers in the Pythagorean triplets: leg, leg, hypotenuse.

4, _4 √3_ 8

30, 40 50

9, 10, √181

110. Graph y = x − 2.

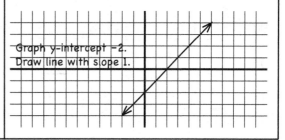

Graph y-intercept −2.
Draw line with slope 1.

Pre-algebra Sample Test B	
111. Given a circle inscribed in a square with side 20 units. Find the area in sq un between the circle and square. A sq = 400 d = 20 r = 10 A circ = 100π $400 - 100\pi$	116. Write the equation of the line shown. $y = -2x - 2$
112. Write the function rule. | x	f(x)
1	−2
0	−7
−1	−12 | $f(x) = 5x - 7$ Note linear relationship: As x increases by 1, y increases by 5. y = mx + b Use (0, −7) y = mx − 7 Use (1, −2) y = 5x − 7	117. Evaluate for x = $\frac{2}{3}$, y = $\frac{5}{6}$. $18x^2 - 36xy + 72y^2 - 11$ $8 - 20 + 50 - 11$ 27
113. Operate. $\frac{7}{15} + \frac{5}{6} + \frac{4}{5}$ $\frac{14 + 25 + 24}{30}$ $\frac{63}{}$ $\frac{21}{10}$	118. Find the area of the figure: a triangle with altitude 4 attached to a parallelogram with altitude 2 attached to a semicircle with diameter 8. b = 8 r = 4 A tri = 16 A para = 16 A circ = 16π $32 + 8\pi$
114. Solve. $4(2x - 1) + 6x - 2 = 3(x - 1)$ $8x - 4 + 6x - 2 = 3x - 3$ $14x - 6 = 3x - 3$ $11x = 3$ $x = $ $\frac{3}{11}$	119. Solve the system by substitution. $x + y = 4$ $x = -y + 4$ $2x + 3y = 11$ $2(-y + 4) + 3y = 11$ $-2y + 8 + 3y = 11$ $y = 3$ x = 1 y = 3
115. Complete. 60^0 1 0^{60} 0 1^{60} 1 6^{-2} $\frac{1}{36}$ -2^4 −16 $(-2)^4$ 16	120. Multiply. $(2x - 3)(4x + 5)$ $8x^2 + 10x - 12x - 15$ $8x^2 - 2x - 15$

CURRICULUM GUIDE

1. ABSOLUTE VALUE

Level 6
1. Solve equations with a variable or integer inside absolute value bars.
2. Evaluate whole number expressions with addition/subtraction and absolute value.
Level 7
1. Solve equations with $x \pm k$ inside absolute value bars.
2. Evaluate whole number expressions with 4 operations and absolute value.
Level 8
1. Solve equations with $ax \pm k$ inside absolute value bars.
2. Find the number of integer values satisfying an absolute value inequality.

2. ADDITION (WHOLE NUMBER)

Level 6
1. Add five 9-digit numbers.
Levels 7, 8–none

3. AGE PROBLEMS

Level 6
1. Find ages of 3 or 4 people for one point in time.
2. Solve age word problems by completing 2-row, 3-column charts with constants.
Level 7
1. Solve age word problems by completing 3-row, 3-column charts with constants.
Level 8
1. Solve age word problems by completing 2-row, 3-column charts with variables.

4. ANGLES (see FRACTIONAL PARTS, POLYGONS, RATIOS)

Level 6
1. Find the complement of a multiple-of-5 angle.
2. Find the supplement of a multiple-of-10 angle.
3. Label angles formed by 2 parallel lines and 1 transversal with vocabulary words.
4. Find a specified angle measure in a figure of 2 parallel lines and 2 transversals.
Level 7
1. Find the complement or supplement of any whole number angle.
2. Operate on the complement or supplement of angles as algebraic expressions.
3. Label statements about angles SOMETIMES, ALWAYS, or NEVER.
4. Find all angle measures in a figure of 2 parallel lines and 2 transversals.
5. Find a missing angle in a figure of intersecting lines.
Level 8
1. Solve word problems given information about angle complements/supplements.
2. Solve word problems given information about angles as algebraic expressions.
3. Find a missing angle in a crook problem.
4. Find a missing angle in a figure of intersecting polygons.

5. <u>AREA</u> (see CIRCLES, RECTANGLES, RHOMBUSES, TRAPEZOIDS, TRIANGLES)

Level 6
1. Calculate the area of an irregular figure on a grid by the addition/subtraction method.
Level 7
1. Calculate areas of polygonal and circular shapes that divide a rectangle on a unit grid.
2. Calculate shaded area shown in figures of polygons and circles.
3. Calculate the area of a triangle or rectangle bounded by the graphs of lines.
Level 8
1. Calculate the area of a parallelogram or trapezoid bounded by the graphs of lines.
2. Calculate the area of a uniform rectangular border.
3. Calculate the area of a regular hexagon.

6. <u>AVERAGES</u>

Level 6
1. Find whole average of an arithmetic sequence with even or odd number of numbers.
2. Find whole average using the rightmost digits method.
3. Find two missing numbers given average, the other numbers, and a max/min rule.
Level 7
1. Find decimal average of an arithmetic sequence with even or odd number of numbers.
2. Find decimal average using the rightmost digits method.
3. Find a new average given the original average and a rule to alter the numbers.
Level 8
1. Find additional or omitted numbers causing a change in a given average.

7. <u>BASES</u>

Level 6
1. Name the number 1 after or 1 before in a given base.
2. Convert numbers in 1-digit bases to base 10.
3. Add or multiply basic facts in 1-digit bases.
Level 7
1. Convert numbers in base 10 to 1-digit bases.
2. Convert numbers in 2-digit bases to base 10.
3. Add or subtract 4-digit numbers in 1-digit bases with regrouping.
Level 8
1. Convert numbers in base 10 to 2-digit bases.
2. Multiply 4-digit by 1-digit numbers in non-ten bases with regrouping.

8. <u>BOUNDARY</u>

Level 6
1. Find the least or greatest product or difference of numbers in given ranges.
Level 7
1. Find the least quotient of numbers in given ranges.
Level 8
1. Find the bounds of a side of a rectangle given P and the max value of the other side.

9. <u>CALENDAR</u>

Level 6
1. Calculate a 3-digit number of days after today given today's day of the week.
Level 7
1. Calculate a 3-digit number of days before today given today's day of the week.
Level 8
1. Calculate the first date in a month given the sum of specified dates.

10. <u>CIPHERS</u>

Level 6
1. Solve addition cipher with two 3-digit addends.
Level 7
1. Solve addition cipher with two 4-digit addends.
Level 8
1. Solve multiplication and division ciphers.

11. <u>CIRCLES</u> (see FRACTIONAL PARTS)

Level 6
1. Complete a chart given radius (whole), diameter, area, or circumference.
2. Find the area of a circle with an even radius inscribed in a square.
Level 7
1. Complete a chart given radius (not whole), diameter, area, or circumference.
2. Find the area of a circle with an odd radius inscribed in a square.
3. Use the area, rather than the radius or diameter, in a proportion.
Level 8
1. Find the area of a square inscribed in a circle.
2. Find the circumference given the area in a form other than $k\pi$ (or find A given C).

12. <u>CLOCKS</u> (see TIME)

13. <u>COMBINATIONS</u>

Levels 6, 7–none
Level 8
1. Calculate the combination of x things taken y at a time.

14. <u>CONES</u> (see CYLINDERS)

Level 6
1. Know that a cone is not a polyhedron.
Level 7
1. Find the volume of a cone given the radius and height.
Level 8
1. Find the volume of a cone given slant height and radius, diameter, or altitude.
2. Find the surface area of a cone given slant height and radius, diameter, or altitude.

15. <u>CONGRUENCE</u> (see POLYGONS, TRIANGLES)

16. <u>COORDINATE PLANE</u> (see PARALLELOGRAMS, TRANSFORMATIONS)

Level 6
1. Plot and label points on the coordinate plane.
2. Name the plotted points on the coordinate plane.
3. Given coordinates of two opposite vertices of rectangle, find the other two.
Level 7
1. Find area and perimeter of rectangles and triangles given vertices as coordinates.
2. Find the point equidistant from four coordinates.
3. Name lattice points that are described distances from given points.
Level 8
1. Given 2 points, find the 3rd point that makes a triangle of a specified area.
2. Find the area and perimeter of trapezoids given vertices as coordinates.
3. Find the coordinates to make one rectangle a given times the area of another.

17. <u>COUNTING</u>

Level 6
1. Draw a 2-set Venn diagram with Neither from a word problem with the overlap given.
2. Construct a cross-tabulation chart from a word problem.
3. Solve a word problem using the Multiplication (Fundamental Counting) Principle.
4. Answer "how many?" using "subtract and add one."
Level 7
1. Solve a word problem drawing a 3-set intersecting Venn diagram.
2. Count the paths on a rectangular grid between 2 points moving only up and/or right.
3. Solve a word problem using the Multiplication Principle with 1 restriction.
4. Count the number of triangles in a picture of intersecting lines.
5. Answer "how many?" perfect squares in a range using generating numbers.
Level 8
1. Solve a word problem drawing a Venn diagram and using algebra.
2. Solve a word problem using the Multiplication Principle with more than 1 restriction.
3. Count the number of multiples in a range with subtraction of the overlap.

18. <u>CUBES</u>

Level 6
1. Find the volume of a cube given its surface area.
2. Find surface area of a cube given its volume.
3. Find the number of cubes that fill a rectangular solid.
4. Find the volume and total surface area of solids comprised of cubes.
Level 7
1. Find the number of unit cubes with 0, 1, 2, or 3 faces painted of a painted cube.
2. Find the face diagonal of a cube given the edge (E), volume (V), or area of a face.
Level 8
1. Find the space diagonal of a cube given the E, V, area of a face, or face diagonal.
2. Solve assorted word problems about cubes.

19. <u>CYLINDERS</u>

Level 6
1. Compute the volume and surface area of a cylinder.
Level 7
1. Compute the volume of a cylindrical wedge.
Level 8
1. Compute the volume between 2 cylinders with concentric bases and equal height.

20. <u>DECIMALS</u> (see SCIENTIFIC NOTATION)

Level 6
1. Multiply a decimal by a whole number using mental math.
2. Rewrite a list of decimals in ascending order.
3. Convert a repeating decimal with an ellipsis into bar notation.
4. State whether a fraction would terminate or repeat if converted to a decimal.
5. Divide a decimal by a power of 10.
6. Solve word problems using decimal division.
7. Convert a fraction (denominator 9, 11, 33, or 99) to a repeating decimal.
8. Solve decimal equations by clearing decimal points.
Level 7
1. Convert fraction (denominator 45, 90, 300, 330, 900, or 990) to a repeating decimal.
2. Convert a repeating decimal with no leading digit or a leadng 0 to a fraction.
3. Subtract or divide a decimal and a whole number using mental math.
Level 8
1. Convert a repeating decimal with a leading nonzero digit to a fraction.

21. <u>DISTANCE-RATE-TIME</u>

Level 6
1. Write an equation to solve DRT problems (whole number without unit conversion).
2. Draw and complete a chart to solve DRT problems (whole constants).
Level 7
1. Write an equation to solve DRT problems (decimal or fraction with unit conversion).
2. Draw and complete a chart to solve DRT problems (decimal or fraction constants).
Level 8
1. Use the Pythagorean Theorem to solve DRT problems.

22. <u>DIVISIBILITY</u> (see DIVISION, FACTORS)

23. <u>DIVISION (WHOLE NUMBER)</u>

Level 6
1. Divide a 6-digit number by a 2-digit number.
Level 7
1. Find the remainder without dividing for a 2-digit divisor ≤ 21 using divisibility rules.
Level 8
1. Find the remainder without dividing for a 2-digit divisor ≥ 22 using divisibility rules.

24. <u>EQUATIONS</u> (see DECIMALS, EXPONENTS)

Level 6
1. Convert from English to an algebraic expression.
2. Add and subtract like terms without distributing.
3. Solve 1st degree 2-step equations with fractional answers.
Level 7
1. Convert from English to an equation.
2. Add and subtract like terms with distributing.
3. Solve 1st degree equations requiring more than 2 steps.
Level 8
1. Solve 2-variable, 2-equation systems with integral answers.
2. Solve quadratic equations in factored form.

25. <u>ESTIMATING</u>

Level 6
1. Estimate the solution of a word problem.
Grade 7
1. Estimate a fraction with many decimal factors to the nearest whole number.
Level 8
1. Estimate the nearest whole square or cube root.

26. <u>EVALUATING</u> (see ABSOLUTE VALUE, EXPONENTS, OPERATIONS–ORDER)

27. <u>EVENS & ODDS</u>

Level 6
1. Know that zero is even.
2. Find the product of 2 consecutive even/odd numbers given their sum.
3. Find the percent of even/odd numbers in a set.
Level 7
1. Solve even/odd word problems with constants.
Level 8
1. Solve even/odd word problems with variables.

28. <u>EXPONENTS</u> (see SCIENTIFIC NOTATION)

Level 6
1. Evaluate expressions with parentheses and positive exponents.
2. Multiply or divide 2 whole numbers times powers of 10 with regrouping.
Level 7
1. Evaluate a positive number with a zero, positive, or negative exponent.
2. Evaluate expressions with parentheses and zero, positive, or negative exponents.
Level 8
1. Evaluate a positive number with a fractional exponent.
2. Simplify expressions using the 3 rules of exponents.
3. Solve equations with variable exponents.

29. FACTORIALS

Level 6
1. Evaluate factorial expressions.
2. Simplify fractions containing factorials.
3. Find the missing factors or divisors for one factorial to become another factorial.
Level 7
1. Find the number of times a given factor occurs in a factorial.
Level 8
1. Solve equations with a variable factorial.

30. FACTORS (see FRACTIONS)

Level 6
1. Complete a factor chart for 5-digit numbers using divisibility rules for 1 through 11.
2. Find the missing digit to achieve divisibility by a given number.
3. Extend the divisibility-by-6 rule to divisibility by two-digit numbers.
Level 7
1. Complete a factor chart for 5-digit numbers using divisibility rules for 1 through 99.
2. Find the sum of specified factors of a number.
Level 8
1. Find the missing digits (multiple answers) to achieve divisibility by a given number.
2. Find the sum of numbers with a specified odd number of factors.

31. FICTITIOUS OPERATIONS (see PROPERTIES)

Level 6
1. Calculate integral values using exponents and absolute value.
Level 7
1. Calculate decimal or fractional values using exponents and absolute value.
Level 8
1. Calculate nested values.
2. Calculate values using algebra.

32. FRACTIONAL PARTS

Level 6
1. Compute the fractional part that one standard measurement is of another.
2. Solve fractional part word problems with a known starting number using bifurcation.
3. Find the fractional part that a central angle is of 360°.
Level 7
1. Compute the fractional part that one decimal number is of another.
2. Solve fractional part word problems with a known starting number using trifurcation.
3. Find the fractional part of fraction #1 that fraction #2 exceeds fraction #3.
4. Find the fractional part that one percent is of another percent.
5. Solve fractional part word problems by forming the fraction "is/of."
Level 8
1. Solve fractional part word problems with an unknown starting number.

33. <u>FRACTIONS</u> (see DECIMALS, FRACTIONAL PARTS, MIXED NUMBERS)

Level 6
1. Add 2 or 3 fractions with 2-digit denominators.
2. Find mentally the LCM (LCD) of two or three 2-digit or 3-digit numbers.
3. Find mentally the GCF of two or three 2-digit or 3-digit numbers.
4. Find the LCM and GCF of two 3-digit numbers using prime factorization.
5. Multiply a fraction times a 2-digit whole number.
6. Write fractions (halves through tenths, not sevenths) as decimals from memory.
7. Simplify fractions with parts described as prime/composite factors.
8. Simplify complex fractions with + or − in the numerator and denominator.
9. Simplify continued fractions less than 1.
10. Rewrite fractions in ascending order by comparing number and size of "pie pieces."
Level 7
1. Find the LCM and GCF of 3 numbers using prime factorization.
2. Simplify complex fractions with x or ÷ in the numerator and denominator.
3. Simplify continued fractions greater than 1.
Level 8
1. Add, subtract, multiply, and divide simple (one-term parts) algebraic fractions.
2. Simplify fractions with numerator and denominator having repeating decimals.

34. <u>FUNCTIONS</u>

Level 6
1. Complete a chart of x- and y-values for functions in the form $y = mx + b$ (m integer).
2. Identify a set or mapping as a function.
Level 7
1. Complete a chart of x- and y-values for linear functions in the form $ax + by = c$.
Level 8
1. Evaluate the functions ABS, SGN, TRUNC, INT, SIGMA, SQR, and SQRT.
2. Identify a graph as a function.
3. Evaluate composition of linear functions.

35. <u>GEOMETRY–PLANE</u> (see ANGLES, AREA, CIRCLES, LINES, PERIMETER, POLYGONS, SYMMETRY, TRIANGLES)

36. <u>GEOMETRY–SOLID</u> (see CONES, CUBES, CYLINDERS, NETS, PRISMS, PYRAMIDS, SPHERES)

37. <u>GRAPHS (and TABLES)</u>

Level 6
1. Compute statistics on data shown in a stem-and-leaf plot.
2. Compute the number of paths from one vertex to another on a directed graph.
Level 7
1. Compute statistics on data shown in a frequency distribution.
Level 8
1. Compute statistics on data shown in a line plot.

38. <u>GREATEST COMMON FACTOR</u> (see FACTORS, FRACTIONS)

39. <u>HISTORICAL MATH</u>

Level 6
1. Identify Euler graphs: figures drawn without lifting the pencil or retracing a line.
Level 7
1. Show Goldbach's Conjecture: even integers greater than 2 are the sum of 2 primes.
2. Verify the first 3 perfect numbers as the sum of their proper divisors.
Level 8
1. Show that primes of the form $4k + 1$ are the sum of 2 squares.
2. Identify the first 6 Fibonacci primes.
3. Identify the first 3 abundant numbers: sum of proper divisors is greater than number.

40. <u>INEQUALITIES</u> (see ABSOLUTE VALUE)

Level 6
1. Solve a 2-step inequality without flipping sign and without simplifying fraction.
Level 7
1. Solve a 2-step inequality with flipping sign and with simplifying fraction.
Level 8
1. Graph on the number line and write the solution to a factored quadratic inequality.

41. <u>INTEGERS</u>

Level 6
1. Add and subtract 2-digit multiple-of-5 integers.
2. Multiply and divide integers using basic facts.
Level 7
1. Add and subtract any 2-digit integers.
2. Multiply and divide any 2-digit integers.
Level 8
1. Find consecutive integers with a given sum or product.
2. Find consecutive even or odd integers with a given algebraic relationship.

42. <u>INTEREST</u>

Level 6
1. Find simple interest given principal, 1-digit percent rate, and time in 1-11 months.
2. Complete a chart for principal that is multiple of $10 and rate that is multiple of 0.5%.
Level 7
1. Find simple interest given principal, decimal percent rate, and time in 13-35 months.
2. Mentally find the principal for a given a rate and one-year interest.
Level 8
1. Find the unknown principal/rate given the rate/principal, interest, and time in years.
2. Find the total value for annually compounded principal at given rates and years.

43. <u>LEAST COMMON MULTIPLE</u> (see FRACTIONS)

44. <u>LINES</u> (see NUMBER LINE)

Level 6
1. Find a segment's midpoint given endpoints with positive coordinates.
2. Graph lines of the form y = mx.
Level 7
1. Find a segment's midpoint given endpoints with negative or decimal coordinates.
2. Graph lines of the form y = mx + b.
3. Find slopes and both intercepts given lines in standard or slope-intercept form.
Level 8
1. Find a segment's endpoint given the midpoint and one endpoint.
2. Graph lines of the form ax + by = c with a and b factors of c.
3. Compute the slope of a line given 2 points.
4. Find the points of trisection of a line segment given endpoints.
5. Label lines as parallel, perpendicular, or neither given two equations.

45. <u>LOGIC</u>

Level 6
1. Complete a truth table for NOT, AND, and OR.
2. Write the converse of a conditional statement, and label it true or false.
Level 7
1. Write the inverse of a conditional statement, and label it true or false.
Level 8
1. Write the contrapositive of a conditional statement, and label it true or false.
2. Use DeMorgan's Laws to rewrite statements.

46. <u>MATRICES</u>

Level 6
1. Add or subtract square matrices.
2. Perform scalar multiplication on square matrices.
Level 7
1. Add or subtract rectangular matrices.
2. Perform scalar multiplication on rectangular matrices.
Level 8
1. Determine if matrices may be multiplied. If yes, give dimensions of the product.
2. Multiply matrices.
3. In a matrix equation, solve for a variable expressed as a matrix element.

47. <u>MEASUREMENT–METRIC</u> (see FRACTIONAL PARTS)

Level 6
1. Convert among metric measures.
Level 7
1. Convert among squared and cubed metric measures.
Level 8
1. Solve word problems having metric unit conversion.

48. <u>MEASUREMENT–STANDARD</u> (see FRACTIONAL PARTS)

Level 6
1. Write a variable number of units as an algebraic expression.
2. Convert whole measures to decimals with a different unit.
Level 7
1. Multiply and divide expressions in related units by a whole number with regrouping.
Level 8
1. Convert square and cubic units.
2. Find the area of rectangles and volume of rectangular prisms with unit conversions.

49. <u>MENTAL MATH</u> (see PLACE VALUE)

Level 6
1. Multiply a whole number by 5 (times by 10, cut in half).
2. Multiply a whole number by 4 (double twice).
3. Divide a whole number by 4 (halve twice).
4. Multiply a 2-digit number by 11 with regrouping.
Level 7
1. Multiply a whole number by 25 (divide by 4, tack on 2 zeros).
2. Multiply a 3-digit or 4-digit number by 11 without regrouping.
3. Divide a whole number by 5 (double, move decimal point 1 place to left).
Level 8
1. Square an odd multiple of 5.
2. Multiply two 2-digit numbers using DPMA.
3. Multiply two 2-digit numbers using the difference of 2 squares.

50. <u>MIXED NUMBERS</u>

Level 6
1. Add or subtract two mixed numbers.
2. Multiply or divide two mixed numbers.
Level 7
1. Add or subtract three mixed numbers.
Level 8
1. Simplify complex fractions containing mixed numbers.
2. Multiply two mixed numbers as binomials.

51. <u>MODULO ARITHMETIC</u>

Level 6
1. Find values in 1-digit mods.
Level 7
1. Find opposites and reciprocals in 1-digit mods.
2. Find values after operating in 1-digit mods.
Level 8
1. Find values in 2-digit mods.
2. Find square roots in 1-digit mods.

52. MONEY

Level 6
1. Find the number of ways two coin denominations can make a given value.
Level 7
1. Find the equal number of coins that make a given value.
Level 8
1. Calculate the percent that given quantities of coins are of a dollar value.

53. MULTIPLICATION (WHOLE NUMBER)

Level 6
1. Multiply a 5-digit number by a 4-digit or 3-digit number.
Levels 7, 8–none

54. MYSTERY NUMBER WORD PROBLEMS

Level 6
1. Use divisibility by 8 and 11 clues to find a 6-digit mystery number.
Level 7
1. Use divisibility by 12 and 15 clues to find a 6-digit mystery number.
Level 8
1. Use divisibility by 18 and 25 clues to find a 6-digit mystery number.

55. NETS

Level 6
1. Name the solid depicted in a net.
Level 7
1. Identify a net as a cube or not a cube.
Level 8
1. Identify the net of an octahedron.

56. NUMBER LINE

Level 6
1. Graph inequalities on the number line.
2. Identify an equation or inequality as a point, line, ray, or segment if graphed.
3. Calculate the whole point that is a fractional part of the distance between 2 points.
Level 7
1. Graph inequalities with AND or OR on the number line.
2. Identify an equation or inequality as an open ray/segment or 2 points.
3. Calculate the decimal point that is a fractional part of the distance between 2 points.
Level 8
1. Find a length given bisection or trisection points of number line segments.
2. Find a length given overlapping number line segments.
3. Find a length given the ratio of lengths of number line segments.

57. OPERATIONS–ORDER

Level 6
1. Evaluate expressions of 5 numbers with parentheses yielding whole numbers.
Level 7
1. Evaluate expressions of 6 numbers with parentheses yielding decimal numbers.
Level 8
1. Evaluate expressions knowing $(-a)^b$ versus $-a^b$.

58. PARALLELOGRAMS

Level 6
1. Find the area of a parallelogram given a picture on a unit grid.
Level 7
1. Find the area of a parallelogram given the coordinates of its vertices.
2. Find the area of a parallelogram using the Pythagorean Theorem to find the height.
Level 8
1. Find the perimeter of a parallelogram given its sides as algebraic expressions.
2. Find all angles of a parallelogram given two angles as algebraic expressions.
3. Solve probability problems about parallelogram angles.
4. Solve word problems with algebraic information about parallelogram angles.

59. PATTERNS–LETTER

Level 6
1. Find the next letter alternating cases, moving forward or backward, and skipping 3+.
Level 7
1. Find the next letter moving forward and/or back with alternating or intertwining skips.
Level 8
1. Find the next letter with alternating/intertwining patterns and both upper/lower cases.

60. PATTERNS–NUMERIC

Level 6
1. Find missing numbers in a sequence using multiplication patterns.
Level 7
1. Find the next term in a sequence using successive differences.
Level 8
1. Find the formula that generates a numeric pattern.

61. PATTERNS–VISUAL

Level 6
1. Find the number of given shapes in a specified nth item of the pattern sequence.
Level 7
1. Find the perimeter in a specified nth item of the pattern sequence.
Level 8
1. Find the 5th image in a pattern given the first four.

62. <u>PERCENTS</u> (see INTEREST)

Level 6
1. Mentally find a multiple of 5% of a whole number yielding a whole number.
2. Compute sales tax (and add) or discount (and subtract) given an original cost.
3. Solve multiple-step word problems FORWARD with discount and tax.
4. Complete fraction/decimal/percent chart for numbers > 1 or repeating decimals.
Level 7
1. Mentally find any % of a whole number yielding a whole number.
2. Solve multiple-step word problems BACKWARD with discount and tax.
3. Find the percent change from one whole number to another.
4. Find a missing variable in problems of the form w% of x is y.
Level 8
1. Find successive percents of a number.

63. <u>PERFECT SQUARES AND CUBES</u> (see COUNTING, FACTORS)

Level 6
1. Find the least natural number n such that kn is a perfect square.
Level 7
1. Find the least natural number n such that kn is a perfect cube.
Level 8
1. Solve word problems about perfect squares and cubes.

64. <u>PERIMETER</u> (see BOUNDARY, RATIOS, RECTANGLES)

Level 6
1. Find the perimeter of a right-angled figure on a unit grid.
2. Given area, equate perimeters of a regular triangle and square to find whole sides.
3. Find the perimeter of attached polygons.
Level 7
1. Find the perimeter of a figure on a unit grid using the Pythagorean Theorem.
2. Given area, equate perimeters of a regular triangle and square to find decimal sides.
Level 8
1. Find the area of a regular hexagon given its perimeter.

65. <u>PERMUTATIONS</u> (see COUNTING)

66. <u>PLACE VALUE</u>

Level 6
1. Convert word form to standard form or vice versa with whole and decimal places.
2. Convert expanded form to standard form with whole and decimal places.
Level 7
1. Arrange 6 digits to form 3-digit numbers with greatest and least positive differences.
2. Evaluate an expression using values determined from a place-value equation.
Level 8
1. Multiply by 101 or 1001 or 10,001 mentally.

67. <u>POLYGONS</u> (see AREA, PERIMETER, RATIOS)

Level 6
1. Label statements about polygons true or false.
2. Label polygons congruent, similar, or neither.
3. Label polygons convex or concave.
Level 7
1. Find the sum of the angles of a polygon.
Level 8
1. Find the measure of an angle of a regular polygon.

68. <u>PRIMES</u> (see FRACTIONS, HISTORICAL MATH)

Level 6
1. Write the prime factorization of 3-digit numbers using exponents.
Level 7
1. Write the prime factorization of 4-digit or 5-digit numbers using exponents.
Level 8
1. Solve word problems about sums of prime numbers.

69. <u>PRISMS</u> (see CUBES, MEASUREMENT–STANDARD)

Level 6
1. Find the volume of rectangular prisms given areas of 3 faces with 2 prime factors.
2. Find volume of a folded rectangular container formed by removing square corners.
Level 7
1. Find face diagonals of rectangular prisms.
2. Find the volume of rectangular prisms given areas of faces with 3 prime factors.
Level 8
1. Find the volume of polygonal prisms given altitude and base dimensions/area.
2. Find the space diagonal of rectangular prisms.

70. <u>PROBABILITY</u>

Level 6
1. Find the probability of getting a math attribute when tossing one standard die.
2. Find the probability of getting a small sum when tossing 2 standard dice.
3. Find the probability of selecting certain cards from a full deck of cards.
4. Find the probability of selecting a point in a shaded area of a square or rectangle.
5. Find the probability that a positive integer in a range is a factor of a given number.
6. Find the probability of drawing an item from a collection.
Level 7
1. Find the probability of a specified event when tossing 3 standard dice.
2. Find the probability of selecting a point in an irregular shaded area.
3. Find a probability when drawing from a deck of cards with overlapping conditions.
Level 8
1. Find the probability of making successive draws without replacement.
2. Find a probability of "at least" an event occurring.

71. <u>PROPERTIES</u>

Level 6
1. Distribute positive or negative constants using DPMA.
2. Label closure, associative, identity, inverse, & commutative properties and ZPM.
Level 7
1. Distribute positive or negative variables using DPMA.
2. Label reflexive, symmetric, and transitive properties of equality.
Level 8
1. Identify properties for fictitious operations.
2. Identify whether sets with given operations are closed.
3. Use DPMA to simplify expressions.

72. <u>PROPORTIONS</u> (see CIRCLES)

Level 6
1. Solve a proportion using equivalent fractions.
2. Solve word problems with direct or inverse proportion and no unit conversion.
Level 7
1. Solve a proportion using cross multiplication, including two solutions.
2. Solve word problems with direct or inverse proportion and unit conversion.
Level 8
1. Solve algebraic proportions with direct, inverse, or joint variation.
2. Find the positive geometric mean between two positive whole numbers.

73. <u>PYRAMIDS</u> (see NETS)

Level 6
1. Identify a pyramid as a polyhedron.
2. Recognize the net of a square pyramid and equilateral triangle pyramid.
Level 7
1. Find the volume of a square pyramid given the base edge and altitude.
Level 8
1. Find the volume of a rectangular pyramid given the base dimensions and altitude.
2. Find the surface area of a square pyramid given the base edge and lateral edge.

74. <u>PYTHAGOREAN THEOREM</u> (see DISTANCE-RATE-TIME, TRIANGLES)

Level 6–none
Level 7
1. Determine if 3 whole numbers form a Pythagorean triplet.
2. Find the length of a diagonal line segment given a picture of 2 attached triangles.
3. Find the area of an isosceles triangle.
Level 8
1. Determine if 3 decimal or radical numbers form the sides of a right triangle.
2. Determine if a triangle is acute or obtuse given 3 sides.
3. Solve N-S-E-W directional word problems using right triangles.
4. Find the distance between 2 points that form a diagonal line segment.

75. <u>RADICALS</u>

Level 6
1. Evaluate the square root of whole number perfect square.
2. Find and operate on whole numbers between two non-perfect-square radicals.
Level 7
1. Evaluate the square root of decimal perfect square.
2. Solve a 2-step radical equation with a whole solution.
Level 8
1. Simplify whole non-perfect-square radicals.
2. Solve an equation of the form: factor times root x = x or factor times x squared = k.

76. <u>RATES</u>

Level 6
1. Find the cost of a number of items given the cost of a different number of the same.
Level 7
1. Find the unit rate given an amount and a cost for the amount.
Level 8
1. Find the less costly unit rate of two items with unit conversions.

77. <u>RATIOS</u> (see NUMBER LINE, PROPORTIONS, RECTANGLES)

Level 6
1. Solve word problems relating the ratio, sum, and product of two numbers.
2. Solve word problems relating a ratio with the total of the parts.
3. Find the ratio of a side of a regular polygon to the perimeter.
Level 7
1. Find a ratio given 5 consecutive points on a line and ratios of their lengths.
2. Find all angles of a non-regular polygon given their ratio.
Level 8
1. Find the ratio of an angle to its complement/supplement given its ratio to the other.
2. Find a ratio of 2 variables from among 6 variables in two given ratios.

78. <u>RECTANGLES (and SQUARES)</u> (see BOUNDARY, COORDINATE PLANE, MEASUREMENT–STANDARD, UNIFORM BORDER PROBLEMS)

Level 6
1. Calculate the perimeter (P) of a square given the area.
2. Calculate the area (A) of a square given the perimeter.
3. Find the maximum area of a rectangle given the perimeter.
4. Find the maximum perimeter of a rectangle given the area.
Level 7
1. Find the perimeter/area of a rectangle given the area/perimeter and one dimension.
2. Find the area/diagonal of a rectangle given the diagonal/area and one dimension.
Level 8
1. Find A/P of a rectangle given the P/A and a relationship between the dimensions.
2. Find the area of a rectangle given the perimeter and the ratio of the sides.

79. <u>RHOMBUSES</u>

Level 6
1. Calculate the area of a rhombus as a parallelogram given a picture or dimensions.
Level 7
1. Calculate the area of a rhombus given 2 diagonals.
Level 8
1. Calculate the area of a rhombus given the perimeter and 1 diagonal.
2. Calculate the perimeter of a rhombus given 2 diagonals or the area and 1 diagonal.

80. <u>SCIENTIFIC NOTATION</u>

Level 6
1. Convert between standard form and scientific notation with positive exponents.
Level 7
1. Convert between standard form and scientific notation with negative exponents.
Level 8
1. Write a multi-operation expression in scientific notation.

81. <u>SEQUENCES and SERIES</u>

Level 6
1. Identify a sequence as arithmetic (label d), geometric (label r), or neither.
2. Find missing terms of arithmetic/geometric sequences with d/r whole and positive.
3. Find the sum of an arithmetic sequence with all terms visible and positive.
Level 7
1. Find missing terms of arithmetic/geometric sequences with d/r decimal or negative.
2. Find the sum of an arithmetic sequence with all terms visible and negative.
3. Find a specified nth term of a repeating sequence.
Level 8
1. Find a specified term of a sequence described in a word problem.
2. Find the sum of an arithmetic sequence with terms not all visible (with ellipsis).
3. Find the nth term of an arithmetic or geometric sequence given the first 4 terms.

82. <u>SETS</u> (see PROPERTIES)

Level 6
1. Identify a number as a member of N, W, Z, Q, and/or R.
2. Perform union and intersection operations on sets of numbers.
Level 7
1. Perform complementation operation on sets.
2. List all subsets of a set.
Level 8
1. Draw Venn diagrams to show the relationship among sets.
2. Perform union and intersection operations on sets of points.
3. Find the number of subsets of a given set.

83. <u>SIMILARITY</u> (see POLYGONS, TRIANGLES)

84. SPHERES

Level 6–none
Level 7
1. Find the volume and surface area of a sphere.
Level 8
1. Convert ratios of radii of 2 spheres to ratios of volume and surface area.
2. Find the number of smaller spheres into which a larger sphere may be re-formed.

85. STATISTICS (see GRAPHS)

Level 6
1. Find five numbers given their statistical values.
2. Complete data in a chart for a circle graph.
Level 7
1. Find seven numbers given their statistical values.
2. Find statistics for perfect squares, perfect cubes, and primes.
3. Find statistics for factors of given numbers.
Level 8
1. Find the algebraic mean of algebraic expressions.
2. Find the numeric mean and range of algebraic expressions given the median.
3. Find an unknown in a list given statistical facts about the numbers.

86. SUBTRACTION (WHOLE NUMBER)

Level 6
1. Subtract two 9-digit numbers.
Levels 7, 8–none

87. SURFACE AREA (see CONES, CYLINDERS, PRISMS, SPHERES)

88. SYMMETRY (see TRANSFORMATIONS)

89. TIME

Level 6
1. Convert whole time measures to decimals in another time unit.
2. Multiply a unit fraction by a quantity of days-hours-minutes.
3. Add hundreds of minutes to a given time with AM/PM conversion.
Level 7
1. Subtract hundreds of minutes from a given time with AM/PM conversion.
2. Find the angle formed by clock hands at quarter past, half past, or quarter to.
3. Find the fraction that an amount of hours is of a span of days.
4. Solve word problems about cycles of clock hands.
Level 8
1. Solve word problems about broken clocks.
2. Find the angle formed by clock hands at any time.

90. <u>TRANSFORMATIONS (ROTATION, REFLECTION, TRANSLATION)</u>

Level 6
1. Identify figures with rotational symmetry and name the number of degrees.
2. Draw a triangle reflected over an axis.
3. Draw a triangle translated right/left and up/down.
4. Find the new point after a point is reflected over an axis and then translated.
Level 7
1. Draw a polygon reflected over a non-axis vertical or horizontal line.
2. Find the new point after a point is reflected over the line y = x and then translated.
Level 8
1. Draw a polygon reflected over the line y = x or y = –x.
2. Find the new point after a point is reflected over the line y = –x and then translated.

91. <u>TRAPEZOIDS</u>

Level 6
1. Calculate the area of a trapezoid given a picture with dimensions.
2. Identify a trapezoid as right, isosceles, or scalene.
Level 7
1. Mentally find the median (midline) of a trapezoid with decimal bases.
2. Given the area, height, and one base of a trapezoid, find the other base.
Level 8
1. Use the Pythagorean Theorem to find the area of an isosceles trapezoid.
2. Use the Pythagorean Theorem to find the perimeter of an isosceles trapezoid.

92. <u>TRIANGLES</u> (see PYTHAGOREAN THEOREM)

Level 6
1. Find the area of a triangle with 3 diagonal sides on a grid.
2. Find missing whole-number sides of similar triangles.
3. Determine whether 3 whole numbers are valid sides of a triangle.
4. Given two sides of triangle, find the range of the 3rd side.
5. Label a triangle given facts about its sides and angles.
6. Identify corresponding parts of congruent triangles by symbols or pictures.
Level 7
1. Find the height of an isosceles right triangle given its area.
2. Find missing decimal-number sides of similar triangles.
3. Determine whether 3 decimal numbers are valid sides of a triangle.
4. Given two sides of triangle, find the range of the perimeter.
5. Find the area of an equilateral triangle.
6. Find a missing side of a 45-45-90 or 30-60-90 special right triangle.
Level 8
1. Find a missing angle, side, or area in a picture of adjacent triangles.
2. Find the number of triangles with whole-number sides given the greatest side.
3. Selecting 3 of 4 numbers, find the probability that they form sides of a triangle.
4. Find the ratio of the sides, altitudes, perimeters, and areas of similar triangles.
5. Find the area of a special right triangle given the hypotenuse or a leg.

93. <u>TRIPLET PROBLEMS</u>

Level 6
1. Solve word problems using a chart without fractions.
Level 7
1. Solve word problems using a chart with fractions.
Level 8
1. Solve word problems using a chart with variables.

94. <u>UNIFORM BORDER PROBLEMS</u> (see AREA)

95. <u>VOCABULARY</u>

Levels 6, 7, 8
1. Complete the blank using words in the vocabulary list.

96. <u>WORD PROBLEMS</u> (see AGE, AREA, TRIPLET, UNIFORM BORDER, WORK)

Levels 6, 7
1. Solve "Fence Post" problems.
2. Solve "Organized List" problems.
3. Solve "How Many?" problems.
Level 8
1. Solve "Mixture" problems.
2. Solve "Coin" problems.
3. Solve "Profit" problems.

97. <u>WORK PROBLEMS</u>

Level 6–none
Level 7–none
Level 8
1. Find the time for two people to do a job together given their separate times.
2. Find the time for one person to do a job given another's time and the time together.

<u>CURRICULUM NOTES</u>

NOTES